George Thomas Bettany, William Kitchen Parker

The Morphology of the Skull

George Thomas Bettany, William Kitchen Parker

The Morphology of the Skull

ISBN/EAN: 9783337307073

Printed in Europe, USA, Canada, Australia, Japan

Cover: Foto ©berggeist007 / pixelio.de

More available books at **www.hansebooks.com**

THE

MORPHOLOGY OF THE SKULL

BY

W. K. PARKER, F.R.S.,
HUNTERIAN PROFESSOR, ROYAL COLLEGE OF SURGEONS;

AND

G. T. BETTANY, M.A., B.Sc.,
SHUTTLEWORTH SCHOLAR, CAIUS COLLEGE, CAMBRIDGE; LECTURER
ON BOTANY IN GUY'S HOSPITAL MEDICAL SCHOOL.

London:
MACMILLAN AND CO.
1877.

[All Rights reserved.]

Cambridge:
PRINTED BY C. J. CLAY, M.A.
AT THE UNIVERSITY PRESS.

PREFACE.

A SKETCH of the history of the skull in the principal types of vertebrates is here for the first time presented. It has been attempted to narrate the facts by means of a consistent terminology, amplifying what Prof. Huxley has admirably developed; but the descriptions involve as few theoretical opinions as possible. The convenience of students has been considered throughout; summaries of nearly every stage, and of each Chapter from the second to the eighth, have been carefully drawn up. By the help of the index the history of individual bones or tracts can be examined comparatively.

Many points of interest have necessarily been omitted in a work treating of anatomy especially in a developmental aspect. It is not expected that the book will be thoroughly intelligible after mere reading. The limitations of space and of illustration will in some cases account for this; but frequently the complexity of the structures described is such that a brief description may easily lack clearness. The student should consult the very much larger series of figures to be found in original memoirs in the Philosophical Transactions and elsewhere[1];

[1] Fowl, Phil. Trans., 1869; Frog, 1871; Salmon, 1873; Pig, 1874; Axolotl, 1877 (in the press); Sharks and Rays, Zool. Trans., 1877. A memoir on the Snake is in preparation.

but personal dissection will in many cases be required for a full comprehension. It is hoped that the book may be found useful as a help to practical work. The types chosen are as accessible as any (the Newt or Salamander, which will do as well, being substituted for the Axolotl). The adult skull should first be mastered as far as possible; and then the earlier stages should be worked out both by ordinary dissection, and by transverse and longitudinal sections.

So far as interpretations are put forward, they are given merely as honest endeavours, not as final judgments. It has been our desire neither to exaggerate the value of what is known, nor to force facts to bear more than legitimate inferences. To assist many students to learn morphology for themselves is far more our object than to persuade them to accept our momentary ideas. Consequently we have not discussed the views of the great anatomists of the days before embryology had illuminated the dark problems of animal structure. Our omission to mention their names is due to no undervaluation of their vast labours and the treasures they have bequeathed to us.

One name, however, will remain ever connected with the insight which realised the bearing on the vertebrate skull of its developmental history. Prof. Huxley, since delivering his Croonian lecture (Proc. Roy. Soc. 1858), has never rested in his efforts to throw light upon this subject; and he has had a very great share in the researches and elucidations which have made this book possible. His papers in various transactions and journals[1],

---

[1] Proc. Roy. Soc.; Proc. Zool. Soc.; Jour. Geol. Soc.; Decades Geol. Survey; Jour. Anat. and Phys.

and his larger works[1], bear the most emphatic testimony to his labours. We express our cordial acknowledgments to Mr F. M. Balfour, M.A., Fellow and Lecturer of Trinity College, Cambridge, for reading many pages of proof-sheets, and especially for his kind help in reference to the first Chapter.

The illustrations are, with very few exceptions, reproductions, mediate or immediate, of Mr Parker's original drawings; some of them have appeared previously in other works. They have been drawn on wood by Mr T. P. Collings, and engraved by Mr J. D. Cooper.

In concluding our difficult task, we cannot but ask indulgence for the errors or imperfections which must necessarily be associated with the first production of a work on what may be called almost a new subject.

[1] *Elements of Comparative Anatomy: Anatomy of Vertebrated Animals.*

LONDON, *September,* 1877.

CONTENTS.

CHAPTER I.
PRELIMINARY EMBRYOLOGY . . . PAGE 1

CHAPTER II.
THE SKULLS OF THE DOGFISH AND THE SKATE . . 14

CHAPTER III.
THE SKULL OF THE SALMON . . 43

Appendix on the Skulls of Fishes 83

    Murænoids 83, Siluroids 84, Ganoids 84, Ceratodus 87, Lepidosiren 88, Chimæra 89, Elasmobranchs 89.

CHAPTER IV.
THE SKULL OF THE AXOLOTL . . 91

Appendix on the Skulls of Urodeles 129

    Proteus 129, Siren 130, Menopoma 131, Menobranchus 133, the Salamandrine Skull 134.

CHAPTER V.
THE SKULL OF THE COMMON FROG . . . 136

Appendix on the Skulls of Anura 176

    Rana pipiens 176, Pseudis 176, Bufo 177, Dactylethra 177, Pipa 181.

CHAPTER VI.

THE SKULL OF THE COMMON SNAKE 187

Appendix on the Skulls of Reptiles 213
    Chelone 213, Lizards 215, Crocodiles 217.

CHAPTER VII.

THE SKULL OF THE COMMON FOWL . 219

Appendix on the Skulls of Birds 262
    Struthionidæ 262, schizognathæ 262, desmognathæ 263, ægithognathæ 264, saurognathæ 265.

CHAPTER VIII.

THE SKULL OF THE PIG . 267

Appendix on the Skulls of Mammalia . . 301
The Human Skull . . . 304

CHAPTER IX.

THE MORPHOLOGY OF THE SKULL 310

The Cartilaginous Skull 310; the Sense Capsules 320; the Arches 326; the Cranial Nerves 332; plan and segmentation of the Cartilaginous Skull 336; the Osseous Skull 343; Evolution 358.

LIST OF ILLUSTRATIONS.

FIG.		PAGE
1.	Head of embryo Dogfish, 11 lines long	15
2.	Head of embryo Dogfish, 11 lines long, median longitudinal section	17
3.	Head of embryo Skate, 1⅓ in. long	20
4.	Head of embryo Dogfish, 1⅛ in. long	22
5.	Head of embryo Dogfish, second stage; basal view of cranium	24
6.	Head of embryo Dogfish, third stage; median longitudinal section	28
7.	Skull of adult Dogfish, side view	36
8.	Skull of Skate, nearly adult	40
9.	Embryo Salmon, about ¾ inch long; side view of head within chorion	44
10.	Embryo Salmon, about ¾ inch long; upper view of head, dissected, the neural tissue having been removed	45
11.	Embryo Salmon, about ¾ inch long; lower view of head, with the arches shining through	47
12.	Embryo Salmon, partly hatched; median longitudinal section of head	49
13.	Embryo Salmon, not long before hatching; under view of head, with arches seen through	50
14.	Embryo Salmon, not long before hatching; lower view of skull dissected, the branchial arches having been removed	52

LIST OF ILLUSTRATIONS.

FIG.		PAGE
15.	Salmon Fry, second week after hatching; upper view of skull dissected	54
16.	Salmon Fry, second week after hatching; transverse section of head, through forebrain and eyeballs	56
17.	Salmon Fry, second week after hatching; side view of skull	59
18.	Young Salmon of the first summer, about 2 inches long; side view of skull, excluding branchial arches	62
19.	Adult Salmon; lateral view of chondrocranium with its ectosteal bones	69
20.	Adult Salmon: median longitudinal section through skull, after removal of jaws and arches	70
21.	Adult Salmon: side view of skull with all bones attached	75
22.	Head of Axolotl, just after hatching, side view	95
23.	Larval Axolotl, about five lines long; upper view of skull, dissected	99
24.	Larval Axolotl, about five lines long; transverse vertical section of head, through eyeballs	100
25.	Larval Axolotl, three-quarters of an inch long; upper view of basis cranii and lower jaw, dissected	104
26.	Young Axolotl, 2¼ inches long; under view of skull, dissected, the lower jaw and gill-arches having been removed	109
27.	Young Axolotl, 2¼ inches long; upper view of skull; lower jaw removed	111
28.	Adult Axolotl; under view of skull, the lower jaw and arches being removed, and also the investing bones on the right side	115
29.	Axolotl, nearly adult; side view of skull	120
30.	Skull of Amblystoma, side view	124
31.	Tadpole four lines long, four or five days after hatching; side view of head	137
32.	Embryo Frog, just before hatching; side view of head, with skin removed	138
33.	Tadpole of Common Toad, one-third of an inch long; cranial and mandibular cartilages seen from above	141
34.	Tadpole about one inch long; view of face and cranial floor from above, the brain having been removed	144

LIST OF ILLUSTRATIONS.   xiii

FIG.		PAGE
35.	Tadpole one inch long; side view of skull dissected	146
36.	Tadpole with tail beginning to shrink; side view of skull without branchial arches	150
37.	Young Frog, with tail just absorbed; side view of skull	155
38.	Young Frog, near end of first summer; upper view of skull, with left mandible removed, and the right extended outwards	158
39.	Adult Frog; side view of auditory region, with semicircular canals and parts of middle ear displayed	161
40.	Adult Frog; upper view of skull with investing bones and lower jaw removed	163
41.	Adult Frog; median longitudinal section of skull, lower jaw removed	165
42.	Adult Frog; upper view of skull	166
43.	Adult Frog; under view of skull	168
44.	Adult (edible) Frog: under view of skull; investing bones removed on right side (Huxley)	169
45.	Adult Frog; side view of skull, dissected.	171
46.	Adult Frog; columella auris	172
47.	Adult Frog; hyobranchial plate	173
48.	Embryo Snake, 1¾ inch long; side view of head with facial arches seen through	190
49.	Embryo Snake, about 1¾ inch long; chondrocranium seen from above, the brain and jaws having been removed	191
50.	Embryo Snake, about 2½ inches long; side view of head, dissected	196
51.	Adult Snake; side view of skull, with jaws removed	206
52.	Adult Snake; under view of skull	208
53.	Adult Snake; skull seen from above.	210
54.	Embryo Chick of the fourth day of incubation: head viewed from below as an opaque object (Foster and Balfour)	220
55.	Embryo Chick, fifth day of incubation; view of cranial structures from above	221

LIST OF ILLUSTRATIONS.

FIG.		PAGE
56.	Embryo Chick, fifth day of incubation; head viewed from below, with skeletal parts seen through	222
57.	Embryo Chick, sixth day of incubation; head seen from below (Huxley)	224
58.	Embryo Chick, seventh day of incubation; side view of skull.	226
59.	Embryo Chick, middle of second week of incubation; under view of skull, with arches removed	232
60.	Embryo Chick, end of second week of incubation; posterior view of cranium	235
61.	Embryo Chick, end of second week of incubation; inner view of hinder part of cranium	236
62.	Embryo Chick, end of second week of incubation; upper view of skull, the brain and parostoses being removed . . .	238
63.	Chick two days old; median longitudinal section of skull, the brain being removed	242
64.	Chick two days old; external lateral view of skull . .	244
65.	Chick two days old; under view of skull, with lower jaw removed	246
66.	Fowl of first winter; median longitudinal section of skull .	250
67—71.	Views of nasal structures of Fowl of first winter . .	252
72.	Fowl several years old; side view of skull . . .	254
73.	Fowl several years old; under view of skull with lower jaw removed.	256
74.	Adult Fowl; side view of columella auris . . .	258
75.	Fowl several years old; hyobranchial apparatus from above .	259
76.	Embryo Pig, two-thirds of an inch long; side view of head and neck	268
77.	Embryo Pig, two-thirds of an inch long; elements of the skull seen somewhat diagrammatically from below . .	270
78.	Embryo Pig, two-thirds of an inch long; palatal view .	272
79.	Embryo Pig, an inch and a third long; median longitudinal section of head, with nasal septum removed . . .	281
80.	Embryo Pig, an inch and a third long; posterior view of a section through the basal region of the skull . .	284

LIST OF ILLUSTRATIONS.

FIG.		PAGE
81.	Embryo Pig, an inch and a third long; side view of mandibular and hyoid arches	285
82.	Embryo Pig, 2½ inches long; under view of skull with lower jaw removed	287
83.	Embryo Pig, 2½ inches long; vertical section of head, showing structures between and beneath the orbits	290
84.	Embryo Pig, six inches long; outer view of occipital and auditory regions	293
85.	The Pig at birth; outer view of auditory capsule, &c.	296
86.	Auditory chain of bones	ib.

THE MORPHOLOGY OF THE SKULL.

CHAPTER I.

PRELIMINARY EMBRYOLOGY.

1. IN the study of the morphology of the skull, it is necessary to take into account all the forms which the skull assumes, from its origin in the embryo to its adult condition. Inasmuch as skeletal elements are apparent in the head at a very early period of development, no student will thoroughly apprehend their nature without some knowledge of the processes by which the body of a vertebrated animal is evolved from its germ. The following summary is intended to refresh the memory of those who have studied embryology practically, or by reading special treatises, and also for the information of others who may not have time for such study.

2. The fertilised and developing ovum of all vertebrates contains a membrane called the *blastoderm* or germinal membrane, which is the rudiment of the future animal. It is produced by a process of segmentation, varying greatly in details, affecting the whole or part of the primary ovum.

3. The blastoderm, at an early stage after its formation, consists of three layers of cells. The upper or external layer is called the *epiblast*; the middle layer, the *mesoblast*; and the lower or inner, the *hypoblast*. The epiblast gives rise (1) to all the epidermis and epidermic appendages of the body, (2) to the epithelial lining of

the cerebro-spinal canal and its derivatives, the ventricles of the brain, (3) to the cerebro-spinal nervous centres, and (4) to various parts of the organs of special sense. The hypoblast is the source of the epithelium of the alimentary canal and all the organs developed as diverticula from it. From the mesoblast the remaining parts of the body are derived, including muscles, bones, connective tissue, and blood-vessels; the generative and urinary organs; the dermis, and the mucous membranes of the alimentary canal and of the organs connected therewith.

4. To produce this diversity of parts, the layers of the blastoderm become thickened in various regions; folds and pits arise in many situations, and form pouches, sacs, tubes, and processes; the mesoblast especially splits into distinct strata. At the same time the cells composing the layers gradually alter their character as they multiply, and, instead of being very much alike, become more and more unlike, and specialised for the functions they have to perform in the growing organism. In most cases the blastoderm sooner or later grows round the mass of nutritive material constituting the yelk, and encloses it[1].

5. A portion of the blastoderm is directly developed into the embryo: among the functions of the remainder is the absorption and transmission of nutriment to the embryo. The locality of the latter is early marked out by the formation of a trench bounding an oval region in the blastoderm. The part enclosed becomes thenceforth the embryo; and the demarcation is made more definite by the trench, as it grows, continually undermining the enclosed parts. As this trench extends inwards, the embryo rides upon the rest of the blastoderm, now called the umbilical vesicle, and its enclosed yelk. The

---

[1] This description will not hold good for Mammals, where there is no proper yelk; but the main facts, independently of those due to the existence of the yelk, are true for them. It is not intended to include the development of Amphioxus in this general view. The account in § 6 requires modification in its application to Osseous Fishes.

head of the creature is from the first distinguishable as being included within the earliest developed part of the trench, at one end of the oval. Afterwards, certain membranous appendages, the amnion and the allantois, arise in the development of several classes of vertebrates: for their description the reader is referred to general embryological works.

6. We have described the formation of a simple longitudinally tubular embryo by gradual constriction of a portion of the blastoderm from the remainder. Coincident with this process is another, by which a second longitudinal tube is formed parallel with and above the first. This arises by the growth of two parallel folds from the upper surface of the embryo along its axis, having a shallow longitudinal groove between them. These *medullary folds*, after growing to a certain height, arch over towards each other and coalesce, for a greater or less distance, so as to form a tube open at both ends; it is the neural tube, in connection with which the cerebro-spinal nervous system takes its rise. The anterior and posterior openings of the tube become closed at a variable period after its formation.

7. The neural tube is necessarily lined with epiblast, since it originates on the upper or epiblastic surface of the embryo. When it is closed, there is an external epiblastic surface as before, and an internal epiblast lining the tube. Between these two strata the growing mesoblast gradually penetrates until it has completely separated the outer from the inner epiblastic layer. The neural canal marks the dorsal region of the embryo; while the parts beneath it belong to the ventral aspect of the body.

8. The original embryonic cavity, or space constricted off by the trench as it grows inwards, is converted into a double tube by a process of splitting, which divides the mesoblast into two layers. The inner of these layers, where it comes into relation with the hypoblast, forms with it a membrane called the

splanchnopleure, or visceral wall: while the outer layer of mesoblast with the adherent epiblast is known as the *somatopleure,* or body-wall. The splanchnopleure grows inwards on the ventral side of the embryo more rapidly than the somatopleure; and it early forms a very distinct alimentary tube, shut off from the umbilical vesicle. A space is thus left all round the digestive canal between the somatopleure and the splanchnopleure: this is the pleuroperitoneal cavity. The somatopleure of either side, growing towards the middle ventral line at a slower rate, meets its fellow eventually and coalesces with it to form the ventral wall.

9. The main organs of the body which surround the alimentary canal are developed, (1) as pouches or tubular protrusions from the wall of the digestive tube in different situations: these tubes becoming branched, and surrounded by differentiated mesoblast, gradually assume characteristic forms, and fill up the larger portion of the original pleuroperitoneal space, which however everywhere surrounds them; the lungs, liver, and pancreas are good representatives of this class; of course all these organs contain hypoblast; (2) as thickenings and developments of the mesoblast, independently of the hypoblast or epiblast; such are the Wolffian bodies, kidneys, and sexual organs.

10. A structure of great importance arises early, along the axis of the embryo, between the neural and the alimentary tube. One or other of the principal germinal layers[1] gives origin to a median longitudinal string of cells, forming an axial rod beneath the floor of the neural tube. Anteriorly this rod, called the *notochord,* does not extend quite so far forwards as does the neural canal. The formation of the limbs, as paired outgrowths from lateral ridges on the sides of the embryo, must not engage our attention here.

---

[1] The controversy in which the subject of the origin of the notochord is at present involved renders this indecision necessary.

11. The neural tube and the rudiments of the nervous system in the anterior region of the body very early manifest differentiation from the corresponding parts behind them. The cephalic end of the tube is dilated in a pyriform manner: this dilatation increases in all dimensions, and its walls show a series of constrictions. Ultimately, out of the fore end of the tube, three cerebral vesicles are formed, one behind another, having their cavities freely continuous with the rest of the tube behind, which is thenceforth distinguished as the spinal canal. The notochord, underlying the neural tube, stops short of its cephalic end, terminating underneath the middle or the hinder cerebral vesicle. The nervous tissue becoming differentiated in the walls of these vesicles constitutes three main divisions of the brain, viz. the fore, the mid, and the hindbrain. The mesoblast ultimately encircles the cranial cavity as it does the spinal canal.

12. As the cerebral vesicles enlarge in all dimensions, their increase is not followed by any high development of the anterior part of the lower or visceral tube. Consequently, at an early period the head consists mainly of the large neural protuberances, and these bend downwards around the anterior end of the notochord and of the alimentary tube. This flexure, taking place about a point beneath the middle or hinder part of the middle cerebral vesicle, is called the *mesocephalic* (less precisely the *cranial*) *flexure*. The first cerebral vesicle comes to occupy an infero-anterior, the second a supero-anterior position. In addition to the definite mesocephalic flexure, there is in many cases a considerable ventral curvature of the body, and especially of the neck. One of the most remarkable things in vertebrate development is the way in which the cranial axis ultimately becomes once more approximately straight; while the ventral appendages of the skull acquire a high degree of development.

13. The first cerebral vesicle enlarges transversely, and the lateral protuberances after a time become con-

stricted from the axial region. Thus arise stalked *optic vesicles*, which at a later period, being met by involuted epiblast from the exterior, are pitted so that the external wall is opposed to the inner concave wall, just as part of a hollow indiarubber ball may be pushed in till it meets the opposite surface of the ball. The stalks of the optic vesicles are subsequently known as the optic nerves. These nerves soon appear to proceed from the under part of the brain; for the constriction by which the optic vesicles are distinguished is much deeper above than below, and consequently the stalk is carried downwards towards the base of the brain.

14. The first cerebral vesicle retains its distinctive appellation of forebrain, and gives origin to the cerebral hemispheres, which are not constricted from the axis by a stalk, and which in higher vertebrates achieve a preponderance over all other cranial structures. The hemispheres themselves bud out anteriorly into smaller bulbs constituting the pair of olfactory lobes, from which the olfactory nerves are subsequently derived. The remains of the primary cerebral vesicle form what is known as the vesicle of the third ventricle, the lateral cavities or ventricles of the hemispheres having been originally considered in cerebral topography as the first two ventricles. From the centre of the floor of the third ventricle, which usually becomes the hindmost of the elements produced from the original first vesicle, arises a funnel-shaped process, the *infundibulum*, which is extended downwards to join the pituitary body situated at the extreme fore end of the alimentary canal.

15. The remaining principal parts of the brain, as related to the primary vesicles, may be briefly mentioned. The middle vesicle gives rise to the optic lobes (corpora quadrigemina of mammals) and to the crura cerebri where they are distinct; its cavity in higher forms becomes reduced to a narrow channel between the third and the fourth ventricles, or the primary cavities of the fore and the hindbrain. In the latter, the roof developes into the cerebellum in front, the floor behind forms the medulla oblongata.

16. The chief cranial nerves other than the olfactory and optic, which have already been referred to, arise on either side of the floor of the hindbrain, and in some cases certainly by outgrowths from it. There are four principal masses, two of which are near the anterior end of the notochord, two near the posterior limit of the head. The first is the rudiment of the trigeminal nerve; it early becomes forked peripherally, one branch passing forwards to the eyeball, the other downwards to the mandible. The second becomes the facial nerve; the third and fourth, the glossopharyngeal and vagus (or pneumogastric). All these nerves, at or near their cerebral termination, possess special ganglia.

17. Three pairs of pouch-like ingrowths of epiblast originate on the outside of the primary membranous cranial cavity, and contribute to form organs of special sense. The mesoblast becomes distinctively aggregated around these pouches, and more or less completely cuts off communication between the pouch and the external epiblast.

18. The anterior of these organs are the nasal sacs, which are always related at first to the forebrain, usually lying under it. Their involution is distinguished by the prominent rim bounding it. This border is in certain cases deficient in its lower part, and when the process which bounds the outer side of the mouth grows over it to unite with the fore lip, this deficiency is converted into a short canal opening on the roof of the mouth; while the orifice of the original involution becomes definitely located on the face. Thus two narial orifices are formed, anterior and posterior, or superior and inferior. The inferior openings may sometimes be carried backwards in the mouth by supplementary plates developed beneath them, closing their access to the anterior part of the mouth, and only allowing communication posteriorly.

19. In its later development, each nasal pouch becomes a cavity lined by a sensory membrane, part of which receives special nerves from the olfactory lobes of the brain. Mostly, the nasal cavities are elaborated in a labyrinthic manner; that is, the surface of the lining membrane is extended by the projection of complex folds into the cavity; and these are supported by corresponding growths of hard parts. The latter are derived from or are continuous with the cranial skeleton, and play a great part in the morphology of the skull.

20. The second pair of involutions, the optic sacs, are usually much larger than the nasal. Their primary position is above and behind the olfactory sacs, at the sides of the midbrain; but this location is frequently modified afterwards, and they become placed at the sides of and often below the forebrain—almost entirely anterior to it in some cases. The optic pouches of epiblast grow inwards until they indent the optic vesicles of the brain, and, pushing their outer part inwards till it comes in contact with their inner surface, give rise to the cup which subsequently constitutes the retina. The epiblastic mass gets completely cut off from the exterior, and forms the crystalline lens. Mesoblastic growths form all the other structures of the eyeballs. The shape of the eye is mainly determined by a strong fibrous membrane, the *sclerotic*, which often becomes cartilaginous, and even acquires ossifications, called sclerotic plates; but inasmuch as the eyeballs must be mobile, the sclerotic plates never enter into combination with the cranial skeleton, and will scarcely receive any mention in the following pages.

21. A cavity called the orbit, open externally and bounded by skeletal elements, is in most cases formed, to lodge and protect the mobile eyeballs. The study of the orbit is of high interest in cranial morphology; but it is rendered comparatively simple by the fact that the sense-organ, with its special skeletal investment, lies complete within the orbit; there is no interlacement of parts or complication of regions. A primary cleft arises between

the optic involution and the maxillopalatine process which bounds the mouth. This opens on the roof of the mouth where no secondary palate is developed, and into the nasal passage where there is such a palate. It is variously called the lachrymal or orbitonasal canal.

22. The third pair of special sense organs, the ear-sacs, begin as large involutions of epiblast at the sides of the hindbrain. The epiblastic pouch speedily becomes hollow, and the cavity early expands to large dimensions, and gives off processes which constitute the auditory labyrinth. Mesoblastic tissue engirths these processes, and forms a hard investment to the whole organ. At first this investment is usually distinct from the proper cranium; but they soon coalesce, and the regions of both become sometimes almost inextricably intertwined.

23. The main part of the cavity of the auditory labyrinth is denominated the vestibule; it has an upward and backward process, the aqueductus vestibuli, towards the situation of the primary involution and the long persistent membranous space or fenestra in that position. Hollow processes arise from the vestibule, anteriorly, posteriorly, and externally; and as they grow, mesoblast penetrates into their concavity, leaving the periphery pervious as a tube, which forms a curve sometimes bearing a close resemblance to a semicircle; hence these three tubes are called *semicircular canals* of the ear. They are present in the same relative situations in all but the lowest vertebrates. One end of each tube dilates where it opens into the vestibule, forming an *ampulla*. Another process of the vestibule, passing forwards and downwards in the ear-mass, is developed into the simple conical *cochlea* of birds, and the spirally-twisted cochlea of mammals.

24. In almost all vertebrates the mesoblastic investment of the otic capsule becomes at least cartilaginous, very often bony: but the cartilage is scarcely ever complete on all aspects. It may be incomplete internally on the cranial side; or fenestræ may be formed in it

externally, where the labyrinth abuts on certain secondary structures found in higher vertebrates.

25. Another organ of special sense, that of taste, is formed by epiblastic involution in the head. It does not present a manifest paired aspect, its involution being part of that of the mouth-cavity or alimentary vestibule, which discharges many functions besides that of taste. The oral involution early takes place in the angle between the downbent cranial vesicle and the fore part of the alimentary tube. The mesoblast outside the latter grows forward ventrally and laterally, surrounding a hollow on the inferior surface of the head, into which the copious ingrowing epiblast is received. The hinder limit of this involution is somewhere about the level of the fore end of the hindbrain. It is at a comparatively late period that a slit places the alimentary canal in communication with the oral cavity.

26. At about the point of junction of these two cavities, a diverticulum of epiblast is given off, passing upwards into the floor of the cranium: this forms the *pituitary body* already spoken of (§ 14, p. 6). The stalk of connection with the cavity beneath is obliterated, and the pituitary body always occupies a definite position in the cranial floor in front of the end of the notochord.

27. The mouth, throat, and neck are parts little differentiated in the embryo when already the neural vesicles have attained a high development. Subsequently they increase very greatly in size, growing both forwards and backwards. The mesoblast does not split at its extreme anterior end to form a visceral wall within the body-wall. A transverse section beneath the fore end of the notochord shows a simple tubular cavity, which is the commencement of the digestive canal. In this region, as it grows more extensive, a series of thickenings arises on each side in a vertical plane, constituting what are called *visceral folds*. In general they are curved somewhat forwards towards the middle line below, and the foremost of

them grows especially forwards to envelope the oral involution of epiblast.

28. The greatest number of these folds which occurs in vertebrates is nine; frequently no more than six appear. The first pair of folds receives the name of *mandibular;* the second, that of *hyoid;* the remainder are named *branchial.* The valleys between the folds usually become perforated in more or less of their extent; thus there is established a series of *visceral clefts,* which may remain open throughout life, affording a channel of communication between the cavity of the throat and the exterior; or some or all of the clefts may become closed at a later stage of development, and so remain during the life of the animal.

29. The boundaries of the mouth are primarily constituted, (1) anteriorly, by the *nasofrontal process* which is the termination of the investment of the head beneath the forebrain, and between the olfactory sacs; (2) superiorly, by the floor of the cranial vesicles, forming the primary *palate:* (3) laterally and somewhat superiorly, by the *maxillopalatine processes* which grow forwards on each side along the line of the primary palate from the upper end of the mandibular folds; and (4) inferiorly by these mandibular folds.

30. The visceral folds, and especially those named branchial, become the basis for the development of gills in varied forms, which may persist throughout life, or only during a portion of embryonic existence. In the higher forms, where no gills are developed, the appearance of the branchial folds is to a large extent evanescent, though not less certain and definite than in the other cases.

31. A brief reference to the vascular system will close this general account, and enable us to proceed to the history of the skeletal elements. The heart is formed in the mesoblast at the anterior region of the pleuroperitoneal cavity. A tube proceeds from its fore end in the mid-ventral mesoblast: this is the *bulbus aortæ.* It

passes along the ventral line into the neighbourhood of the visceral folds, and gives off a series of branches, which surround the fore end of the alimentary canal, one passing outwards and upwards along each visceral fold to gain the dorsal side of the throat. These channels or *aortic arches* unite to form a pair of tubes corresponding to that from which they originated ventrally; and these dorsal tubes pass backwards right and left of the notochord at some distance from it, uniting sooner or later to constitute the primitive dorsal aorta.

32. The aortic arches never quite maintain their primitive arrangement in adult life. They most nearly retain it in the permanently branchiate forms. But one or two of the anterior arches always become disconnected with the dorsal tube, and by means of secondary ramifications supply the head and brain with blood. One pair of these vessels, the internal carotid arteries, will be mentioned in succeeding pages, as passing into the cranial cavity in a definite position. Other portions of the aortic arches are differentiated in abranchiate conditions or forms of higher vertebrates, and furnish blood-vessels to internal respiratory organs: and finally but two pairs, or one pair, or one aortic arch, may be left in the adult in continuity with the dorsal aorta.

33. The skeletal elements with which we have to do arise entirely as developments of the mesoblast. Very early the mesoblast at the sides of the neural tube behind the head is divided by transverse partitions into a longitudinal series of more or less definite segments, the *protovertebræ*. These extend backwards as far as the end of the neural canal, but stop short anteriorly at the level of the posterior end of the third cerebral vesicle. The slightly differentiated mesoblast grows upwards so as to surround the neural canal, and also invests the notochord.

34. When chondrification takes place, a continuous cartilaginous investment surrounds the notochord. This cartilage is ultimately segmented into vertebral pieces, so

that the lines between adjacent vertebræ are intermediate between the dividing lines of the protovertebral segments; thus each vertebral body corresponds with the contiguous parts of two protovertebræ. Lateral vertebral cartilages become gradually extended round the spinal canal on each side so as to form neural arches, which at a later period coalesce on the dorsal aspect of the canal.

35. By the time that the vertebra is complete in cartilage the notochord usually begins to degenerate, especially where it is encircled by the vertebral bodies or centra. In these situations it may become perfectly obliterated. Very frequently remnants of the notochord persist in the intervertebral spaces.

36. Ossification commences in the vertebræ in the region of the centrum immediately surrounding the notochord. Other centres of ossification arise, one on either side of the neural canal. Where ribs are developed they grow outwards from the region of the vertebral centra. Some of the ribs may extend completely round the body-wall so that the corresponding pairs meet ventrally; and a junction is effected between several succeeding medio-ventral elements, to form a cartilaginous sternum. The costal arches ossify by one or two centres on each side; the sternum often by several paired centres.

37. No protovertebræ are formed by the side of the notochord in the cranial region. Much of the mesoblast investing the brain and constituting the visceral folds and other processes belonging to the head, undergoes a gradual but early transformation into cartilage, forming definite skeletal tracts before intercellular substance has appeared. It is the object of the following pages to trace in several types of vertebrates the rise and history of these cartilaginous tracts and of the bones developed within or in proximity to them.

CHAPTER II.

THE SKULLS OF THE DOGFISH AND THE SKATE.

38. THE eggs of Elasmobranch fishes, of which the principal are known as Sharks and Rays, when deposited, are enclosed in a strong horny capsule or "purse," secreted from the oviduct. This capsule is pillow-shaped, the corners being pointed in the Rays, and produced into long tendril-like processes in the Shark and Dogfish. The embryo remains enclosed in the purse until about six months after oviposition, and during this period the most important metamorphoses take place.

First Stage.

39. In embryos of the lesser Spotted Dogfish (*Scyllium canicula*), from eight lines to an inch in length[1], the head and the branchial region are proportionately large and conspicuous, and external gills are present; the body is slender, tapering to a long thread-like tail. The embryo is extremely active, and has attached to it a yelk-sac of about three-quarters of an inch in diameter.

40. The neural tube in the head is completely closed, but the covering of the third or posterior cerebral vesicle is very thin. The latter vesicle (Figs. 1 and 2, $C\,3$) lies directly above the anterior end of the notochord: the second vesicle ($C\,2$) is immediately in front of the notochord, forming the foremost rounded part of the head; while the first vesicle ($C\,1$) is beneath the second, and is

---

[1] These embryos correspond in most respects with stages M and N, described and figured by Mr Balfour (*Journ. Anat.* Vol. x. p. 563, and Plate xxv.).

totally below the level of the notochord as well as anterior to it. Thus the mesocephalic or cranial flexure is already fully established.

41. Each sense capsule is seen in a very rudimentary condition as an infolding of the epiblast of the side of the head. These infoldings are at present subequal in size. The nasal sacs (Fig. 1, *Na*) are situated upon the inferolateral surface of the head, external to the first cerebral vesicle. The young eyeballs (*E*) are almost vertically over the nasal sacs: above each eyeball is a noteworthy elevation of the skin, the *supraorbital band*. The ear-sacs (*Au*) are on a higher level and more posterior, flanking the hinder part of the third cerebral vesicle.

Fig. 1.

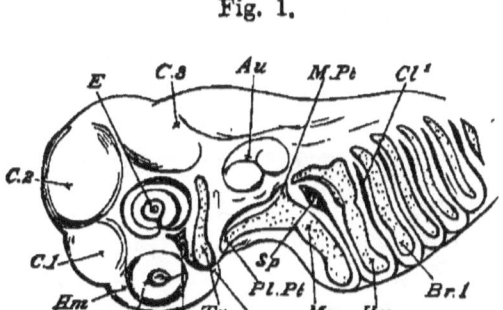

Head of embryo Dogfish, 11 lines long. *Tr.* trabecula; *Pl. Pt*, palatopterygoid; *M.Pt*, metapterygoid region; *Mn*, mandibular cartilage; *Hy*, hyoid arch; *Br.*1, first branchial arch; *Sp*, mandibulo-hyoid cleft; *Cl*1, first branchial cleft; *Lch*, so-called lachrymal cleft; *Na*, olfactory rudiment; *E*, eyeball; *Au*, auditory mass; *C* 1, 2, 3, cerebral vesicles; *Hm*, hemispheres; *f.n.p*, nasofrontal process.

42. On the under surface of the head is a large square mouth, bounded in front by the median antero-inferior projection of the head, called the *nasofrontal process* (*f. n. p.*), lying between the nasal sacs; behind, by the rudimentary lower jaw (*Mn*); and laterally by a process passing forward from the upper end of each half of the mandible (*Pl. Pt*). Behind the mouth on each side are seen six clefts (*Sp., Cl.*) curving downwards from the neural towards the hæmal region of the neck roughly

parallel to one another. The clefts on each side correspond, but they do not meet below.

43. Thick and somewhat prominent ridges intervene between the visceral clefts; and from the surface of each ridge or arch there arises a series of long filamentous external gills. The heart is plainly to be seen through the transparent skin in the middle line beneath, between and below the posterior arches just described. Behind the heart is the umbilicus, and on either side of the umbilical region the rudiment of a pectoral fin projects.

44. The notochord in the cephalic region (Fig. 2, *n.c.*) lies beneath the neural axis, and is curved somewhat downwards[1], at the same time slightly tapering. It does not extend so far forwards as the third cerebral vesicle does, but ends above the middle of the mouth. In embryos of the Dogfish at this stage and also in Pristiurus embryos the anterior end of the notochord has been seen submoniliform in outline, presenting from five to seven beadings in the distance between its anterior extremity and the middle of the auditory region.

45. The pituitary body (*py*), lying behind the first or inferior cerebral vesicle, and connected with it by the infundibular process, is closely in front of and a little below the notochord; it is the lower boundary of a space formed by the curvature of the neural axis, resembling the concavity of the hook of a crozier. This space is filled with delicate gelatinous tissue; it is the transitory "middle trabecula" of Rathke.

46. Flat bars of nascent cartilage, the *parachordals*[2], are found in the hinder part of the cranial floor on either side of the notochord, extending backwards into the neck for three or four times the length of their intracranial portion, without showing any vertebral segmentation. The inner edge of each parachordal is grooved to embrace

---

[1] This downward curvature is considerably greater at an earlier stage; see Balfour, *Journ. Anat.* Vol. x. Pl. xxiv. *G*, *H*, and *I*.

[2] Fig. 5, p. 24, which shows these structures in the next stage, may be consulted for many points in this description.

the notochord partially; but they are not identified with its sheath, which is already cartilaginous. The parachordals extend further forwards internally, where they reach the middle of the beaded region of the notochord; externally, they pass nearly to the fore end of the auditory masses. In front they abut unconformably, at an angle of about 120°, upon the hinder part of the trabeculæ. The parachordals are crescentically scooped and bevelled where they lie between the auditory capsules, and the latter rest upon the bevelled edge: they are wider behind this region, having a straight external edge.

Fig. 2.

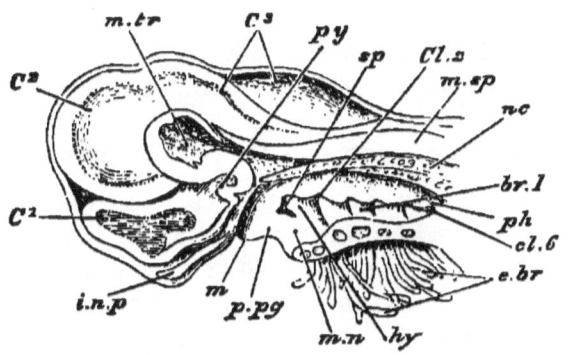

Head of embryo Dogfish, 11 lines long, median longitudinal section. C 1, 2, 3, cerebral vesicles; *nc.* notochord; *py.* pituitary body; *m.tr.* middle trabecula; *i.n.p.* internasal plate; *m.sp.* medulla spinalis; *m.* mouth; *ph.* pharynx; *sp.* spiracle; *cl.* visceral clefts; *mn.* mandibular, *hy.* hyoid, *br.* 1, first branchial, arches; *e.br.* external branchiæ.

47. Before the parachordals can be plainly made out, a pair of broad flat bars is manifest in the hinder part of the down-turned cranial floor, behind and below the first two brain-vesicles, and partly below the third. Their plane is directed forwards and downwards from the middle of the base of the third vesicle to the middle of the first. These *trabeculæ* (Fig. 1, *tr*) embrace the pituitary body and infundibulum by an arcuate inner margin, so as to leave between the two bars an oval space equal in width to themselves. In front they very nearly meet behind the nasal sacs, where they project bluntly towards one another. From this inner point their boundary sweeps

crescentically outwards, and then again bends inwards in front of the auditory capsules. The trabeculæ are expanded where they abut against these capsules, wedging in for a little distance between them and the parachordals. Posteriorly the trabeculæ lie on the fore end of the parachordals, forming a slight elevation directed backwards, behind the pituitary region: they closely embrace the fore part of the beaded portion of the notochord.

48. In front of the trabeculæ and between the nasal sacs there is a median styloid tract of condensed tissue, broad behind, filling up the angle where the trabeculæ almost meet, and then growing forwards as a pointed rod between the nasal sacs and beneath the fore-brain. Before the end of this stage the ear-sacs begin to acquire a cartilaginous covering: but excepting in the regions which have been described, the boundaries of the cranial cavity are entirely membranous and very diaphanous.

49. Each visceral arch has a cartilaginous axis, forming a more or less arcuate rod (Fig. 1, *mn, hy, br*); and these cartilages will in future be referred to as the mandibular, hyoid, and branchial bars or arches. The upper ends of the branchial bars point directly inwards: but each of the two anterior arches sends forwards from its upper extremity a considerable flat process. The process from the mandibular arch is so large as to be really a continuation of it; and it occupies the thick ridge which forms the lateral boundary of the mouth on either side: it is the *palato-quadrate* plate.

50. There is very little upward expansion of the mandibular arch where it bends forwards; a little later, ligamentous fibres connect the region of the bend with the side wall of the cranium immediately in front of the auditory capsule and below the exit of the trigeminal nerve. The palato-quadrate plate of either side grows towards its fellow, and ultimately meets it below the level of the eyes and behind the nasal sacs: the two become connected by ligament, and a deep groove is left between the bars and the nasal sacs. The anterior process of the

hyoid arch applies itself along the lower side of the auditory capsule posteriorly. The branchial arches remain simple throughout this period.

51. During the growth of the processes of the first two arches, the upper part of the cleft between them becomes much wider, and is finally almost triangular, with the base above. The succeeding clefts also become wider above than below. The posterior edge of each arch developes a valance-like ridge or fold tending to cover the cleft behind it.

The subsequent growth of these folds closes in the clefts very much, leaving only the adult branchial slits. Such folds entirely close most of the visceral clefts in abranchiate vertebrates. The fold on the hyoid arch is homologous with the opercular fold in osseous fishes.

52. In the youngest embryos at this stage a series of rounded papillæ, the rudiments of external gills, are found on all the visceral arches except the last. Later on, these papillæ develope into long filaments, each containing a single capillary loop. About ten of these spring from the hyoid and each of the branchial arches; the mandibular has four, much shorter than the others, arising from its upper posterior edge in front of the first cleft. More internally on the posterior edge of each arch bearing external gills, is a series of cog-like projections, the rudiments of the internal gills: they form double series on the first four branchial arches.

53. In embryos of a Skate (*Raia maculata*), about an inch and a third in length, (Fig. 3), taken from the purse seven weeks after oviposition, development has not proceeded farther than in the Dogfish already described. The extra length is due to the tail: in other respects it does not differ markedly from the larval Dogfish. In the whole cranial region there is no difference which requires special notice.

54. The mandibulo-hyoid cleft (*sp*) early fills up below, and is pear-shaped above. The first branchial

cleft (between the hyoid and the first branchial arch) is similar to the preceding in its upper region, but retains the lower slit-like part. All the arches except the last branchial bear long spatulate external branchial filaments; there are six or seven to each arch except the mandibular, which has only four. In an earlier stage the rudiments of the branchiæ are seen as very numerous minute buds arranged in series. Certain of these shoot out to become the external transitory gill-filaments; the others afterwards develope into the permanent gills.

Fig. 3.

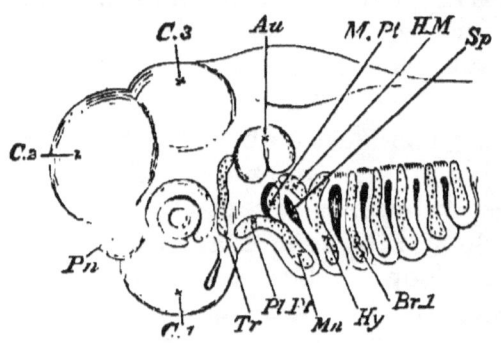

Head of embryo Skate, 1¼ in. long. *Tr.* trabecula; *pl. pt., mn.* palato-mandibular arch; *m. pt.* metapterygoid cartilage; *h. m.* hyomandibular; *hy.* remainder of hyoid arch; *br.* 1. first branchial arch; *pn.* pineal gland; *sp.* mandibulo-hyoid cleft or spiracle.

55. The mandibular and hyoid arches are more markedly bifurcate above than in the Dogfish, the anterior forks, however, being much the larger. The palato-quadrate region (*pl. pt.*) has the same relations as in the Dogfish: a separate cartilage (*m. pt.*) arises in the hinder fork of the arch, in front of the first cleft. It is placed in front of and below the auditory capsule, and is the rudiment of the spiracular cartilage.

56. Behind the first cleft, the anterior process of the hyoid arch is developed as a separate cartilage, lying below the auditory mass: it is the *hyomandibular* element (*h. m.*). The hinder fork of the arch is loosely connected with the ear-sac behind, and it grows inwards towards

the axial parts in a pointed manner; it diverges upwards and backwards from the ventral part of the arch (*hy*), almost as much as the palato-quadrate diverges forwards from the mandible proper.

57. The branchial arches (*br*) in the greater portion of their extent, laterally and below, are simple unsegmented arcs; but a separate little curved *pharyngobranchial* element early developes in each, above and mesiad of the main bar, in the pharyngeal roof.

58. The primary basal elements of the skull are already clearly manifest in these young embryos, the parachordals forming the base of the hinder cranial region, and the trabeculæ of its anterior portion. The latter relation is somewhat masked by the mesocephalic flexure; the trabeculæ appearing to be behind rather than below some parts of the brain. But they lie truly in the cranial floor, as is very evident after the flexure has passed away. The mandibular arch is seen in these types more distinct from other parts than in many vertebrate forms: its shape is already specialised almost to the adult condition. The origin of the hyomandibular element separately in the Skate is noteworthy as compared with its segmentation from the main hyoid bar elsewhere. In other respects these fishes show to great perfection the separate elements of which the skull is composed.

Second Stage.

59. In embryos of the Dogfish from 12 to 15 lines in length the external gills are two or three times as long as in the preceding stage; the four filaments on the mandibular arch are only one-third the length of those on the other arches. The mesocephalic flexure remains, although rapidly diminishing, and the "middle trabecula" is not absorbed; the brain has become much more complex.

60. The trabeculæ (Figs. 4 and 5, *tr.*) are now fully chondrified, but they are still distinct from one another. They are relatively longer, and the intertrabecular space is

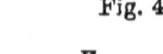

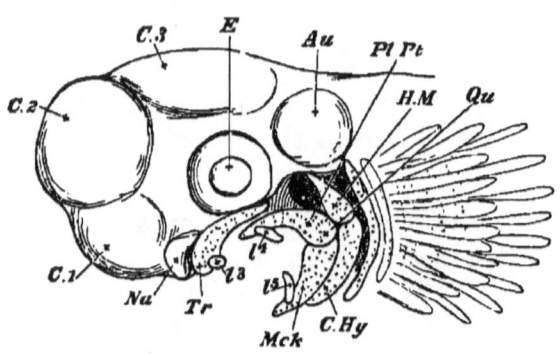

Head of embryo Dogfish, 1¼ in. long. *Tr.* trabecula; *pl. pt, qu.* palato-quadrate bar; *mck.* meckelian cartilage; *h.m.* hyomandibular; *c.hy.* ceratohyal; *l* 3, 4, 5, labial cartilages; the branchial arches and external gills are indicated behind the hyoid arch.

oblong, with rounded angles. Externally the trabecular edge is straightened, and so also is the anterior margin, which extends furthest forwards in the middle line. Postero-externally a crest rises from the hinder half of the trabeculæ in the cranial wall, nearly reaching to the auditory mass, and arching back over the exit of the trigeminal nerve. The hinder part of each trabecula has coalesced with the corresponding parachordal (*pa. ch.*); the trabecula at the junction forms a low posterior clinoid wall, which is not continuous across the notochord, but which reaches to the auditory capsule and fuses with it by its thick outer extremity. The angle where the trabeculæ and parachordals unite is fast lessening, and they will soon lie in one plane as a continuous basicranial bar on each side.

61. The parachordals (Fig. 5, *pa. ch.*) are well chondrified, as also is the distinct tubular tract surrounding the notochord; the parachordals have not united. The apex of the notochord is flat, the beaded appearance being

lost; it is wedged in between the two halves of the clinoid wall just described, and the hinder face of the pituitary body. The auditory capsules (*au.*) are chondrified; they are elongated oval in shape, showing enlargements over the course of the semicircular canals. Most parts of the roof and walls of the cranium are still membranous.

62. The internasal element is not fully chondrified; it is in the same condition as the nasal roofs, but distinct from them. Where it lies between the nasal sacs its relations are substantially as before; but anteriorly it has sent forward three prenasal lobes, a median rounded piece projecting forwards as the *prenasal* element (*pn.*), and a lateral pair of curved bifoliate lobes growing round and applying themselves to the antero-internal face of the nasal sacs below: these are the *cornua trabeculæ* (*c. tr.*).

63. The palato-mandibular arch has undergone transverse segmentation just at its greatest bend: so that the arch now consists of an upper and anterior portion answering to the *palatopterygoid* and *quadrate* regions of other vertebrates (Fig. 4, *pl. pt.*, *qu.*), and of a lower and hinder portion, the meckelian or mandibular cartilage. The metapterygoid ligament is further developed, but retains its primitive relations; while a stronger ligamentous union arises, below the remains of the first cleft (the *spiracle*), between the posterior edges of the quadrate and articular regions of the mandible, and the front edge of the hyomandibular and ceratohyal (see following paragraphs); the upper portion of this ligament may be called *symplectic*, the lower mandibulo-hyoid. In this way by the weakness of the direct attachment of the jaws to the cranium, and their suspension from the hyomandibulars, their great mobility is provided for.

64. The cartilages of the upper and lower jaw just mentioned are simple finger-shaped cartilages, those of each pair becoming rounded towards the middle line, approaching one another, and being united by ligament. The palato-quadrate is the larger; it is at first directed

forwards, then somewhat inwards, and finally almost abruptly inwards so as to become nearly transverse where

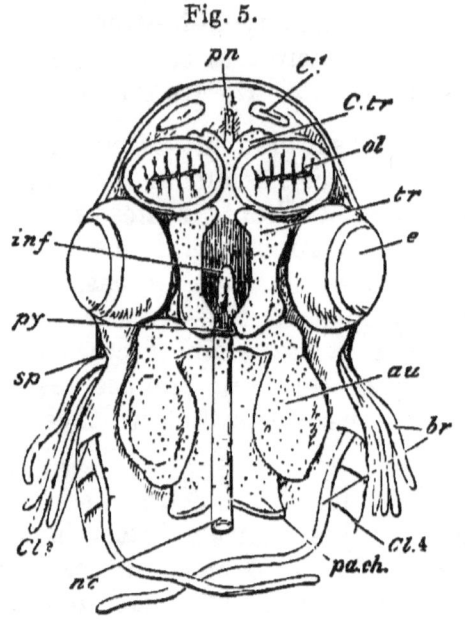

Fig. 5.

Head of embryo Dogfish, second stage; basal view of cranium from above, the contents having been removed. *C*1, forebrain; *ol.* olfactory sacs; *au.* auditory capsule; *nc.* notochord; *py.* pituitary body; *pa.ch.* parachordal cartilage; *tr.* trabecula; *inf.* infundibulum; *c.tr.* cornua trabeculæ; *pn.* prenasal element; *sp.* spiracular cleft; *br.* external branchiæ; *cl.* 2, 4, visceral clefts.

it lies behind the nasal sac and exactly beneath the anterior termination of the trabecula. The hinder or quadrate region is somewhat compressed, and lies in the cheek above its convex articular condyle. The meckelian cartilage is thick and solid where it receives the condyle in a concave surface, and curves but slightly in proceeding forwards and inwards to its rounded termination in the chin.

65. The upper part of the hyoid arch, including its forward process, has been segmented off in a position parallel with the articulation of the jaws. This *hyomandibular* element (*h. m.* Fig. 4) developes an oblong articu-

lating surface in relation to the infero-lateral region of the auditory mass at the middle of its length. Below this facet the cartilage bends forwards so as to be concave on its hinder margin, while the projecting angle in front is attached to the quadrate region of the upper jaw by the symplectic ligament. Below this again the hyomandibular turns a little backwards, and bears a notch and a posterior angular process for articulation with the simple arcuate lower piece, the *ceratohyal* (*c. hy.*). Between the ventral ends of the ceratohyals there intervenes a flat shield-shaped *basihyal* cartilage.

66. The branchial arches are divided into upper and lower pieces in a manner generally similar to the arches in front, forming *epibranchial* and *ceratobranchial* segments; and above each epibranchial a small separate cartilage is developed, turning inwards and backwards in the roof of the pharynx: this is the *pharyngobranchial*.

67. Morphological advance has been signalized in this stage by opposite processes in the cranial and the facial regions. In the former the elements which were primitively distinct, the trabecular and the parachordal, have coalesced: but the palato-mandibular and the hyoid bars have undergone segmentation, and in them the essential relations of the adult condition have already been established. The nasal region gives as yet only a mere indication of its future cartilaginous skeleton: while the cranium itself remains membranous for the most part.

Third Stage.

68. Embryos of the Dogfish, from an inch and a half to two inches long, have undergone great metamorphosis. The external gills are still long, but the internal gills are functional. The mesocephalic flexure is lost; so that the buccal cavity is not now behind the first cerebral vesicle, but lies below it. (See Fig. 6.)

69. The parachordal cartilages (*pa. ch.*) have grown round the notochord, above and below, enclosing it completely, and have coalesced with its cartilaginous sheath; thus they unite with each other to form the *basilar plate*. The cranial part of the notochord is soon reduced to a mere posterior cone of gelatinous tissue in the hinder part of this plate. Behind, it bears a rounded knob on each side for articulation with the first vertebra; in front, although it is fused with the trabecular plate, an emargination in the middle line is very plain below, the cartilage becoming very much thinner.

70. Seen from below, the basilar plate is broad and subquadrate, generally convex, with a groove in the middle line. It extends widely outwards so as to support the auditory capsules entirely, and come into relation with the hyomandibular articulation.

71. The lozenge-shaped sub-pituitary space is now thinly floored by cartilage which is confluent with the trabecular plate all round, and rises into a low posterior clinoid ridge behind, directed a little forwards. The subpituitary cartilage is pierced in its widest part by the internal carotid arteries (*i. car.*), which, however, lie close together. The so-called "middle trabecula" is a mere fissure between the mid-brain and the hind-brain.

72. The trabeculæ have coalesced and now form a broad flat trabecular plate. This extends most widely outwards behind, where a short curved process is sent out beyond the prominent antero-external angles of the basilar plate, being not yet perfectly confluent with the latter[1]. More anteriorly the trabecular plate not only forms the floor of the cranial cavity, but also partially supports the eyeballs. Each antero-external angle is curved outwards and somewhat backwards, being loosely connected with the

---

[1] The imperfect confluence of these regions, combined with an exaggerated growth external to the auditory capsules, produces a large fenestra in certain forms, as *Carcharias glaucus*, and *Lamna cornubica*, bounded by an extensive curved band of cartilage.

palatine cartilage of the same side. This angle may be termed the *antorbital* or ethmopalatine process.

73. In front of the ethmopalatine region, the trabecular plate, which has been gradually rising towards the level of the nasal roof, is suddenly narrowed, so that it has a nearly transverse anterior margin except in the middle line, where it is confluent with the internasal element. This internasal plate, which is not very wide, sends forwards in the middle line from its transverse anterior margin a short linear prenasal rostrum.

74. The internasal plate gives off on its upper aspect a double ascending lamina, the moieties of which soon diverge outwardly as broad wings, and coalesce with the inner wall of each nasal sac. Each of these *nasoseptal laminæ* bears a small sigmoid cornu (*c. tr.*) in front, passing inwards towards the prenasal cartilage. The roof and outer walls of the nasal sacs (*na.*) have very largely chondrified, forming dome-shaped capsules; and in addition to the coalescence of each olfactory cartilage with the corresponding nasoseptal lamina, it is confluent with a cartilaginous tract developed in the supraorbital band (§ 41, p. 15).

75. No cartilaginous laminæ have arisen in the multiplicate nasal membranes. The opening through which each olfactory crus reaches the corresponding nasal cavity is large, and is situated in the inner side wall of the dome behind. Owing to the outspreading of the nasoseptal wings, the opening is nearly horizontal at this stage; subsequently it becomes slanted upwards and forwards.

76. Very much of the cranial roof and walls is chondrified, the elements being more or less distinct, and more or less fused with the cartilages already mentioned. There is a flat cartilaginous cranial roof, or *tegmen cranii* (Fig. 6), extending from the supraoccipital or posterior cranial region (*s. o.*) to the level of the fore part of the eyeballs, and on either side is the curved supraorbital tract, continued anteriorly to a junction with the olfactory capsules. The front edge of the cranial roof is crescentic and

incomplete, exposing the olfactory crura as they diverge to their destinations.

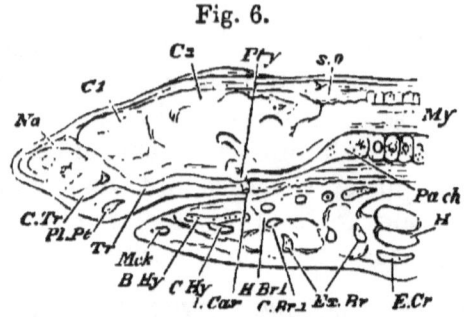

Fig. 6.

Head of embryo Dogfish, third stage; median longitudinal section. *C*1, forebrain, *C*2, midbrain, *my.* myelon; *na.* olfactory sac; *pa.ch.* parachordal cartilage; *s.o.* supraoccipital cartilage; *pty.* pituitary body; *i.car.* internal carotid artery; *c.tr.* cornua trabeculæ; *pl.pt.* palato-pterygoid cartilage; *mck.* mandibular cartilage; *b.hy.* basihyal; *c.hy.* ceratohyal; *h.br.*1, first hypobranchial; *c.br.*1, first ceratobranchial; *ex.br.* extrabranchial cartilage; *h.* heart; *e.cr.* epicoracoid cartilage; behind *pa.ch.* vertebral centra.

77. Between each trabecula below and the supra-orbital band above, there is a cartilaginous sphenoidal wall continuous with both, extending from the auditory to the nasal regions. The supraorbital ridge forms a partial roof for the orbit, while the outer edge of the trabecula furnishes a partial floor. The optic nerve pierces the sphenoidal wall at its anterior third. The trigeminal nerve passes out in the primary fissure between the sphenoidal wall and the auditory sac; when they have coalesced, this fissure becomes crescentic, and may be called the trigeminal foramen. The nasal branch of the ophthalmic nerve (from the trigeminal) passes forwards in the angle between the sphenoidal wall and the preorbital part of the supraorbital band where it goes to join the olfactory capsule. The facial nerve emerges at the junction between the auditory mass and the trabecular plate, and curves upwards and backwards round the former.

78. The cartilaginous capsules of the auditory sacs are complete on both the inner and outer sides except where nerves and vessels pass. The capsules have coalesced with

the supraorbital band in front and above, with the supra-occipital cartilage and the tegmen cranii on the inner side above, and with the basilar plate below. The semicircular canals are prominent and large, and the surrounding cartilage is correspondingly bulged outwards. Over the junction of the anterior and posterior semicircular canals an unchondrified tract is seen on the upper surface of the skull. The foramina for the glossopharyngeal and the vagus nerves are one behind the other, behind the auditory capsule, near the postero-external angle of the basilar plate.

79. The palato-quadrate bar has become much larger; and it is very like the articulo-meckelian in form and size. The moieties of each jaw meet in the middle line, where a strong ligament unites them. At the symphysial ends the rods are terete and incurved; proximally they are flat and sinuous. The articular region is gently scooped to work on the convex quadrate condyle.

80. The enlarged hyoid arch has undergone no important change: the hyomandibular is articulated with the auditory capsule and the edge of the basilar plate below the horizontal semicircular canal. Between the ventral ends of the right and left hyoid bars there is a broad, flat, *basihyal* plate of cartilage, (*b. hy.*), which is like a keystone to the arch. One more flat keystone occurs behind, between the two last branchial arches; but the hypobranchial pieces of the second and third arches turn backwards to reach the front edge of this *basibranchial*. The first hypobranchials are but small lobes (*h. br.* 1), and do not extend backwards.

81. The branchial arches are very similar, each consisting of four pieces. Their *hypobranchials* partially floor the throat cavity; their *cerato-* and *epibranchials* form its lateral boundaries which are very convex outwardly. The upper segments or *pharyngobranchials* are turned backwards in the roof of the throat. The fourth pharyngobranchial is forked externally, the hinder fork being applied to the fifth arch so as to be continuous with its epibran-

chial; there is a fifth ceratobranchial, from which no hypobranchial is cut off.

82. Two other series of cartilages, called *labials* and *extrabranchials*, have now to be described. Three labial cartilages are related to the anterior and outer edges of each nasal capsule. The first is pyriform and is already furrowed above by two or three slime-glands; the second and third are found in valvular folds of the nasal opening: the second is crescentic, and its lower horn partially floors the nasal aperture; the third is ear-shaped, and lies on the outer edge of the capsule. A fourth labial, lanceolate in shape, underlies each palato-quadrate bar on the upper edge of the angle of the mouth; whilst a fifth labial, converging towards the fourth at the angle, is upon the external surface of the mandible. Similar cartilages, the *extrabranchials* (*ex. br.*), appear outside each of the first four branchial arches. They are pointed above and broader below; the last is very small.

83. In an embryo of *Raia maculata* four inches long, taken from the egg-pouch three months after oviposition, the tail is twice as long as the body proper, and the yolk-sac is still of the size of a small walnut. Yet the skull has assumed the adult form and structure even more completely than in the Dogfish. On either side the flabelliform fin has grown forwards and united with the cheek, and consequently all the branchial clefts appear entirely on the ventral surface, while the mandibulo-hyoid cleft is seen dorsally as the spiracular opening. From the gill-slits the external branchiæ still project as long filaments. The nasal apertures are on the ventral surface, and the head is produced in front of them as a flattened rostrum, very distinct from the cranial convexity. The eyeballs lie in their sockets on the upper surface, bounded in front, internally, and behind, by the prominent supraorbital ridges.

84. The general character of the metamorphosis is very similar to that of the Dogfish. The prenasal rostrum

is much larger, and the first labials lie on either side of it in front, parallel with it. The internasal plate, continuing the gradual elevation of the anterior part of the cranial floor, is much broader than in the Dogfish, being as wide as the cranial floor itself; and instead of giving rise to distinct nasoseptal wings, it simply rises and becomes thickened on each side, and is continuous with the nasal roofs. No fenestra arises in these cartilages nor is there any development of cornua in front. The olfactory crura pass into the back part of the nasal roofs through a large fenestra on each side in front of the fore-brain. A small antorbital plate is carried on the outside of the base of each nasal capsule, where it forms the antorbital boundary.

85. A large part of the cranial roof is unchondrified in front of the supraoccipital region. This median fontanelle is elliptic in shape, and limited anteriorly by a transverse bar of cartilage, and laterally by the supraorbital tracts. In front, the cranial cavity is quite unclosed by cartilage, as in the Dogfish. Instead of the former continuous parachordal mass in the basicranial and cervical regions, there is an occipital segmentation, dividing an occipital cranial tract from a cervical cartilaginous tube, which is perforated at the sides by a series of fissures for the cervical nerves.

86. There is now a complete lunule of cartilage in the anterior wall of the spiracular opening. It is attached to the side wall of the cranium, below the antero-external or *sphenotic* process of the auditory capsule; and it bears a comb-like pseudobranchia. Below, the spiracular cartilage is attached by ligament to the hindmost part of the palato-quadrate bar.

87. The upper jaw is directed quite transversely, instead of being concave behind, as in the Dogfish; the two palato-quadrate cartilages are closely in apposition in the middle line. The hinder jaw is similarly transverse, the mandibular cartilages being parallel to the palato-quadrate. The emarginate nasofrontal process of integument leaves

the anterior jaw largely uncovered in the middle line. The small valvular nasal openings with the labial cartilages enclosed in the valves are situated on either side of the nasofrontal process.

88. The hyomandibular cartilage articulates with the auditory capsule beneath its supero-external (*pterotic*) ridge; below, it is entirely detached from the rest of the hyoid arch, turning forwards behind the spiracle, which it guards, as the spiracular cartilage does in front. The hinder end of the palato-quadrate is strongly fastened by ligament to the lower end of the hyomandibular.

89. The growth of opercular skin-folds between the branchial arches has left only small sigmoid external openings to the five branchial sacs. They are portions of the lower parts of the original clefts, and they converge towards the middle line from before backwards. A special ear-shaped opercular flap covers each of the openings. At this period the hyoid and branchial arches carry two sets of branchiæ in full function; the external are at their highest development; and the internal branchial laminæ are perfect, though small.

90. The epihyal segment of the hyoid arch is attached by an *interhyal* ligament behind the hyomandibular to the postero-external angle of the skull (pterotic ridge). The whole hyoid arch is smaller than the succeeding branchial arches—a great contrast to the Dogfish. The hypohyal is styliform, and widely separated from its fellow; indeed it is attached by ligament to the first hypobranchial. (Compare Dogfish, § 80, p. 29.)

91. The branchial arches are almost bent double, in accordance with the flattened shape of the animal. The pharyngo- and epibranchials are above, the ceratobranchials external and inferior. As in the Dogfish, there is no separate fifth pharyngobranchial. A single series of cartilaginous rays proceeds externally from each arch; these are pedate at their outer ends. The first hypobranchial is elongated and slender, meeting and coalescing with its fellow in the middle line. The second, third, and

fourth hypobranchials are small. The fifth arch has its hypobranchials fused into a broad flat plate which lies between the last three branchial arches, having two diverging cornua in front.

92. There are no extrabranchial cartilages, nor are there any oral labials. One rostral pair, and three nasal pairs in the valves of the nasal openings, are the only representatives of labials found in this Skate; the internal labial related to the nostril is the largest, and projects backwards over the upper jaw.

93. It is of high interest to note the perfection of the cranial structures here described, when the embryos are still so small in proportion to their adult size, and also the simplicity of the changes by which they are developed from the primary elements. The chondrification of mesoblastic tissue, in continuity with the parts originating earliest, comprises most of the history of the proper cranial and capsular skeleton. The facial arches have attained the special arrangements found in this group by a method of segmentation, essentially resembling that which will be seen to occur in every type examined. The prenasal and the antorbital pieces are to be carefully borne in mind, for they throw light upon many facts to be afterwards noticed. The labial and extrabranchial cartilages and the branchial rays are structures which are more specially characteristic of the Elasmobranchs, but they are charged with lessons for the student who accurately notices their relations.

Skull of the adult Dogfish.

94. The changes which have to be described are not morphologically very important. The olfactory capsules have become relatively larger, and more closely approximated. The incurved cornua projecting from the nasoseptal laminæ have disappeared. The first labials, instead of being wholly anterior to the nasal region, extend backwards over the nasal roofs, and are pitted by numerous slime-glands. The second labials have partly

coalesced with the anterior edges of the nasal domes, and with the adjacent antero-external angles of the internasal cartilage. There is a considerable space between the superficial extrabranchials and the deeper branchial arches. Between the corresponding bars of each series, numerous cartilaginous rods are developed, situated in the septa between the successive gill-pouches.

95. The adult skull may now be described in detail. It is entirely fibrous and cartilaginous, excepting that there are closely-set superficial calcifications in tesseræ on most of the cartilages. There are no investing or "splint" bones; but as a cartilaginous skull this is the most perfect and complete of its kind.

96. The chondrocranium is very flat both above and below; perforated behind by the foramen magnum, and open in front between the hinder boundaries of the nasal capsules. The latter are large domes attached to the antero-external angles of the cranium. Thus the olfactory region of the skull is much wider than the interorbital. A strong concave supraorbital ridge partly roofs each of the large eyeballs; it forms a distinct antorbital plate, running down towards the antorbital process of the cranial floor behind the olfactory capsule, and at its opposite extremity coalesces with the front region of the auditory cartilage, where a blunt sphenotic process projects. Internal to each supraorbital ridge is a considerable groove, in the front part of which the supraorbital foramen opens. Mesially the cranial roof or tegmen cranii is convex.

97. The surface of the otic mass exhibits three well-marked prominences for the semicircular canals, anterior, posterior, and external; and mesiad of the elevation for the anterior canal is a small opening, which is the remnant of the primitive involution of the ear-sac. There is also a strong external *pterotic* ridge, horizontally placed (*pt. o.*).

98. Viewing the skull from behind, the foramen magnum is seen, somewhat triangular in shape, bounded

at the basal angles by the well-developed occipital condyles. Between these is a slight elevation, marking the original situation of the notochord. The lateral regions of the cranial cartilage extend widely outwards, the upper edge, as seen from behind, being longer than the lower, which is constituted by the basilar plate. The greater portion of each lateral mass of cartilage is periotic; but the foramen magnum is completely encircled by an occipital ring not due to the ear-capsule. Laterally, the basilar plate furnishes the lower part of the facet for the hyomandibular.

99. The inferior surface of the cranium is an oblong tract, of which the hinder half is the basilar plate, extending under the ear-capsules, and the fore part is due to the trabeculæ. A notch under the front of the auditory region marks the separation between the basilar plate and the hinder angle of the trabecula. Farther forwards the trabecular plate is narrower; but behind the nasal region it sends out an antorbital process on either side, to which the ethmopalatine ligament is attached. Between the olfactory sacs the trabecular cartilage is much narrowed, but it grows upwards and forwards to coalesce with the adjacent part of each nasal roof. On each side of this *ethmoidal* region the olfactory fibres pass through a large membranous fenestra down into the nasal sacs. These have their folds of membrane pinnately arranged, and containing no cartilage. The prenasal rostrum ($p.\ n.$) projects forwards and upwards for a short distance beyond the internasal region. The inner wall of each olfactory capsule presents a large unchondrified space or fenestra; it is in the region originally formed by the nasoseptal wing.

100. The long first labials ($l.\ 1$) meet in the middle line anteriorly to any other skeletal element, and curve backwards over the outer part of the olfactory capsules. The second labials ($l.\ 2$) are partly confluent with the front of the capsules, and with the anterior angles of the internasal plate at the base of the rostrum. A third labial

(*l.* 3) lies in the skin-fold protecting the outer edge of each nasal aperture.

101. Laterally, the skull presents continuous cartilage, embodying olfactory capsule, supraorbital bar, antorbital and postorbital (sphenotic) processes, auditory investment, and basilar plate. The hinder part of the skull is very little higher than the front region. The rounded olfactory boss and the smaller antorbital prominence are succeeded by the concave orbital wall; while, behind, the ear-cartilage with its sphenotic process (*sp. o.*) and pterotic ridge (*pt. o.*) forms the outstanding buttress of the cranial edifice.

102. In this skull, having no bones or distinct cartilages, the nerve foramina must be carefully studied as

Fig. 7.

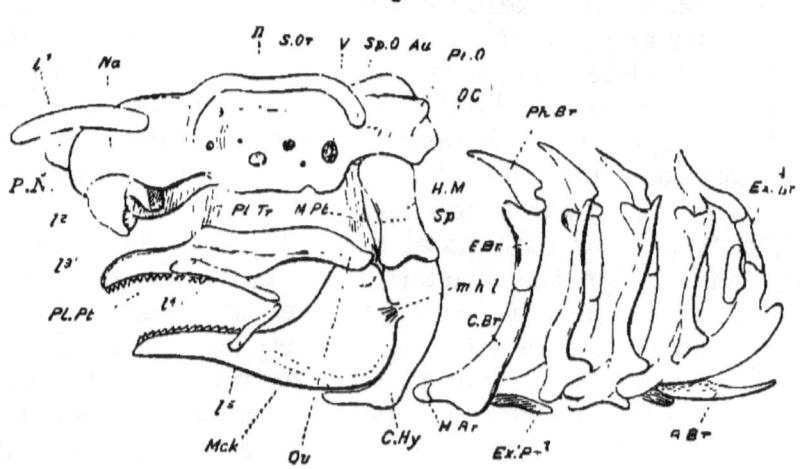

Skull of adult Dogfish, side view. *O.c.* occipital condyle; *au.* periotic capsule; *pt.o.* pterotic ridge; *sp.o.* sphenotic process; *s. or.* supraorbital ridge; *na.* nasal capsule; *p.n.* prenasal cartilage; *II.* optic foramen; *V*, trigeminal foramen; *pl.pt., qu.* palato-quadrate arcade; *m. pt.* metapterygoid ligament (including a small cartilage); *pl. tr.* ethmo-palatine or palato-trabecular ligament; *mck.* lower jaw; *sp.* spiracle; *h.m.* hyomandibular; *c.hy.* ceratohyal; *m.h.l.* mandibulo-hyoid ligament; *ph.br.* pharyngobranchial; *e.br.* epibranchial; *c.br.* ceratobranchial; *h. br.* hypobranchial; *b. br.* basibranchial; *ex.br.* extrabranchial; *l*1, 2, 3, 4, 5, labial cartilages; the dotted lines within *mck* indicate the basihyal.

regional landmarks. On the upper surface there is a small foramen for the supraorbital nerve in a groove between the fore part of the supraorbital ridge (*s. or.*) and the frontal part of the cranial roof. The olfactory fenestra has already been described. The foramen for the optic nerve (Fig. 7. *II.*) appears low down in the middle region of the lateral (sphenoidal) wall of the cranium. The trigeminal foramen (*V.*) is similarly placed in the hinder part of the sphenoidal wall, and the orifice for the facial is a little behind and below this. The external foramina for the glossopharyngeal and vagus nerves are on the posterior aspect of the skull, the former low down and external, marking the original separation between auditory mass and basilar plate; the latter higher up, and at the side of the foramen magnum in the line of junction of the occipital and auditory regions.

103. Examining the inner surface of the skull, the pituitary body is seen to be guarded in front and behind by small cartilaginous ridges in which some bony deposit takes place. These markedly resemble the anterior and posterior clinoid ridges bounding the "sella turcica" of Mammalia; and the resemblance is strengthened by the entrance of the internal carotid arteries on either side of the pituitary body. With such facts may be coupled the likeness of the occipital articulation with the atlas to that of Mammalia.

104. The palato-mandibular cartilages, constituting the main part of the oral apparatus, are loosely swung from the sides of the basis cranii by two short ligaments. The upper jaw, or palato-quadrate arcade, is attached by the ethmopalatine ligament to the antorbital region; by the metapterygoid ligament (*m. pt.*) to the postorbital trabecular angle behind the trigeminal foramen, and in front of the spiracle (*sp.*); and by the symplectic ligament to the hyomandibular behind. The metapterygoid and the symplectic ligaments are very closely connected, and some fibres of the former are attached to the hyomandibular cartilage. The hinder or quadrate end of the upper jaw (*qu.*) forms a condylar surface for articulation with the lower jaw or articulo-meckelian element.

105. The upper jaw is narrow behind, broad and convex outwardly in the middle, and bends inwards to be united with its fellow by a strong ligament. The lower jaw is very similar to the upper; it is deep and outwardly convex behind, for attachment of the oral muscles. Teeth are borne on the mucous membrane of the fore and inner part of both jaws. The fourth and fifth labials (*l.* 4 and 5), set on the upper and lower jaws obliquely, make an acute angle by the approach of their pointed hinder ends, considerably within the angle of the jaw. When the mouth is opened the angle between them becomes obtuse. There is a small cartilage in the anterior wall of the spiracle.

106. The large phalangiform hyomandibular (*h. m.*), the upper segment of the hyoid arch, is articulated with the side wall of the cranium beneath the pterotic ridge. The spiracle is in front of it, while distally it is connected with the quadrate region by the symplectic ligament. The stout and elongated ceratohyal (*c. hy.*) is articulated with the distal facet of the hyomandibular, and is locked within the everted postero-inferior edge of the mandible, being strongly bound to it by the mandibulo-hyoid ligament (*m. h. l.*). At its ventral end the ceratohyal is bilobate, and articulates by its anterior lobe with the broad basihyal plate, which it also slightly underlies behind.

107. The succeeding four arches have supero-transverse pharyngobranchials (*ph. br.*), which are in contact with their fellows in the middle line beneath the vertebral column; lateral epibranchials (*e. br.*) and ceratobranchials (*c. br.*), and infero-median hypobranchials (*h. br.*) directed backwards. Each epibranchial has a small forward spur at its upper end. Normally the pharyngobranchials are turned somewhat backwards; but when the pharynx is distended they may be directed transversely or even a little forwards. Each pharyngobranchial has at its lower end a backward spur. The fifth branchial arch is much smaller than the rest, having its pharyngobranchial confluent with the fourth, and no hypobranchial. The median

basibranchial (*b. br.*) lies between the last three arches, in contact anteriorly with the hinder hypobranchials, and with the fifth pair of ceratobranchials behind.

108. There is a vascular plexus (pseudobranchia) on the anterior side of the spiracular opening. The five gill-clefts are the external openings of as many pouches, the anterior and posterior walls of which bear transversely plaited branchial laminæ. Internally each pouch opens into the pharynx. In the partition between each pouch is a series of cartilaginous rods, or branchial rays, most of them simple, extending from the hyoid and first four branchial arches towards the skin. Those in relation with the hyoid are pectinately branched; three of them are in the hyomandibular region.

109. Four extrabranchial cartilages (*ex. br.*), parallel to the ceratobranchials, are superficial to the corresponding sets of branchial rays. They are pedate below, and very greatly resemble the scapulo-coracoid cartilages situated immediately behind them (see Fig. 6, *e. cr.*). The fifth branchial arch has no gill-pouch behind it, no rays, and no extrabranchial.

Skull of the adult Skate.

110. The most notable change in the appearance of the skull is produced by the great elongation of the prenasal rostrum (Fig. 8, *p. n.*), which has the first labials (*l.* 1) on either side of its fore extremity. The lateral portions of the internasal plate where they adjoin the nasal capsules are continued into the rostrum, so that the cranial floor is continuous with a long concavity on the hinder part of the rostrum. All traces of the external branchiæ have disappeared; and cartilaginous branchial rays, about a dozen in the anterior wall of each gill-sac, extend from the arches outwards and backwards. The antorbital plate has become a thick backwardly and outwardly curved rib-like bar, pointed behind, lying outside and above the upper jaw.

111. The following are, briefly, the chief points in which the Skate's skull differs from that of the Dogfish:—

Fig. 8.

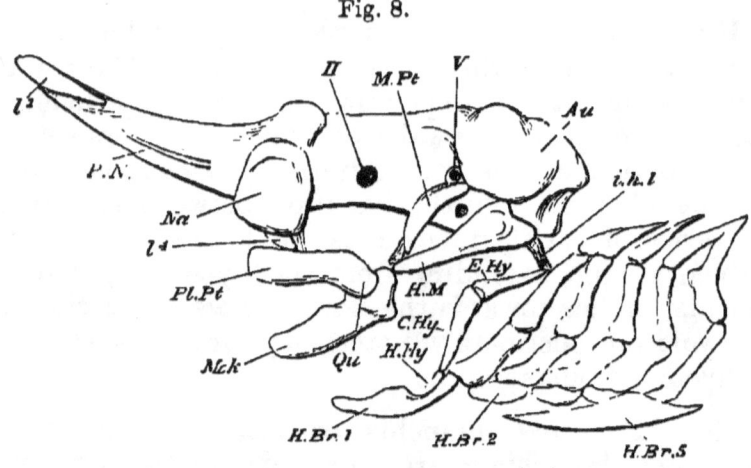

Skull of Skate, nearly adult. *Au.* periotic capsule; *na.* olfactory capsule; *pn.* prenasal cartilage; *II, V.* in the side wall of the cranium, optic and trigeminal foramina; *pl.pt., qu.* palato-quadrate bar; *mck.* mandibular cartilage; *m.pt.* metapterygoid; *h.m.* hyomandibular; *i.h.l.* interhyal ligament; *e.hy.* epihyal; *c.hy.* ceratohyal; *h.hy.* hypohyal; *h.br.* 1,2,5, hypobranchials; above them are seen the cerato, epi, and pharyngobranchials; *l*1,4, labials.

the presence of a supero-median cranial fontanelle, and of a large prenasal rostrum; the second labials not coalescing with the olfactory capsules; the fourth labials becoming involved in the inner fold of the nasal openings, so that three pairs of labials protect these orifices; the distinctness and size of the antorbital cartilage; the large size of the cartilage (*m. pt.*) in the anterior wall of the spiracle[1]; the form and forward direction of the hyomandibular, and its dissociation from the rest of the hyoid arch, except by a posterior and superior interhyal ligament attached to a distinct epihyal; the smaller size of the hyoid arch, and its close connection with the first branchial; the absence

---

[1] The presence of one or more spiracular cartilages is constant in Sharks; a small one is found in the Dogfish.

of a basihyal, and the junction of the first hypobranchials across the middle line; the coalescence of the fifth hypobranchials (*h. br.* 5) to form a broad median basibranchial plate; the absence of the fifth pair of labials, and of the extrabranchial cartilages.

112. SUMMARY. The history of these skulls is singularly uncomplicated; but this does not necessarily indicate that the type is a low one. Paired basal elements arise at first, the trabeculæ in the anterior portion of the cranial floor, and the parachordals behind. By their union, and by the chondrification of the greater part of the cranial wall in continuity with them, a simple cartilaginous cranium is formed, deficient in one or more regions above. The ear-cartilages are impacted in the cranial mass behind, and the nasal capsules have their own skeletal defence in front, coalescing with the cranial cartilage as well as with an internasal element which is at first distinct from the trabeculæ, but afterwards fused with the cranial floor. The small prenasal spur found in the Dogfish is represented by the great rostrum of the Skate; while the antorbital region is characterised in the latter by the development of an antorbital rod extending backwards.

113. The skeleton of the jaws and gills possesses from its origin substantially its adult relations; and no part of it is at any time fused with the cranium proper. The upper and lower jaws are primarily continuous, although the bars are sharply bent; but they are speedily cut in two. In the Skate an upper and posterior region of the mandibular arch chondrifies early to form the spiracular cartilage.

114. The suspension of the jaws is effected by the intermediation of an element of the hyoid arch; but in the Dogfish the hyomandibular is the segmented upper extremity of the bar, while in the Skate it arises separately in the antero-superior region of the arch. The stout ceratohyal is in the Dogfish directed forwards and closely related to the lower jaw and the distal end of the hyoman-

dibular; while in the Skate the hyoid arch is feebler, but is connected directly with the skull and the proximal end of the hyomandibular. It has an epihyal as well as a ceratohyal piece, and is in relation with the first branchial arch.

115. The branchial arches commence as simple rods, afterwards segmented into a series of pieces, the arches of opposite sides being connected by basal cartilages; the Skate is distinguished by the separate origin of its upper branchial elements. Branchial rays are found in the partitions between the gill-pouches.

116. Two series of cartilages occurring less generally in other types than in those which have been described are named labial and extrabranchial. These are developed between the primary skeletal elements and the skin, and enter into relation with the rostrum, the nostrils, the jaws, and the gill-clefts.

CHAPTER III.

THE SKULL OF THE SALMON.

First Stage: Unhatched embryos, with simple facial arches.

117. THE embryos here described have their neural tube open both anteriorly and posteriorly. The long tape-like body lies on a yelk-sac, with the umbilicus very little constricted, so that its rim reaches to within a short distance behind the mouth. The involutions to form the sense-organs are widely open; yet the primary elements of the skull can be distinguished, though merely by being composed of more consistent tissue than the surrounding parts. The rods or bars to be spoken of are formed of the very small mother-cells of hyaline cartilage.

118. The enlargements of the neural tube to form the cerebral vesicles are very slight, not constituting any obvious bulging; but the three vesicles are distinct, the first being largest, and having the landmark of the pituitary body (Fig. 10, *py.*) projecting backwards from its postero-inferior surface. There is no mesocephalic flexure of the neural tube. A little behind the pituitary body is the pointed anterior end of the notochord (*n. c.*), lying just below the middle of the second cerebral vesicle. There is no clear distinction between the outer layer of epiblast and that which limits the neural tube.

119. Almost at the anterior extremity of the embryo, on its under surface, the olfactory sacs appear as small pits surrounded by a circular ridge of epiblast; the extreme open end of the neural tube is immediately above them.

The involutions for the eyeballs are much larger, and unclosed; they are behind the olfactory sacs, and lie quite

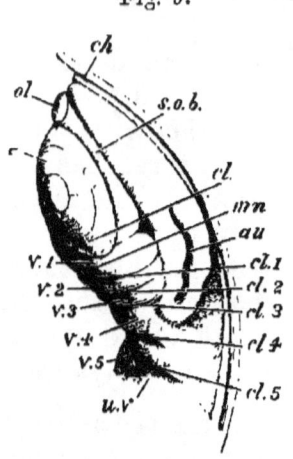

Fig. 9.

Embryo Salmon, about ¾ inch long; side view of head within chorion.

ch. chorion; *ol.* olfactory sac; *e.* eyeball; *au.* auditory mass; *s.o.b.* supraorbital band; *u.v.* attachment of umbilical vesicle; *cl.* lachrymal cleft; *v.* 1, 2, 3, 4, 5, visceral arches; *mn.* first or mandibular arch; *cl.* 1, 2, 3, 4, 5, visceral clefts.

on the inferior surface. The rudiments of the ear-sacs (*au.*) are by the sides of the third cerebral vesicle, and are consequently more dorsal than and posterior to the eyeballs; their line of involution is open, and longitudinally placed. So large are the eye- and ear-sacs relatively to the skull, that they somewhat overlap one another.

120. The side wall of the head arching over the eyeball is conspicuously thickened, forming the *supraorbital band* (Fig. 9, *s. o. b.*). Below and behind the eyeball, is a thickened suborbital arch, which extends forwards to the olfactory sacs of the same side. Between the anterior extremities of these arches is a space lined with thickened epiblast, the future mouth-cavity (Fig. 11, *m.*). This is bounded in the middle line and behind by the meeting of another pair of arches, the foremost of the visceral series.

THE SALMON: FIRST STAGE.

The latter curve forwards to the middle line below the auditory region, the successive arches having between them slits or clefts (*cl.*), which penetrate into the primitive alimentary canal. It is only at a later period that the mouth has a communication with the same cavity. The umbilical sac is attached in front to the head, in the space between the hindermost visceral arches; the heart (Fig. 11, *h.*) is situated on the wall of the fore part of the alimentary canal, above or within the same region.

Fig. 10.

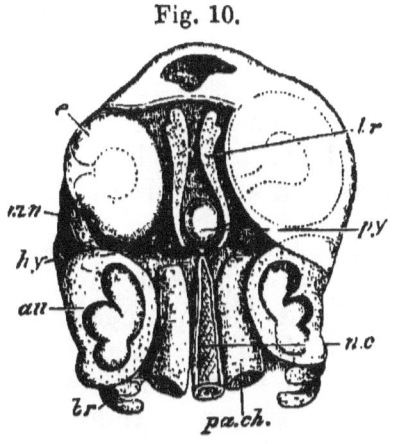

Embryo Salmon, about ¾ inch long; upper view of head, dissected, the neural tissue having been removed.

tr. trabecula; *pa.ch.* parachordal cartilage; *n.c.* notochord; *au.* auditory capsule, showing dilatations for semicircular canals; *py*, pituitary body; *e.* eyeball; *mn.* mandibular, *hy.* hyoid, *br.* branchial, arches. The anterior aperture remaining in the neural tube is shown in the skin in front of the trabeculæ.

121. The notochord ends pointedly at about the level of the front end of the ear-sacs, behind the pituitary body. On either side of it are *parachordal* bars (*pa. ch.*) cut off squarely in front, a little behind the apex of the notochord (Fig. 10). Laterally, the parachordals are flanked by the auditory involutions, and by the mesoblastic investment which is forming a partial girdle of nascent cartilage round them. The inner part of the ear-cavity shows the several dilatations for the semicircular canals (Fig. 10). The nascent cartilage does not appear in the floor or on

the lower outer side of the sac below the line of involution. A little later, the auditory masses unite with the parachordals at two points on either side, one anterior and the other posterior; thus a *fenestra* is formed in the base of the skull (see Fig. 14, *f. s. o.*).

122. The axial *prechordal* region of the cranial floor is occupied by the *trabeculæ* (*tr.*), bowed outwards on either side of the pituitary body. They are small and pointed behind, where they curve towards the notochord, without touching it; and more expanded in front, where they approach one another, and lie under the fore part of the cranial cavity, and in the roof of the mouth. The intertrabecular space is not merely coextensive with the pituitary body, for the trabeculæ lie in the floor of the forebrain, and support it on either side.

123. A *palatine* bar is distinctly formed in the subocular band; it is simple and arcuate, reaching in front to the nasal sac, and behind nearly to the auditory capsule, where it overlaps the posterior extremity of the supraorbital ridge. The palatine rod is from the first clubbed at the fore end, and there solidifies earliest; it passes back into a fine point behind, much later in solidifying. The *mandibular* arch is at first curved but slightly forwards below, and may be nearly transverse, although embryos of the same age vary greatly in this respect. Its upper extremity curves inwards beneath the fore part of the ear-sac, towards the apex of the notochord. It soon becomes distinguishable into three regions; (1) the *metapterygoid* apex, (2) a median swelling directed forwards, the *orbital process*, and (3) the lower or *meckelian*, elongated forwards as a spatulate expansion, but not yet meeting its fellow (Fig. 13, *mt. pg., or. p., mk.*).

124. The *hyoid* arch (*hy.*) has an enlarged and incurved apex, which lies underneath the middle of the ear-sac. Its lower end becomes curved forwards behind the mandibular region. The two hyoids do not meet, but are before long separated only by a small median segment,

THE SALMON: FIRST STAGE.

the *basihyal* (Fig. 13, *g. h.*). There are five pairs of *branchial* rods (*br.*) behind the hyoid, not meeting in the

Fig. 11.

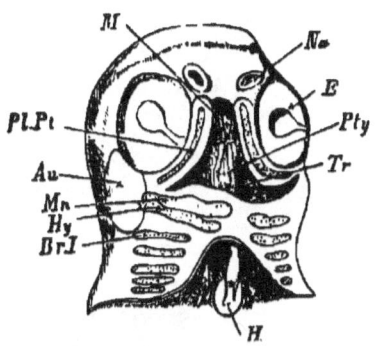

Embryo Salmon, about ¾ inch long; lower view of head, with the arches shining through.

Na. nasal sacs; *e.* eyeball; *au.* ear-mass; *tr.* trabecula; *pl.pt.* palatine or subocular bar; *mn.* mandibular, *hy.* hyoid, *br.* branchial, arches; *pty.* pituitary body; *m.* mouth; *h.* heart.

middle line. They become progressively smaller from the front to the hinder one, and are all sigmoid in shape except the last, with their apices incurved over the cavity of the throat.

125. Most of the primary elements of the skull are present in this early embryo, which has not yet undergone the mesocephalic flexure. The parachordal and trabecular regions are definitely marked out, and the periotic tissue is commencing to chondrify and to unite with the parachordals. The palatine, mandibular, hyoid, and five branchial bars, are manifest in a simple unsegmented condition; but the mandibular piece is already becoming differentiated into regions.

126. In order to simplify the description, no reference has been made to the remarkable asymmetry of Salmon embryos, and the great variations in the degree and kind of asymmetry. The head is usually twisted about its axis so that only one eyeball, which may be either right or left, is visible

in an upper view. The rudimentary postoral arches (which share the general asymmetry at first) are distinct before the clefts between them are complete. The arches differ considerably in their degree of development; the earliest distinct are the mandibular and hyoid bars, then the branchials, next the trabeculæ, and most imperfect, the palatines. The chondrification of the latter is very tardy. They enclose a soft pith even in the third stage. With this may be compared their long suppression in the Frog.

Second Stage: The metamorphosis of the hyoid arch.

127. The series of changes now to be noticed takes place in embryos while still within the egg, or only just hatched. The neural tube is completely closed, the cerebral vesicles are greatly enlarged and very prominent, and their growth has produced a mesocephalic flexure (see Fig. 12). Thus the second cerebral vesicle is a large boss on the upper surface of the head, while the first vesicle is below and in front. Behind this is the eyeball, then the visceral arches; but the mandibular arch is produced below (Fig. 13, *mn.*), forming the prominent hinder boundary of the mouth. The nasofrontal process (*n. f. p.*) has grown wide between the nasal sacs, separating the suborbital arches, so that the oral cavity is broad and quadrate. An opercular skin-fold is passing backwards from the upper part of the hyoid region. Dermal papillæ, the rudiments of the gills, are noticeable on all the branchial arches except the last.

128. The notochord remains relatively in the same position as at first; but the parachordals are thicker in front, and unite with the trabeculæ. Behind, they extend farther, and are pointed; they also tend to invest the notochord more closely. Laterally, their union with the ear-cartilages becomes more and more marked, diminishing the fenestra in the line of union with the periotic mass (Fig. 14). The ear-capsules get much more extensively chondrified; and at the line of junction between them and the parachordals there is a thickening and

upward growth, which tends to form side walls and a roof to the cranial cavity. The semicircular canals produce

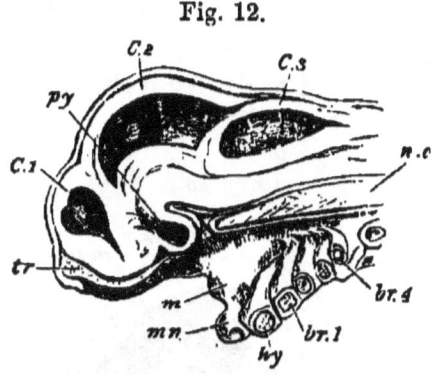

Fig. 12.

Embryo Salmon, partly hatched; median longitudinal section of head.

C 1, 2, 3, fore-, mid-, and hindbrain; *py.* pituitary body; *n.c.* notochord; *tr.* trabecula; *m.* mouth; *mn.* mandibular, *hy.* hyoid, *br.* branchial arches.

bulgings in the ear-cartilage, and a small space remains unchondrified over the posterior canal (*epiotic fenestra*).

129. The mesocephalic flexure causes the trabeculæ to run downwards as well as forwards; anteriorly, however, they curve upwards again, below the fore end of the cranial cavity (Fig. 12, *tr.*). Their hinder-hooked extremities bend inwards more and more towards the parachordals, and finally are completely fused with them. The time of this union is very variable; sometimes it occurs before hatching, in other cases it is not perfect in the second week after that event.

130. In front of the pituitary region the trabeculæ become approximated along their whole length, and then gradually fused; and at the same time a median longitudinal ridge arises. This ridge increases in prominence anteriorly, where the trabeculæ lie in the nasofrontal process, and are laterally expanded, growing towards the subocular or palatine arches; thus the ethmopalatine connection is being formed (Fig. 14). The fore extremi-

ties of the trabeculæ develope a transverse ridge beneath, in the roof of the mouth; above, they diverge from the median crest, and form a flat spatulate lamina on each side, partially overarching the olfactory sac. The remainder of the cranial investment is membranous; parachordals, trabeculæ, and auditory cartilages constitute the whole of the chondrocranium.

Fig. 13.

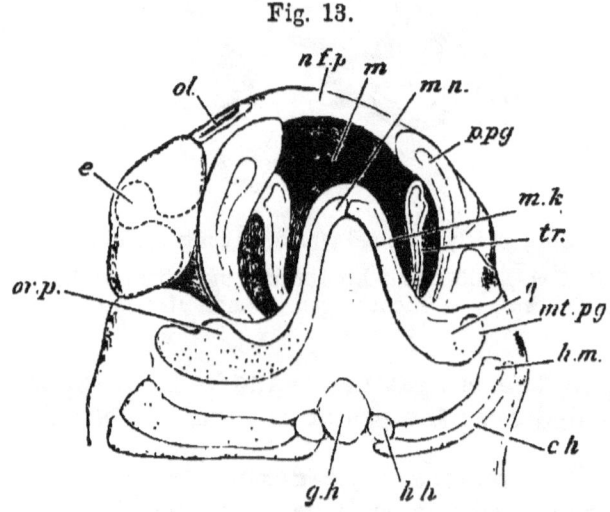

Embryo Salmon, not long before hatching; under view of head, with arches seen through.

N.f.p. nasofrontal process; *ol.* olfactory sac; *e.* eyeball; *m.* mouth; *tr.* trabecula; *p.pg.* palatine cartilage; *m.* mandibular arch; *mk.* meckelian region; *q.* quadrate region; *mt.pg.* metapterygoid region; *or.p.* orbital process; *g.h.* glosso- or basihyal; *h.h.* hypohyal; *c.h.* ceratohyal; *h.m.* hyomandibular.

131. The subocular or palatine bar remains little developed in this stage, forming the lateral wall of the mouth; anteriorly it is in contact with the ethmopalatine process of the trabecula. Behind, there is a cleft between it and the eyeball, passing down into the palate. The mandibular arch undergoes more modification; the inferior meckelian part becomes bent at a large angle with the proximal portion, and meets its fellow in the middle line (Fig. 13).

This flexure generally takes place forwards, but it may be directed backwards, so that very diverse appearances may be presented by the facial arches of different embryos, owing to this cause and also to the asymmetry of the creatures; but the morphological result is the same.

132. The proximal region of the mandibular arch becomes broadened, and an upper segment is cut off in a peculiar manner so as to form a ball-and-socket articulation. The ball is on the upper, the socket in the lower piece; and the junction is defended by an angular process of the latter, running up behind. Furthermore, the whole arch travels considerably downwards from the axial parts and becomes less closely attached to the cranial wall.

133. Before the segmentation of the mandibular cartilage, the most striking division takes place in the hyoid arch. In addition to the median basihyal mentioned in § 124 (p. 46) a small globular piece, the *hypohyal*, is cut off from the inferior end of each hyoid bar; while the rest of the arch is split into two along its entire length, so that one bar lies directly behind the other and in contact with it. The anterior of these is the *hyomandibular*, the posterior the *ceratohyal* (Fig. 13, *hm., c.h.*).

134. At a later period, concurrently with the travelling downwards of the mandibular arch from the cranial mass, the hyomandibular diverges from the ceratohyal below, remaining in contact with it above. The hyomandibular being thus turned forwards, reaches the mandibular arch, and becomes applied to its upper segment and the proximal part of its lower segment (Fig. 14).

135. A further change of relation is subsequently effected. As the throat increases in size, the hyomandibular remains closely apposed to the auditory cartilage, while the ceratohyal is found lower and lower in position, as if it had slid down half way along the hinder edge of the hyomandibular (see next stage, Fig. 17). At this point the upper end of the ceratohyal becomes fixed by an

interhyal ligament, in which cartilage is subsequently developed. The hyomandibular broadens above and developes

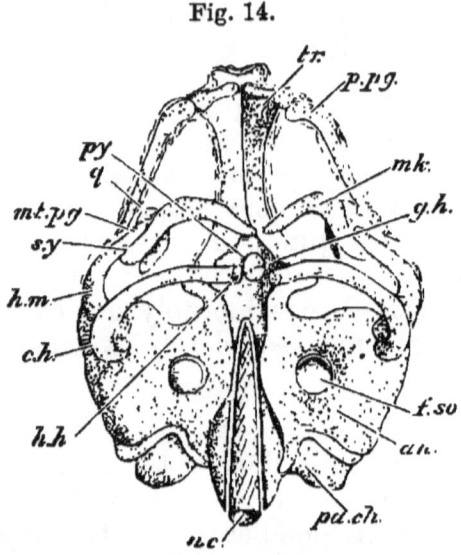

Fig. 14.

Embryo Salmon, not long before hatching; lower view of skull dissected, the branchial arches having been removed.

Tr. trabecula, with transverse ridge in roof of palate; *p.pg.* palatine bar, near ethmopalatine conjunction; *py.* pituitary space; *n.c.* notochord; *pa.ch.* parachordal cartilage which has coalesced with *au.* auditory capsule, except where a fenestra *f.s.o.* is left; *mk.* meckelian cartilage; *q.* quadrate region; *mt.pg.* metapterygoid region; the line should be prolonged to the tract *inside* the meckelian bar; *h.m.* hyomandibular turned forwards at its ventral end, and applied by its symplectic tract *sy*, to the quadrate region of the mandibular arch; *c.hy.* ceratohyal; *h.h.* hypohyal; *g.h.* basi- or glossohyal.

a backward spur; below, it remains styliform. The mandibulo-hyoid cleft becomes gradually obliterated.

136. The branchial arches are not as yet segmented like the arches in front. They fuse in the middle line below, all but the last pair; and then four small median azygous ventral pieces, the *basibranchials*, are cut out. The upper end of these arches curve inwards most elegantly over the sides and roof of the throat. The hinder (fifth) branchial pair remains much smaller than the rest.

137. The important period of development included in this second stage witnesses many significant changes. The mesocephalic flexure being fully established, the trabeculæ unite with the parachordals, and the latter, in close union with the otic cartilages, form lateral and superior occipital growths. The anterior portions of the trabeculæ unite and send up a median crest, the rudiment of the interorbital and nasal septa; while lateral and anterior trabecular growths establish the ethmopalatine and the cornual regions. But the most intense interest attaches to the modifications by which the mandibular and hyoid arches are segmented, and their segments take up the relative positions essentially characteristic of Osseous Fishes. The preponderance of growth in particular regions effects all that to outward observation looks like the shifting of parts. It is, we hope, made plain how the mandibular elements, removed to a distance from the axial parts, derive their *suspensorium* from the arch behind, which is remarkably segmented for that purpose. The formation of definite hypo- and basibranchial pieces is to be noted at this period: the principal branchials do not yet present any segmentation.

Third Stage: Salmon-fry of the second week after hatching.

138. The head has become much larger, and the mesocephalic flexure is lost; a longitudinal section shows the parts of the brain lying approximately in a straight line on the flattened cranial floor. The well-developed brain fills the proportionally large cranial cavity; the auditory protuberances are very marked at the sides. The articulation of the meckelian rod with the proximal mandibular element is still farther removed beneath the level of the cranial floor, and the mandible has grown forwards underneath the fore part of the head. No ossifications in cartilage appear as yet; but parostoses[1] have arisen in

---

[1] By parostosis is meant an ossification in subcutaneous fibrous tissue; it may reach to and involve the perichondrium of a cartilaginous tract;

more than one region, the basicranial splint, or *parasphenoid*, being especially noticeable. In many points there are agreements between the cartilaginous cranium of the Salmon at this stage and the cartilaginous and osseous cranium of the adult Polypterus; see Traquair *On the Cranial Osteology of Polypterus* (*Journ. Anat.* Vol. v. p. 166).

139. The occipital region of the chondrocranium has progressed, but is yet far from perfect. The notochord

Fig. 15.

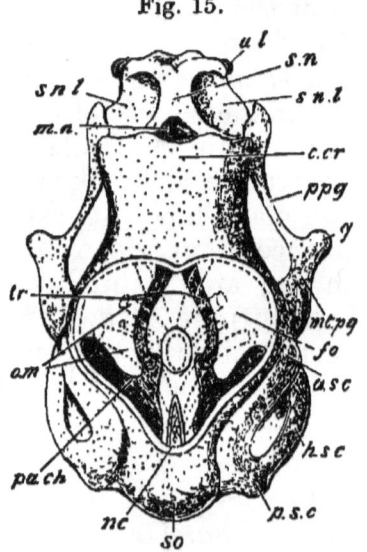

Salmon fry, second week after hatching; upper view of skull dissected, the brain being removed, and the inferior parts somewhat displayed.

Pa.ch. marks the anterior limit of the parachordal cartilage; *nc.* notochord, lying in the posterior basicranial fontanelle produced by the recession of the parachordals; *o.m.* the orbital muscles diverging from beneath the cranial floor in the notochordal region; *so.* supraoccipital tract; *a.sc.* anterior, *h.sc.* horizontal or external, *p.s.c.* posterior semicircular canals; *tr.* trabeculæ; *t.cr.* tegmen cranii, or cartilaginous roof of the cranium; *fo.* superior fontanelle, heart-shaped; *s.n.* nasal septum; *s.n.l.* subnasal lamina; *mn.* mesonasal cavity; *u.l.* upper labial cartilage; *p.pg.* palatopterygoid bar; *q.* quadrate condyle; *mt.pg.* metapterygoid region of mandibular arch.

---

such a bone may be denominated a membrane bone. Ectostosis signifies ossification arising between perichondrium and the superficial cells of cartilage, which gradually becomes absorbed and replaced by bone. Endostosis is the direct calcification of cartilage.

is retracted so as to occupy only the hinder half of the postpituitary (basilar) region, and the parachordal masses have receded from one another, leaving a considerable oval space (*posterior basicranial fontanelle*) between them anteriorly; while posteriorly they do not yet unite with each other in any portion of their length, although they invest the notochord closely behind (Fig. 15).

140. Laterally, the parachordals do not become broad, as in the Dogfish and Skate; they form a floor for the hindbrain, and externally are fused with the periotic cartilages. Nevertheless, there is a distinct notch in front between the otic and parachordal elements; and the latter also project backwards behind the ear-cartilages. Each of these hinder processes is divided into two lips; these are the primordial articulating surfaces of the cranium with the first vertebra. Lateral occipital cartilage now exists, largely fused with the ear-capsules, but rising free on the backward projections of the parachordals. Above, they have passed inwards on each side, and coalesced to form the supraoccipital plate (*s.o.* Fig. 15).

141. The ear-cartilages are not greatly modified; the semicircular canals are large proportionately, and may be distinctly seen through the cartilage. On the under surface the fenestra formerly mentioned (§ 121 p. 46) in the line of junction between the periotic and the parachordal masses is much smaller relatively, being carried outwards by the latter growing underneath the former. The cartilage beneath the bulging vestibule is very thin, and a large otolith within it is well seen from outside. The upper and outer eminence of the ear-capsule (the *pterotic* ridge) is prominent, overhanging the articulation of the hyomandibular.

142. The hinder division of the trigeminal nerve escapes from the cranium by a foramen where the fore part of the parachordal is fused with the auditory cartilage; its anterior (ophthalmic) division passes over the notch between these two elements. The foramen for the

facial nerve (Fig. 18, 7 *a.*) is behind that for the trigeminal (5) on the inferior aspect of the skull; while the glosso-pharyngeal and vagus nerves (8) emerge in the same line near the posterior part of the basis cranii. The orbital muscles (*o.m.*) pass forwards from their basicranial attachment to the orbit on each side of the pituitary body.

143. The hinder portion of the trabeculæ, in the region of the pituitary body, does not show much change: slender arcuate rods enclose a somewhat triangular inter-trabecular space, much larger than the pituitary body; and posteriorly they lie above and unite with the fore extremity of the parachordals. Anteriorly there is great change; the brain-case has become elevated, and the forebrain no longer lies directly on the trabeculæ, but on a membranous floor which forms the upper and inner boundary of each orbit, while a median vertical inter-

Fig. 16.

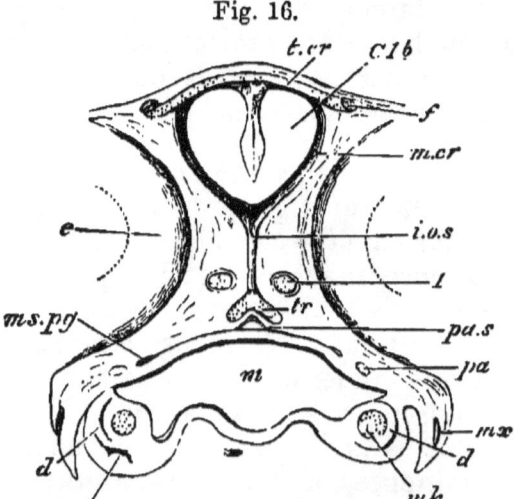

Salmon fry, second week after hatching; transverse section of head, through forebrain and eyeballs.

T. cr. tegmen cranii; *m. cr.* membranous cranial investment; *C1b.* forebrain; *i.o.s.* interorbital septum; *1,* olfactory crus; *tr.* coalesced trabeculæ, with superior ridge; *pa.* palatine cartilage; *mk.* meckelian cartilage; *m.* mouth; *e.* eyeball.

Bones: *f.* frontal; *pa.s.* parasphenoid; *ms.pg.* mesopterygoid; *mx.* maxillary; *d.* dentary; *ar.* articular.

orbital septum connects this floor with the trabeculæ (Fig. 16, *i.o.s.*). The latter are not horizontally placed, but arched upwards so as to unite at rather more than a right angle, and with a sharp edge upwards (Fig. 16, *tr.*) The trabeculæ become broader where they coalesce, and at the front of the orbit they curve suddenly outwards on each side, forming the ethmopalatine processes, to which the fore ends of the palatine rods are connected (Fig. 15, *p. pg.*). The trabecular plate in its most anterior portion is broad and quadrate below, flooring the nasal sacs and ending with a broad edge, truncated at the external angles, and showing very little distinction of free trabecular cornua.

144. The anterior part of the trabecular crest (the nasal septum) has increased in height; its moieties diverge in front and behind, so as to form crescents. The anterior fork becomes shallow towards its termination; the posterior is more elevated, and unites with the ethmopalatine process and with the cranial roof, to be presently described. In the diverging angle between the posterior forks of the internasal cartilage there is a median blind membranous pouch (Fig. 15, *m. n. c.*), beneath which is the trabecular plate. The olfactory crura pass out of the braincase under the hinder part of the forebrain, and run in the orbit horizontally on either side of the interorbital septum, finally gaining a level a little above the coalesced trabeculæ in the ethmoidal region, where they perforate the postnasal wall.

145. The fore part of the cranial roof has become chondrified. This cranial roof, or *tegmen cranii* (*t. cr*), is broad and quadrate, and convex above, covering the forebrain and the anterior part of the midbrain. It is continuous with the posterior forks of the nasal cartilages in front, and shelves down laterally to the ethmopalatine region, forming *lateral ethmoidal* or *antorbital* plates. There is no cartilage in the side wall of the cranium in the region of the tegmen; but each of its posteroexternal angles is continuous with a slender rod of carti-

lage developed in the supraorbital band, which curves outwards and backwards to join the fore part of the auditory cartilage (Fig. 17. *s. or.*). Thus there is a large heart-shaped membranous fontanelle in the roof of the chondrocranium, (*fo.* Fig. 15), bounded in front by the tegmen (which being produced somewhat backwards in the middle line causes an emargination in the fontanelle), laterally by the supraorbital bars and the ear-masses, and posteriorly by the occipital cartilage.

146. No part of the chondrocranium is ossified in this stage: but several parostoses have appeared. The chief of these is the azygous *parasphenoid* (Fig. 16, *pa. s.*), a flat bone lying underneath the cranial floor, extending from the nasal nearly to the hinder part of the auditory region. It is slightly forked both in front and behind. A delicate *supraethmoid* bone overlies the internasal septum, and there is a styloid *frontal* ossification over each lateral edge of the tegmen. Just beneath the front extremity of the trabecular plate is a rudiment of the *premaxillary* on each side.

147. The first important modification to be noticed in the facial parts is the coalescence of the primitively distinct elements, the palatine and the proximal mandibular segment. The palatine or suborbital cartilage joins a forward growth from the mandibular, so as to become expanded behind as well as in front, remaining slender in the middle. The much enlarged upper mandibular segment forms the fore part of the wall of the pharynx below the level of the cranial cavity, and is no longer directly connected with the skull. Its chief relation to the chondrocranium is derived from the arch behind, the upper and forward portion of which is closely applied to it. Thus there is now but one continuous cartilage on each side, extending from the ethmopalatine trabecular process in front to the hyomandibular behind (Fig. 17). Two pairs of small nodular cartilages, the *upper labials* (l_1, l_2) are found in front of the palatine cartilages, underneath the fore part of the trabeculæ. The elongated articulo-meckelian rod, or cartilage of the lower jaw, articulates with the condyle on the proximal mandibular piece, and is produced into a thick angle behind the condylar hollow.

Fig. 17.

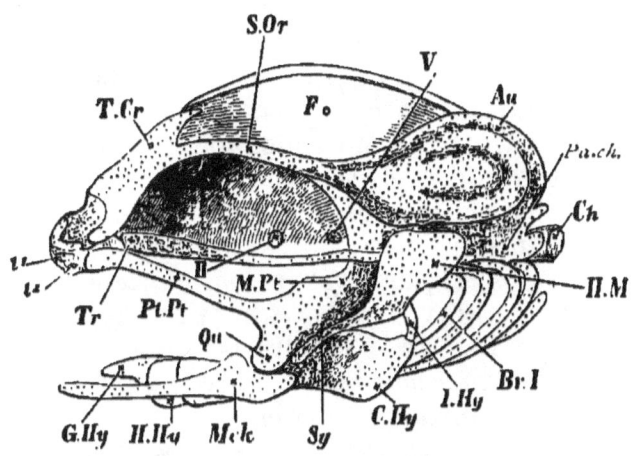

Salmon fry, second week after hatching; side view of skull; parostoses, eyeballs, and nasal sacs removed.

T.Cr. tegmen cranii; *S.Or.* supraorbital band; *Fo.* superior fontanelle; *Au.* auditory capsule; *Pa.Ch.* parachordal cartilage; *Ch.* notochord; *Tr.* trabecula; above the trabecula, the interorbital septum is seen, passing into the cranial wall above and reaching the supraorbital band; *II.* optic foramen; *V.* trigeminal foramen; *l.* 1, 2, labial cartilages; *Pl.Pt.* palatopterygoid bar; *M.Pt.* metapterygoid tract; *Qu.* quadrate region; *Mck.* meckelian cartilage; *H.M.* hyomandibular cartilage; *Sy.* symplectic tract; *I.Hy.* interhyal; *C.Hy.* ceratohyal; *H.Hy.* hypohyal; *G.Hy.* glossohyal; *Br.* 1. first branchial arch.

148. Bony rudiments are now to be found in relation with various parts of the mouth. The first is the *palatine*, an anterior ectosteal lamina on the palatine cartilage; behind this, internally and superiorly, is another ectosteal lamina, the *mesopterygoid;* the proper *pterygoid* is embracing the lower edge. In the connective tissue outside the pterygoid region is a parostosis, the commencing *maxillary* (Fig. 16, *mx.*). In the lower jaw there is a *dentary* parostosis (*d.*) lying all along the outer edge; and near the condyle, on the under edge and inner face, is an ectostosis, the *articular* (*ar.*).

149. The hyoid arch has grown considerably, but its essential relations remain the same as in the last stage (§§ 134, 135). The hyomandibular (*h.m.*) is broad above, where it articulates with the side wall of the ear-capsule, and styliform below (Fig. 17, *sy.*), where it is carried forwards in contact with the hinder edge of the mandibular arch, and actually becomes wedged between the quadrate condyle and the angular process of the articular. The interhyal cartilage (*i. hy.*), which connects the hyomandibular with the remainder of the hyoid arch, is attached to the middle of the hinder edge of the hyomandibular, and curves inwards below, so as to carry the arch forwards within the two moieties of the lower jaw. There is a large *ceratohyal* cartilage (*c. hy.*), flattened above, and knoblike anteriorly (ventrally), where the *hypohyal* (*h. hy.*) fits on to it. The two hypohyals are in close contact, and bear the median *basihyal* or glossohyal (*g. hy.*), which projects forwards into the tongue.

150. The branchial arches are still unossified, but are completely segmented, the pharyngobranchials or upper segments approaching one another beneath the floor of the skull and the anterior vertebræ. The principal lateral pieces are the epibranchial above and the ceratobranchial below. The basibranchials have begun to fuse together.

151. In this most important phase the skull of the Salmon is elaborately constructed as to cartilage, and is yet of a shape and structure very distinct from its adult condition: it answers considerably to the chondrocranium of several Ganoid types. The occipital ring is complete superiorly but not inferiorly; for the basilar plate is not yet constituted, and a very noticeable space separates its moieties anteriorly, where the notochord no longer extends. The fore part of the skull has undergone an extraordinary development, but is destined to be much modified by later growth. Yet in certain distinctive features the adult condition is substantially attained,

namely, in the elevation of the brain considerably above the trabecular floor, in the formation of a cartilaginous cranial roof connected on each side with the periotic investment, in the development of lateral ethmoidal or antorbital, and mesonasal growths, which are excellent landmarks in the study of the skull.

152. The junction of the palatine bars with forward expansions of the upper mandibular segment is the chief event to be noticed in regard to the appendicular pieces. The relations of the hyomandibular element are further perfected, and the branchial arches are segmented. Some membrane ossifications have appeared in proximity to the axial parts; the parasphenoid beneath most of the cranial floor, the premaxillaries in the front part of the roof of the mouth, the supraethmoid over the nasal septum, and the frontals at the sides of the cranial tegmen. The palatine, pterygoid, and mesopterygoid ossifications are especially related to the palatopterygoid cartilages, while the maxillary bones arise externally to them. The dentary and articular bones are found in the lower jaw; the hyomandibular, hyoid and branchial pieces are still unossified.

Fourth Stage: Young Salmon of the first Summer.

153. During the growth of the young Salmon from an inch and a quarter to two inches and a quarter in length, many of the characters which remind us of Ganoid fishes are effaced, while those which are specially Telostean become manifest. The cartilaginous skull is much more massive, and the *tegmen cranii* covers the brain to a large extent; furthermore, many of the cartilage-bones (*ectostoses*) are present. The fore part of the skull has developed so greatly that the nasal sacs appears only as small recesses on either side of it. The tooth-bearing *vomer* is now found in addition to the parasphenoid, being the inferior splint of the fore part of the skull. The yelk-sac is entirely taken into the abdomen, and the tissues generally have become much more perfect and solid.

154. There is no great change to be noted in the cartilages of the hinder part of the skull. The supra-occipital cartilage extends forwards as far as the level of the junction of the anterior and posterior semicircular canals of the ear. The interior cranial surface of the auditory capsule is incompletely chondrified, the ampullæ and the greater portions of the arches of the anterior and posterior canals, as well as the vestibule itself, not being separated by cartilage from the cranial cavity.

Fig. 18.

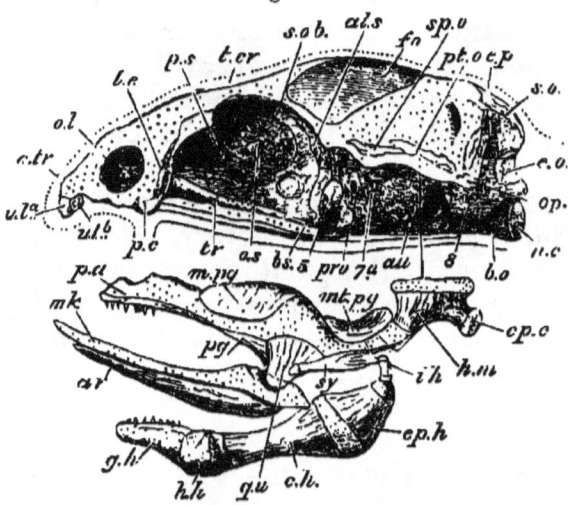

Young Salmon of the first summer, about 2 inches long; side view of skull, excluding branchial arches. The palato-mandibular and hyoid tracts are detached from their proper situations, a line indicating the position where the hyomandibular is articulated beneath the pterotic ridge.

Ol. olfactory fossa; *c.tr.* trabecular cornu; ul^a. ul^b. upper labial cartilages; *p.s.* presphenoid tract; *t.cr.* tegmen cranii; *s.o.b.* supraorbital band; *fo.* superior fontanelle; *n.c.* notochord; *b.o.* basilar cartilage; *tr.* trabecula; *p.c.* condyle for palatine cartilage; 5, trigeminal foramen; 7*a.* facial foramen; 8, foramen for glossopharyngeal and vagus nerves; *mk.* meckelian cartilage; *op.c.* opercular condyle.

Bones: *e.o.* exoccipital; *s.o.* supraoccipital; *e.p.* epiotic; *pt.o.* pterotic; *sp.o.* sphenotic; *op.* opisthotic; *pro.* prootic; *b.s.* basisphenoid; *al.s.* alisphenoid; *o.s.* orbitosphenoid; *l.e.* ectethmoid or lateral ethmoid; *pa.* palatine; *pg.* pterygoid; *m.pg.* mesopterygoid; *mt.pg.* metapterygoid; *qu.* quadrate; *ar.* articular; *h.m.* hyomandibular; *sy.* symplectic; *i.h.* interhyal; *ep.h.* epiceratohyal; *c.h.* ceratohyal; *h.h.* hypohyal; *g.h.* glosso- or basihyal.

155. Ossification arises around the notochordal sheath, laying the foundation of the *basioccipital;* and this bony deposit extends into the surrounding cartilage. A *supraoccipital* ectostosis (Fig. 18, *s. o.*), and a pair of ossicles, the *exoccipitals* (*e. o.*), in the cartilage bounding the foramen magnum laterally, complete the hindmost group of bones in the investment of the brain.

156. Several centres of ossification are present in connection with the auditory cartilage. There is one on its antero-inferior margin, behind and above the main part of the trigeminal nerve, and perforated by its posterior branch; this is the *prootic* (*pr. o.*); a second, the *sphenotic* (*sp o.*), is above the ampulla of the anterior semicircular canal; a third, over the ampulla and arch of the horizontal canal, is the *pterotic* (*pt. o.*); a fourth, the *epiotic* (*ep.*), is above the arch of the posterior canal; and a fifth (*op.*), over the ampulla of the same canal, is the *opisthotic.* The sphenotic, pterotic, and opisthotic form a series on the outer edge of the auditory capsule, overhanging the articulation of the hyomandibular; while the epiotic extends inwards at right angles to this line at its posterior extremity. A remnant of the primordial fenestra between the ear-capsule and the basilar cartilage still persists.

157. The anterior region of the skull has become remarkably modified. The trabeculæ are now separated only in a very small space surrounding the pituitary body. Anteriorly to this they have completely fused and become massive, and have sent up a median ridge in the interorbital septum, occupying its lower part, below the level of the optic foramen. The upper part of the same septum is also chondrified continuously with the upper part of the anterior cranial wall or median ethmoidal region, forming a presphenoidal spur (*p. s.*). Part of the lateral cranial wall in the interorbital region is also cartilaginous, in continuity with the tegmen cranii and the supraorbital tract; while a further portion is chondrified in connexion with the fore part of the auditory

capsule, and the hinder part of the supraorbital bar. The tegmen cranii is thicker and extends farther back over the midbrain.

158. An ectosteal lamina is found in the fore part of each side wall of the cranium, extending downwards on either side so as to press upon the presphenoidal cartilage, and upwards to embrace, by a grooved edge, the downgrowth from the tegmen cranii: this is the *orbitosphenoid* (*o. s.*). The cartilage of the cranial wall in front of the ear-capsule above the optic and trigeminal nerves undergoes ossification, and forms the *alisphenoid* (*al. s.*). Inferiorly it comes into relation with an ectostosis in the hinder part of the trabecular crest, lying in the interorbital septum and projecting backwards in the middle line into the pituitary space. Thus there is formed an almost vertical prepituitary *basisphenoid* bone (*b. s.*), Y-shaped, so as to abut against the lower edge of each alisphenoid. Immediately behind this bone the pituitary body descends to reach the parasphenoid. There is no cartilaginous floor at this point: on either side of the pituitary body one of the internal carotid arteries enters the cranial cavity. The basisphenoid is partly furnished by direct ossification of the membranous interorbital septum.

159. The sloping ethmoidal region is high mesially where it is confluent with the vertical internasal septum and the cranial roof, and descends gradually on either side. An *ectethmoid* ectostosis arises in the lateral ethmoidal region (*l. e.*) near the palatine articulation. The internasal cartilage is high and massive behind, and rapidly descends to form the somewhat triangular beak. The olfactory sacs open laterally by a wide aperture narrowed by membrane, and are otherwise completely surrounded by cartilage. Thus the trabecular floor, comparatively narrow in the interorbital region, is broadly expanded in the nasal tract. There are no ectostoses to be noticed here; but the vomer underlies the internasal cartilage, and bears teeth: the supraethmoid plate is its counterpart above.

160. The following are the bones related to the chondrocranium at this stage. Ectostoses: posteriorly, basioccipital, exoccipitals, supraoccipital; in the auditory region, prootic, sphenotic, pterotic, epiotic, opisthotic; near the pituitary body, basisphenoid, alisphenoids; related to the forebrain, orbitosphenoids; in front of the brain-cavity, ectethmoids. Parostoses: superiorly and in front, parietals and frontals, supraethmoid; inferiorly, parasphenoid, vomer; in front, premaxillaries. The form and relations of the parasphenoid have become very complicated, owing to the growth of the orbital muscles between it and the cranial floor behind, the formation of the basitemporal wings, and of a median crest between the trabecular moieties anteriorly. The fully-developed parasphenoid will be described later.

161. Ossification has advanced greatly in the facial bars, concurrently with modification in the shape of the cartilages. The cartilage in the pterygoid region has become broader and more massive than in the hinder part of the arch. Cartilage has also extended in front of the ethmopalatine connection, so as to constitute a prepalatine region. The ectostoses are, antero-externally, the long styliform *palatine* (*pa.*), bearing teeth; behind this, a small *pterygoid* below (*pg.*), and a much larger *mesopterygoid* (*m. pg.*) internally and above; and in the hinder region, abutting on the hyomandibular, a *metapterygoid* (*mt. pg.*) on the cranial side, and a *quadrate* (*qu.*) in the neighbourhood of the condyle for the lower jaw. In the latter the *articular* bone has increased very much in length. The condylar surfaces, both on the quadrate and articular cartilages, are unossified. The only parostoses to be noticed are the *maxillary* in the edge of the upper jaw, and the *jugal* on the posterior end of the maxillary, the *dentary* flanking the meckelian cartilage outside, and a very small *angular* on the angular process.

162. The hyomandibular cartilage remains of moderate size, and is related as before to the pterotic ridge. There is a broad *hyomandibular* ectostosis above, which

extends partially into a cartilaginous knob on the hinder edge, bearing the opercular bone (*op. c.*). The lower styliform portion of the hyomandibular cartilage acquires a *symplectic* ossification (*sy.*), which fits into the grooved hinder edge of the quadrate, thus firmly clamping one arch to the other. Several parostoses have arisen in the opercular flap; but they will be sufficiently described in the adult skull.

163. The ectostoses in the remaining part of the hyoid arch are the *interhyal* above, the *epiceratohyal* (*ep. h.*) in the upper part of the undivided cartilage formerly named ceratohyal, and the large ceratohyal in the lower part of the same cartilage; finally, the small hypohyal (arising by two ossific centres), and the basi- or glossohyal, which now bears teeth. The branchial arches are completely formed; but their description may be reserved until the adult skull is dealt with.

The Skull of the adult Salmon.

164. The most important differences between the Salmon's skull when a few weeks old and when adult may be briefly referred to, preliminary to a full description. The cranial roof and side walls become almost completely cartilaginous, except where the cartilage is replaced by ectostoses, or where nerves pass out. The cartilage of the cranial roof, however, is covered by large parostoses, conspicuously by the frontals. The precranial region of the skull has greatly increased in relative length, and forms a massive cartilaginous beak, extending considerably beyond the nasal sacs. No new ectostoses arise in it. In the base of the brain-case it is noteworthy that the two orbitosphenoids unite, forming a floor to the anterior part of the cranial cavity, above the presphenoidal cartilage. Another most interesting fact is the growth of the prootics into the basilar cartilage, and their union to form part of the base of the cranium.

165. The mouth is almost completely margined by parostoses, which are mostly dentigerous: the ectostoses

lie within. The orbit has nearly an entire ring of fibrous bones. The hyomandibular and metapterygoid regions are much more massive than they were, increasing the depth of the framework of the head, and throwing the articulation of the lower jaw farther from the axial parts. The opercular skin-fold developed from the upper part of the hyoid arch is provided with a complete series of membrane-bones; while a similar membrane proceeding from the lower part of the same arch has acquired a large number of branchiostegal rays. The same thing in miniature has happened to the branchial arches.

166. The cranio-facial apparatus of the Salmon is very complicated. Cartilage is largely persistent in it, and of the various bones some are but little connected with the cartilage, some closely invest or encircle portions of it; while others are massive, although directly continuous with cartilaginous tracts by their edges. The bones in the chondrocranium are not very large, and the axial structures, cartilages, and ectostoses constitute but a small portion of the entire mass, though the size and outward extension of the membrane-bones applied to the axial parts make the latter appear more considerable than they are. The perfection of the jaws and gill-arches is attained by a large and highly complex apparatus with many parts curiously combined. Ectostoses are reinforced by investing bones laid down in fibrous tissue; yet much cartilage remains.

167. The cartilaginous skull, with its intrinsic ossifications, consists of an elongated brain-case or cranium, having openings where nerves pass out, or where the brain is continuous with the spinal cord; of a strong buttress on either side of the brain-case behind, enclosing the organ of hearing; and of a precranial mass, nearly as long as the cranial box, in which the organs of smell are embedded. Thus the auditory region of the chondrocranium is quadrate as seen from above or below, and is connected with the precranial mass by a narrow isthmus, at the sides of which the orbits lie. In front of the isthmus the fore

part of the cartilage is roughly trihedral, with the base behind, its lateral parts forming the antorbital walls. Anteriorly the cephalic cartilage ends in two short blunt processes or cornua, with an emargination between.

168. On its hinder aspect the skull is lozenge-shaped, the pterotic bones occupying the lateral angles, and the supra- and basioccipital the upper and lower angles. Between the basioccipital and each pterotic are an exoccipital (*e. o.*) and an opisthotic (*op. o.*). The epiotics (*ep. o.*) are on the upper edge, between the supraoccipital and either pterotic. Much cartilage remains in the ear-capsule and above the foramen magnum, which is relatively small, and almost entirely surrounded by the exoccipitals.

169. Each exoccipital contributes a supero-lateral facet to the single occipital condyle, so that the latter is threefold, derived from the basioccipital and two exoccipitals. Between the concave surfaces of the condyle and of the first vertebra (which are lined with articular cartilage), there is a mass of pulpy tissue, which is a notochordal remnant; but cartilage and bone have closed up the cavity in the base of the skull, in which the notochord was formerly contained. The pterotics, the epiotics, and the supraoccipital have rounded bosses upon them which project backwards for the attachment of muscles.

170. Inferiorly, the median region of the skull projects as a thick double ridge, considerably below the periotic masses; thus the latter overarch a large concavity on each side, which is walled in externally by the hyomandibular and other facial elements. The median ridge referred to includes a bifurcated downgrowth from the basilar cartilage, enclosing a longitudinal channel in which the orbital muscles take origin; they diverge right and left from the anterior end of the channel, on either side of the basisphenoid. The two laminæ are considerably ossified by the basioccipital behind, and by the prootics in front (*b. o., pr.o.*, Fig. 20): while underneath, the channel is floored by the parasphenoid parostosis (*pa. s.*) The under

surface of the concavity at either side of the skull, beneath the overarching periotic mass, shows the prootic bone

Fig. 19.

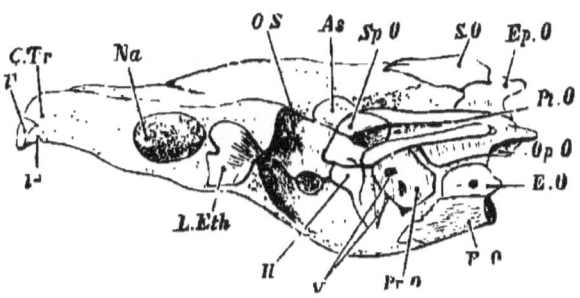

Adult Salmon; lateral view of chondrocranium with its ectosteal bones; the parostoses, jaws, and branchial arches having been removed. The continuous cranial cartilage is dotted.

C.Tr. trabecular cornu; l^1. l^2. labials; *Na.* nasal fossa; *II.* optic foramen; *V.* trigeminal foramina in prootic bone.

Bones: *B.O.* basioccipital; *E.O.* exoccipital; *Op. O.* opisthotic; *Pr.O.* prootic; *Pt.O.* pterotic. bearing a cartilaginous facet for the hyomandibular, and overarching the concavity in which the preceding bones are seen; *Ep.O.* epiotic; *S.O.* supraoccipital; *Sp.O.* sphenotic; *As.* alisphenoid; *O.S.* orbitosphenoid; *L.Eth.* ectethmoid.

anteriorly; behind this, externally, the pterotic, and the small opisthotic; internally, the exoccipital. The extreme forwardly projecting outer angle is formed by the sphenotic (Fig. 19).

171. On the external lateral margin of the auditory region there is the prominent ridge of the pterotic bone, continued forwards by the sphenotic: underneath the projecting ledge is the linear longitudinally extended articular surface for the hyomandibular. The inner portion of the ear-cartilage, between the basioccipital and prootic below, and the supraoccipital above, is quite unossified, and to a considerable extent the labyrinth and canals are even unclosed by cartilage on their cranial aspect.

172. The prootics have become considerable bones, extending upwards for some distance in the side wall of the cranium, and also invading the basilar cartilages even

to the middle line, so as to unite more or less with each other. They grow down on each side in the wall of the basicranial channel, but do not ossify it to its greatest depth. The trigeminal nerve (5) passes out of the cranium in front of the prootic, deeply notching it; its posterior division enters a canal in the same bone, through which it reaches the exterior. The facial nerve penetrates the prootic in front of its middle region, and passes directly outwards and downwards. The auditory nerve perforates the prootic by a large foramen behind the facial nerve; the intracranial foramen for the glossopharyngeal and vagus (9 and 10) is in the same line, between the prootic and the exoccipital; its external orifice is in the exoccipital.

173. The anterior part of the investment of the cranial cavity is of more massive cartilage, much less

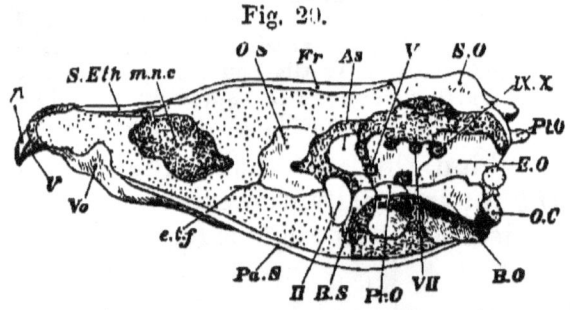

Fig. 29.

Adult Salmon: median longitudinal section through skull, after removal of jaws and arches.

O.C. occipital condyle; *e.t.f.* ethmo-trabecular fissure; l^1. l^2. labials; *m.n.c.* mesonasal cavity; *II.* optic foramen; *V.* trigeminal foramen; *VII.* foramen for the facial; *IX. X.* foramen for the glossopharyngeal and vagus.

Bones: *B.O.* basioccipital, *Pr.O.* prootic, each bone having a descending lamina in the wall of the basilar canal, which is floored by the parasphenoid, *Pa.S.*; the exoccipital, *E.O.*, and the prootic, enter largely into the side wall of the cranium; *Pt.O.* posterior extremity of pterotic, seen outside the cranium; *S.O.* supraoccipital; *B.S.* basisphenoid; *A.S.* alisphenoid; *O.S.* orbitosphenoid; *Fr.* frontals; *S.Eth.* supraethmoid; *Vo.* vomer.

ossified than the hinder part, and the cavity itself is considerably smaller proportionally. Both roof and floor are very thick. The interorbital septum is almost entirely cartilaginous, extending from just beneath the cranial floor to a level as low as the basilar crests enclosing the orbital muscles: it includes the original trabeculæ and their highly developed crest. It is a thick dividing wall rather than a mere septum.

174. From the level of the prootics in the basilar floor, down to the base of the hinder end of the interorbital septum, there is placed a curved basisphenoid bone (Fig. 20, *b. s.*), with an arm diverging on either side into the lateral cranial wall. Thus the bone is Y-shaped. The pituitary body occupies the space immediately behind; it is let down between the arms of the Y, and behind the long leg, to rest only upon the parasphenoid bone. The line of the basisphenoidal arms is continued by the alisphenoids (*a. s.*), which are considerable and thick bones occupying a large portion of the lateral cranial wall between the orbitosphenoids and the prootics, and having the sphenotics on their outer flank, though separated by a tract of cartilage.

175. There is only one cartilage-ossification in the cranial wall in front of the basisphenoid: this is a bone which may be called orbitosphenoid (*o. s.*), but which ossifies more than the orbitosphenoidal region. It contains in a hinder cavity a portion of the forebrain, having the olfactory crura lying on the floor of the cavity, and passing forwards in the substance of the bone. Thus the orbitosphenoid forms the floor and sidewalls and a small part of the roof of the anterior part of the cranial cavity, having beneath it the interorbital septum; and it also ossifies a considerable portion of the hinder ethmoidal region. The optic nerves pass out behind and below the orbitosphenoid on either side, in a gaping chasm between it and the basisphenoid (2).

176. The cranial roof is formed of a thick mass of cartilage, which instead of being gently arched as in previous stages, is trihedral, having a sharp median longi-

tudinal ridge or *culmen cranii* above, and rests laterally upon the orbitosphenoid and the alisphenoids. The tegmen cranii is complete as far as the supraoccipital bone, which extends forwards to about the middle of the auditory tract. In the alisphenoidal region the tegmen has sharp external angles; behind this it is somewhat contracted between the auditory masses, and leaves a pair of small oval parietal fontanelles in the cranial roof just in front of each antero-external angle of the supraoccipital.

177. The ethmoidal, internasal, and prenasal regions of the skull have attained an extraordinary development, quite overshadowing the original nasal septum and sacs. The trabecular cornua (Fig. 19, *c. tr.*) have grown out in front of the nasal cavities, and then coalesced; and behind these prenasal parts, and yet considerably in front of the brain-case, is a great *mesonasal* cavity (Fig. 20, *m. n. c.*), filled with connective tissue and fat, surrounded on all sides by cartilage, with the exception of a small opening on the upper surface of the skull, and irregular breaks in the cartilage inferiorly in the roof of the mouth. This space appears to be formed by a splitting of the original nasal septum into two inner nasal walls with an interval between; it is covered partly by the forward growth (both median and lateral) of the ethmoidal region in continuation of the tegmen cranii, and partly by a retral growth of the upper and back part of the cornua or prenasal cartilage. The olfactory crura diverge to their destination in the hinder region of this cavity.

178. The ethmoidal region, constituting the base of the great precranial pyramid, is the widest and deepest part of it; it is chiefly formed of massive cartilage, extending outwards on each side below to give a rounded ethmopalatine condyle for the palatine cartilage, and passing backwards above as great antorbital wings, largely ossified by the ectethmoid ("prefrontal") bones (Fig. 19, *l. eth.*), which however have a forward and downward extension to the border of the nasal pits.

179. The latter recesses have a high internal wall, which graduates into low anterior and posterior walls shelving downwards to the outer edge of the trabeculo-nasal floor. In front of this are the solid coalesced trabecular cornua, separating again anteriorly and slightly diverging as rounded bosses, which are tipped by the remains of the upper labial cartilages (l_1, l_2), which are united with them by ligament.

180. The cartilaginous palatal roof forms a pointed arch with its angle upwards, and sides approximately straight. From being obtuse behind, in the presphenoidal region, the angle becomes acute in front: it is filled in by parostoses. The cornua are rounded bosses with a groove between them below, in front of these sigmoid ridges. The labial cartilages attached to the cornua project downwards, and on them the premaxillae are moulded.

The palatal roof is bulged and irregularly perforated beneath the mesonasal cavity; and in front of, and external to this is seen, on each side, a sigmoid cartilaginous ridge passing from within outwards and backwards to the edge of the cartilaginous palate. These growths correspond in situation with the more transverse ridges on the fore end of the palatal aspect of the trabeculae in the second stage, (§ 130, p. 50), which subsequently disappeared, to reappear under this form.

181. There are five parostoses, very unequal in size, especially related to the cranial roof. The two parietals (*pa.*) are small, ridged, subarcuate bones, separated by the anterior part of the supraoccipital, articulating with it, and also with the frontal and epiotic of their own side, in front and behind. The much longer and broader pterotic is external to them, a tract of cartilage intervening.

182. The frontal bones (*fr.*) are very large, extending from the supraoccipital and the middle auditory regions to the internasal tract, and not merely lying on the comparatively narrow cranial roof, but stretching far outwards and downwards over the huge orbit. They only meet in their hinder half; being anteriorly separated by the median

ridge of the cranial roof and ethmoidal cartilage. The frontals are much the widest in their middle portions where they extend to the extreme outer limit of the orbits. A great part of the middle and posterior regions of the frontals is pitted by excavations for fatty tissue.

183. Anterior to the frontals, and partly overlapping them by a backward projection on each side, is a considerable median supraethmoid bone (*s. eth.*), broad and quadrate in front, covering all the internasal region and the back part of the cornua. The proper nasal (*na.*) is a small ossification on each nasal roof, external to the supraethmoid in its middle region.

184. The parasphenoid (*pa. s.*) is the great parostosis of the cranial floor: it extends from a little distance behind the palatal ridges mentioned in § 180, almost to the occipital condyle. Its anterior half is elongated-oval in shape: its hinder half is narrowed, but sends up a lamina on either side, outside the lateral walls of the subcranial canal for the orbital muscles. This lamina is continuous in the postorbital region with a basitemporal wing passing in front of the prootic. The anterior portion of the parasphenoid bears a median superior keel, and is arched correspondingly with the palatal roof. The vomer is an oblong azygous bone, the hinder three-fourths of which underlies the parasphenoid; it is armed with sharp recurved teeth anteriorly, where it sends up rather a high crest between the trabecular cornua, and grooving the cartilage in the region of the nasal septum (Fig. 20, *vo.*).

185. The bones and cartilages which remain to be described are principally related as boundaries to the mouth and throat. There are, besides, many subcutaneous bones, one series encircling the orbit, and others developed in skin-folds connected with the hyoid arch. The position of the investing bones contributes to unify these various appended structures with the tracts beneath; but there are two important primary connections of the deeper parts,

namely, the ethmopalatine, where a boss on the lateral ethmoidal region of the skull is tied by ligament to the palatine cartilage; and the hyomandibular, where the longitudinally extended upper edge of the hyoid arch articulates with the periotic cartilage underneath the projecting eave of the pterotic bone (*pt. o.*).

186. From the ethmopalatine connection a massive bar, which is for the most part cartilaginous, extends forwards and backwards. As it passes backwards it becomes

Fig. 21.

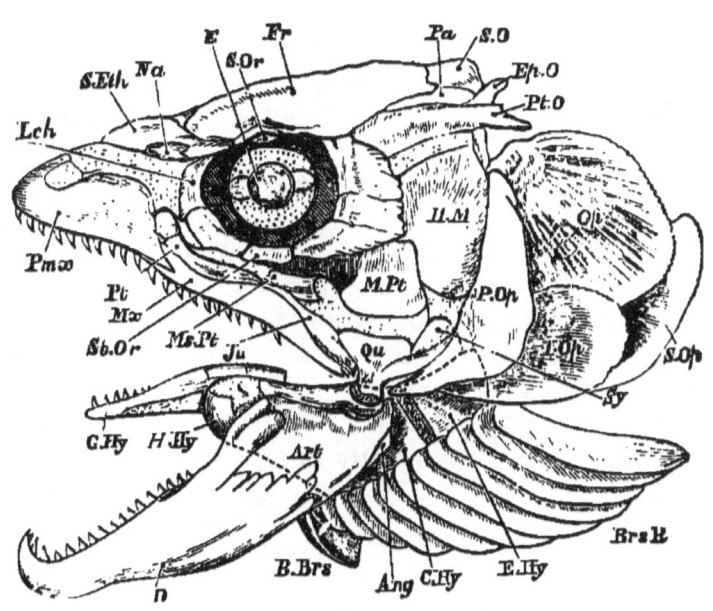

Adult Salmon: side view of skull with all bones attached; cartilaginous parts dotted.

Cartilage bones: *S.O.* supraoccipital; *Ep.O.* epiotic; *Pt.O.* pterotic; *Pt.* palatine; *Ms.Pt.* mesopterygoid; *M.Pt.* metapterygoid; *Qu.* quadrate; *Art.* articular; *H.M.* hyomandibular; *Sy.* symplectic; *E.Hy.* epiceratohyal, partly covered by *P.Op.* and *I.Op.*; *C.Hy.* ceratohyal, hinder part; *H.Hy.* hypohyal; *G.Hy.* glossohyal.

Membrane bones: *Pa.* parietal; *Fr.* frontal; *S.Eth.* supraethmoid; *Na.* nasal; *Lch.* lachrymal; *Sb.Or.* suborbital; *S.Or.* supraorbital; *Pmx.* premaxillary; *Mx.* maxillary; *Ju.* jugal; *P.Op.* preopercular; *Op.* opercular; *I.Op.* interopercular; *S.Op.* subopercular; *D.* dentary; *Ang.* angular; *B.Brs.* basibranchiostegal; *Brs.R.* branchiostegal rays.

much broader, and considerably extended vertically, so as to descend to a level much lower than that of the cranial floor. At the outermost inferior angle of this palato-quadrate tract is a condyle for the mandible. The extended hinder edge of the bar has a peg from the arch behind embedded in it in its lower part, while it overlaps the same arch above. The portion of the palatine cartilage in front of the ethmopalatine connection is considerable, and bears a rounded boss looking upwards to the edge and base of the skull, and articulating with the overlying maxillary bone.

187. The dentigerous palatine (*pl.*), a rod-like bone, extends to the forward end of the bar: it has ossified most of the cartilage in front, but merely invests it behind; only one-third of the length of the bone is behind the ethmopalatine conjunction. The mesopterygoid bone (*ms. pt.*) invests the cartilage behind the palatine, overlapping the latter by a pointed end for half its length on the inner side of the bar, and passing back and broadening to overlap the quadrate and metapterygoid. It also overlaps the proper pterygoid below, and above clings round the upper edge of the cartilage and covers it externally for some extent. The pterygoid embraces the lower edge of the cartilage from the palatine to the quadrate, and does not occupy much space on either the inner or the outer surface, but most on the inner.

188. The metapterygoid (*m. pt.*) is a squarish bone occupying the postero-superior angle of the palatoquadrate bar, and reaching to about half the height of the hyomandibular (*h. m.*), which it overlaps externally. There is a tract of cartilage between the lower edge of this bone and the upper edge of the rather small triangular quadrate (*qu.*), whose inferior angle bears the cartilaginous mandibular condyle, and whose posterior edge is grooved for the peg from the next arch.

189. The investing bones related to these regions belong to the upper jaw. Each premaxillary (*pmx.*) is a

splint of the corresponding trabecular cornu, passing backwards outside the prepalatine cartilage. It has an elevated prenasal plate, and a considerable dentary region. Its line is continued backwards by the smaller maxillary (*m.x.*), which is overlapped by the premaxillary in front, and by the jugal (*ju.*) behind; it also abuts upon the quadrate. The facial plate of the maxillary is small and little elevated; the oval margin is dentigerous in its whole extent. The jugal is a lanceolate bone not entering into the edge of the jaw, but l ing above, on the maxillary and the quadrate.

190. The connective tissue encircling the orbit behind the ectethmoid, above the palato-quadrate bar, below the frontals, and in front of the hyomandibular, has given rise to about seven membrane-bones, which form a nearly complete ring. These orbital bones have some extension within the rim of the orbit and an expansion on the face, which is more marked in the case of the hinder ones. The frontals partly interrupt the ring above, and enter into its circumference.

191. The lower jaw is an elongated bar, very massive behind, rising high in the coronoid precondylar region, having a projecting postcondylar or angular process, and a strong hook on each dentary bone at the symphysis. The proximal part of the cartilage has been ossified by the articular (*art.*), but anteriorly this bone lies as an outside splint on the meckelian cartilage, which extends beyond it almost to the symphysis. The dentary (*d.*) embraces the fore end of the cartilage both externally and internally, but farther backwards it appears principally on the outside, overlapping two-thirds of the articular, and terminating by an upper and a lower fork. It is dentigerous on the anterior three-fourths of its upper edge, but not behind, where its upper fork mounts on the coronoid process of the articular. A small angular (*ang.*) lies on the angle of the jaw beneath the condyle, but the postcondylar process is formed by the articular. The symphysis of the lower jaw

is of small extent; the moieties are united by fibrous tissue.

192. The hyomandibular extends downwards from its elongated horizontal pterotic articulation behind the postorbitals and the metapterygoid. It is narrower as it descends: on its posterior edge in the upper third is a bony process covered by a knob of cartilage, the opercular condyle. The cartilage below the hyomandibular bone is a thick prominent elbow, which turns forwards at a little more than a right angle into the small peg-like symplectic bone (*sy.*), fitting into the groove on the postero-internal face of the quadrate.

193. The inferior division of the hyoid arch is attached to the upper and anterior between its two ossifications, at the elbow just described. A small partially ossified interhyal cartilage intervenes, and the lower bar is directed forwards into the base of the tongue in the inter-mandibular space, being mainly composed of massive cartilage, flat proximally and thick distally. Its proximal third is ossified by the epiceratohyal (*e. hy.*), its distal two-thirds by the ceratohyal (*c. hy.*); the fore extremity of the latter, covered by cartilage, fits into the posterior end of the smaller hypohyal mass (*h. hy.*), which has two distinct ossifications. An azygous basi- or glossohyal (*g. hy.*) passes forwards into the tongue, having the hypohyals on either side behind, and being joined in the middle line posteriorly to the first basibranchial. The long glossohyal is ossified superiorly, and bears a row of teeth which project in the middle of the floor of the mouth.

194. The opercular membrane is a great semi-oval skin-flap extending backwards from the hinder edge of the hyomandibular and quadrate regions, and covering the upper part of the gill-arches. Four large membrane-bones are developed in this flap. The anterior, the preopercular (*p. op.*), occupies nearly its whole vertical extent, and is crescentic in shape, with two horns and a broad median

region; the lower horn almost reaches the postcondylar process of the mandible. The preopercular is burrowed in a radiating manner by mucous glands; and to its upper horn is attached a small supratemporal bone, one of the series of mucous bones which is continued by the scales of the lateral line. The bones of the orbital ring and the nasals also belong to this category.

195. The principal and largest opercular (*op.*) is subquadrate with very rounded angles; it is overlapped in front by the upper horn of the preopercular, and has a cup on its upper anterior angle, for the opercular knob of the hyomandibular. There is a postero-inferior subopercular (*s. op.*), lying within the opercular, which overlaps its upper edge, and also within the next bone to be described, which is outside its forward region. The interopercular (*i. op.*) overlaps the lower anterior angle of the opercular and the anterior portion of the subopercular. It is itself overlapped largely by the preopercular, and tied to the angle of the lower jaw by a strong ligament. Both sub- and interoperculars are very thin bones, of elongated oval shape.

196. The lower division of the hyoid arch has attached to its under edge on the outside a series of twelve flat subfalcate bones, the branchiostegal rays (*br. s. r.*). The posterior is the outermost and largest, and overlaps the next, and so on to the front and smallest. A continuous elastic membrane encloses the whole, passing from the hinder edge of one ray to the fore edge of the one behind it. Where the folds of opposite sides meet in front there is an azygous membrane bone, or basibranchiostegal, (*b. br. s.*), small anteriorly, with a high median crest and a broad hinder basal plate.

197. The five branchial arches form a series in the walls of the throat. The first four bear gills, and are bent backwards at a right angle, the upper portions being about two-thirds the length of the lower and anterior divisions. The upper parts of the arches are also bent

inwards under the hinder region of the cranial floor and the anterior vertebræ: and they present two divisions, a superior and smaller pharyngobranchial and a larger epibranchial, with a fibrous joint between. The fourth pharyngobranchial is attached to the lower hinder surface of the third. All four pharyngobranchials are more or less ossified and project almost directly inwards; they bear no teeth in this type. The forward epibranchials are the larger, and are shaft-like bones tipped with cartilage at both ends; they bear an upper external pointed process. The fourth epibranchial is short and flattened.

198. The lower divisions of the branchial arches are attached to the upper by very perfect hinge-joints. From this joint the long shaft-like osseous ceratobranchial is directed forwards; and the three first arches have a smaller hypobranchial piece segmented off and considerably ossified. The fifth branchial arch has only the lower division developed, and ossified as ceratobranchial; it bears teeth on its inner edge, and is the inferior pharyngeal bone of Cuvier. The bones of the arches are solid in the middle and usually surround a cone of cartilage at either end.

199. The first and fourth branchial arches have projecting from their antero-external margin about a score of small denticular bones, curving inwards; the second and third arches have two series of these, one directed forwards and the other backwards. These little bones are so arranged as to form a colander through which the water is strained as it pours through the branchial clefts. The postero-external face of each arch is grooved from the top of the epibranchial to the bottom of the hypobranchial. In this groove lie the branchial vessels; from it the gill laminæ project.

200. The gill arches of opposite sides are bound together mid-ventrally by a continuous series of bones and cartilages. Each basibranchial, like the basihyal, runs

forwards in front of its arch. There is no segmentation of the cartilage except behind the third basibranchial: and the cartilage behind this point is unossified, and bears both the fourth and the fifth branchial arches. Each of the three solid basibranchial bones is continuous by a considerable tract of cartilage with the next.

201. SUMMARY. Considering first the history of the cartilaginous elements, we find parachordals and trabeculæ in the floor of the cranium very early, and the union of trabecula with parachordal is complete before the parachordals have united with one another. The latter quickly fuse with the adjacent otic cartilages, and an occipital ring is completed in continuity with these masses. The parachordals during growth become separated by a considerable distance in their postpituitary region, and the trabeculæ bend outwards very definitely where they bound the pituitary space. In front they are speedily confluent, and become developed beyond the brain-case into lateral ethmoidal and prenasal or cornual regions, between which the nasal capsules lie.

202. The formation of a median interorbital septum above the coalesced trabeculæ elevates the fore part of the brain. The septum is at first membranous, but afterwards chondrifies continuously with the trabeculæ. The precranial cartilage grows to a height corresponding with that of the septum, and differentiates into massive lateral ethmoidal, subnasal, internasal, and prenasal regions. Upper labial cartilages become connected with its cornual terminations.

203. The supraorbital tract is early chondrified, and is confluent in front with the lateral ethmoidal region, and behind with the auditory capsule. A strong cartilaginous cranial roof is developed from before backwards until it joins the supraoccipital growth, and only leaves a pair of small membranous fontanelles in the parietal region. The side walls of the cranial cavity anteriorly to the otic masses are not extensively chondrified.

204. The mandibular, hyoid, and branchial arches arise as simple rods on each side; and a palatine or subocular bar is at the same time developed, having a definite attachment to the lateral ethmoidal region. The postoral arches undergo a segmentation by which they are divided into upper and lower portions. The inferior piece of the mandibular arch is elongated to constitute the axis of the lower jaw; the upper division grows forwards to join the palatine bar, forming a continuous palato-quadrate tract. It gradually descends from its proximity to the cranial wall, and is finally supported at a much lower level by the upper division of the next arch.

205. The segmentation of the hyoid arch occurs throughout its entire length, the anterior tract becoming superior, and the posterior inferior. The upper part is the suspensorium of the lower jaw: it is articulated by an elongated surface with the otic cartilage, and is applied in front to the hinder edge of the upper mandibular segment. The remainder of the hyoid arch, and also the branchial arches, become more or less segmented, but not remarkably modified. Median basal pieces connect the bars of opposite sides.

206. The proportion of the chondrocranium which is replaced by osseous deposit is but small. The occipital ring has four bones, but they are separated by cartilage. The feeble basi- and alisphenoids, with a peculiar massive orbitosphenoid produced by the junction of a pair of bones, are the other chief cranial ossifications in cartilage, besides those related to the ear-capsules. Each of these acquires five bony centres, of which the pterotic is important as overhanging the hyomandibular articulation, while the prootic encroaches largely on the proper cranial floor. The intrinsic ossifications of the facial bars are very distinct, their history is simple, and recapitulation is unnecessary.

207. The subdermal or membrane bones are very numerous, the principal appearing sooner than the cartilage-bones. Very important elements occur in relation to the

base and roof of the cranium; they are azygous beneath, and mostly paired above. The extent and relations of the parasphenoid demand especial attention. The precranial, palatine, and mandibular cartilages have series of bones related to them as splints; while the large orbits are partly protected by the outward extension of bones related to the cranial roof, and partly by special circumorbital bones. The great opercular and branchiostegal membranes have their own osseous skeleton, while the branchial arches possess many small tooth-like parostoses.

APPENDIX ON THE SKULLS OF FISHES.

208. The skulls of most osseous fishes are constructed substantially like that of the Salmon: cartilage may persist to a greater or less extent, and anchylosis of bones may take place. In Siluroid and Cyprinoid fishes, however, an interorbital septum is not formed. The substance of the ethmoidal cartilage may become partially ossified by median (Cod) or paired (Pike) bones. In the *Murænoids* the trabeculæ become merely narrow bands of cartilage in the orbital region, arched forwards and upwards upon the parasphenoid. Anteriorly there is a median vertically-crested ethmoidal bone, deeply grooved beneath to receive a high crest from the dentigerous vomer. On either side of this crest lies a distinct simple trabecular rod. The olfactory nerves as they pass to the nasal sacs are enclosed by a pair of thick separate ectoethmoidal cartilages; above the membranous olfactory capsule is a small nasal bone. There are no premaxillaries; the vomer, dentigerous all the way, runs to the anterior extremity of the beak; the maxillaries are large and dentigerous. The hyomandibular is very large, with two distinct heads widely separated: its distal part is directed downwards and forwards; the symplectic is distinct. Between these two bones the rest of the hyoid arch is attached by the interhyal ligament. The epi- and ceratohyals are about equal in size, and there is no hypohyal. The apex of the suspensorial part of the mandibular arch is unossified, and is let down to a position opposite the interhyal ligament, being embraced by splint-like processes of bone from the hyomandibular. There is a rod-like quadrate, and in front of it a minute tri-

angular process of cartilage—all the pterygoid cartilage that exists: above it is a styloid palato-pterygoid bone. The lower jaw possesses, besides dentary and articular, a small thick coronoid on the inner face, wedged between the two larger bones.

209. The *Siluroids* resemble in important respects the holostean Ganoids; many of them exhibit a remarkable combination of ganoid dermal plates with strongly ossified cranial cartilage. They agree with certain Ganoids, and differ from other Teleosteans in having no subopercular. In the Siluroid *Clarias capensis* the two dentigerous premaxillaries occupy a large part of the margin of the fore face, and behind them is a crescentic tooth-bearing plate, the anterior part of the single vomer. The maxillaries are minute, and bear a long cirriform filament; above them each nasal capsule has an ossification comparable to a septo-maxillary (chap. v.). The nasal sacs are very small; so are the orbits, which are placed very far forwards. The anterior part of the face is covered by a large broad dermo-ethmoid. Even in the adult, small unclosed spaces persist between the frontals and in the supraoccipital region. The palatine is a distinct rod-like bone attached to the lateral ethmoidal region, carrying the maxillary, and lying partly above and within the pterygoid and the mesopterygoid. The quadrate comes far forwards, and there is no metapterygoid. By comparing this skull with a Salamandrine form many instructive points of agreement are discovered.

The Skull in Ganoids.

210. The holostean *Ganoids* have their chondrocranium ossified essentially like the Teleosteans, certain bones being sometimes missing. The single jugular plate of bone developed between the mandibular rami of Amia, and the two plates found in Polypterus in the same region, are structures not plainly comparable with anything in the Salmon. In other respects the student will find little difficulty in interpreting these skulls.

In *Polypterus*[1], for example, large basal and superior fontanelles persist in the chondrocranium, which is ossified by an occipital ring of bones, by otic masses representing opisthotic and epiotic bones conjoined, by sphenoidal walls united in the cranial floor behind the

[1] See Traquair, "On the Cranial Osteology of Polypterus," *Journ. Anat.* vol. x. p. 166.

pituitary body, and also sending bony plates inwards in the anterior part of the cranial floor, which meet but do not anchylose; by sphenotics or "postfrontals," lateral ethmoids ("prefrontals"); and a median ethmoid in the extreme anterior part of the cranial cartilage, sending back an upper spur which is parosteal. There is no interorbital septum. The basioccipital ossification includes at least one vertebral centrum, the first neural arch being articulated with it above. The posterior margin of the sphenoid is slightly notched by the foramina for the hinder divisions of the trigeminal nerve; near the middle of its lower margin is the optic foramen, above and behind which the motor nerves of the eyeball and the first branch of the trigeminal emerge. The principal roofing bones of the skull are paired parietals, frontals, and nasals, which become more or less anchylosed in old specimens. The septo-maxillaries and median ethmoid appear slightly on the surface. The parasphenoid extends underneath almost the whole occipital region, and reaches forwards nearly to the extremity of the palate: it has remarkable basitemporal processes, and a minutely denticulated anterior region. The premaxillaries and maxillaries have considerable palatal plates. The palato-quadrate cartilages are ossified by small metapterygoids and quadrates, large pterygoids and mesopterygoids, and very small palatines; the vomers continue the line of the pterygoids, within and parallel to the palatal plates of the premaxillaries and maxillaries. The hyomandibular is rod-like, bent almost at a right angle in the region of the opercular condyle, and it has a small additional ossification at its upper end. There are large preoperculars or squamosals, considerable operculars, and small suboperculars. The superficial interspaces between the bones named are filled by many small dermal and membrane bones, which are more or less modified representatives of the lateral line series of scales.

In *Amia*[1] there is a chondrocranium perfectly complete, covered by a shield of suturally united dermal bones. The occipital cartilage bones are all present, two neural arches being articulated with the hinder part of the basioccipital. The basisphenoid is represented by a pair of ossifications on either side of the pituitary fossa; there are large alisphenoids, having a descending plate: the orbitosphenoid is simple. The ear-capsule has a complete set of bones, with the exception of the pterotic. The parasphenoid has

[1] See Mr Bridge's valuable account in *Jour. Anat.*, Vol. XI.

basipterygoid wings extending to and propping up the sphenotic. The supracranial dermal shield consists of a median dermo-occipital, flanked by a pair of supra-epiotics; of large paired frontals; a pair of nasals; and an anterior supraethmoid which partly underlies the nasals. The circumorbital bones are highly developed; there are two large postorbitals which extend backwards by the side of the supra-epiotics. The palatine series of bones is normal. Each premaxillary has two ascending processes surrounding the nostril and uniting above it, and *beneath* the nasal bone. The maxillary has a jugal behind it. The suspensorium and the branchial arches need not be remarked upon; but the lower jaw is singularly interesting. The meckelian cartilage is largely persistent, ossified proximally by four very distinct centres, three of which enter into the condylar surface, and distally by a mento-meckelian. The splenial is in five distinct pieces, one being larger, but all dentigerous. The other bones are dentary, angular, and surangular. The opercular set of bones is complete, with the addition of a supratemporal.

211. In the *Sturgeon* there is a very solid cartilaginous skull, overlaid by dermal bones more or less representing the superficial bones in Teleosteans, and having a large parasphenoid at its base. The precranial region is of great extent, forming a massive beak, surrounded by very generalised dermal bones. The mouth is swung from a large hyomandibular, which has a small sheath of bone near the top, and swells out below into a broad solid plate of cartilage. To this is loosely jointed a symplectic piece, largely ossified, and turned forwards, bearing the upper and lower jaws. The broad metapterygoid passes towards the middle line in the roof of the mouth, above and behind the continuous palato-quadrate cartilage which is partially ossified as pterygoid and palatine: sometimes a small quadrate bone is found near the condyle. The margin of this palatine tract is flanked by the edentulous maxillary, which reaches to the quadrate articulation, and bears a styloid jugal. In old specimens there is a separate mesopterygoid. In the lower jaw the meckelian cartilage persists, surrounded by the dentary, and sometimes ossified anteriorly by a mento-meckelian: there is also a small angular. The lower division of the hyoid arch starts from the hinder end of the symplectic, and has interhyal, ceratohyal, and hypohyal segments.

The Skull of Ceratodus.

212. In *Ceratodus*[1], the massive cartilaginous cranium is produced laterally and inferiorly into a large palato-suspensorial process, continuous behind with the ear-capsule. From the latter there projects backwards a thin wide cartilaginous plate roofing the branchial chambers. The persistent meckelian cartilage is articulated with the condyle on the suspensorium at its farthest outward extension. A small cartilage identified by Prof. Huxley as hyomandibular is found on the hinder side of the suspensorium, firmly bound to it at its junction with the cranium proper; and immediately in front of its anterior edge the posterior division of the facial nerve emerges from the cranium. It has a symplectic process embedded in the strong hyosuspensorial ligament, by which the main hyoid bar is fastened to the suspensorium. There are cartilaginous representatives of branchiostegal rays. Each of the anterior four branchial arches consists of a long ventral and a short dorsal piece of cartilage; there are rudiments of basibranchials. The fifth arch consists of a single piece of cartilage on each side, curved forwards and united with the principal piece of the fourth arch, both above and below. Two pairs of small labial cartilages have been found defending the posterior narial orifices.

213. The cartilaginous cranium is covered by an osseous shield consisting of two median bones, one behind the other, occupying almost the entire length of the skull, and of two pairs of lateral bones, one outside the other, extending over the hinder two-thirds of the head, and lying principally above the otic cartilage. The hinder and larger part of this osseous shield lies upon the large temporal muscles which entirely cover the chondrocranium. The outer bone on each side appears to be homologous with the preopercular of Teleosteans and the squamosal of higher vertebrates: it has an elongated process running down nearly to the quadrate articulation. A fibrous band stretching between the antorbital process and the ventral end of the suspensorium contains three suborbital bones. The cranial floor is supported by a large parasphenoid which extends backwards to a point just beyond the attachment of the third pair of ribs, several anterior vertebræ being, as in the

[1] Prof. Huxley's description is here principally followed (*Proc. Zool. Soc.* 1876); see also Günther, *Phil. Trans.* 1871.

Sturgeon and other Ganoids, confluent with the skull, so that the hinder boundary of the latter has to be imagined between the exits of the pneumogastric and of the first spinal nerve. Just in front of this boundary there lies on either side, deep in the substance of the cartilage, a hollow cone of bone (exoccipital). The parasphenoid is narrow behind and lozenge-shaped anteriorly, with two pointed basitemporal wings at the junction of the two regions. In front of these processes it is conterminous with the two elongated palato-pterygoid bones, which extend inwards from the region of the condyle for the mandible, to meet each other in the middle line for some distance in front of the parasphenoid. The fore part of each palato-pterygoid has anchylosed to it a large dentary plate lying obliquely, and bearing prominent ridges on its external region. Vomers are represented merely by the bases of the pair of anterior trenchant teeth planted on the ethmoidal cartilage, meeting each other in the middle line and diverging outwards and backwards at a right angle to one another. A downward process from the inner and anterior of the lateral bones of the cranial roof meets the fore part of the palatopterygoid. There is no interorbital septum. The meckelian cartilage is sheathed by dentary, splenial, and angular elements, the splenial bearing a dentary plate corresponding to the large palatal one above described. In addition to the preopercular or squamosal already mentioned, there is a principal opercular and a smaller rod-like interopercular. The only other ossification to be noted is the ensheathing bone of the large hyoid bar.

214. The skull of *Lepidosiren* presents many features of resemblance to that of Ceratodus, especially in its cartilaginous parts; no tract, however, identifiable with the hyomandibular appears to have been clearly made out. There are two exoccipitals, a large parasphenoid extending partially beneath the vertebral column, a single roofing bone from the occipital to the ethmoidal region (called parieto-frontal by Prof. Huxley), a pair of nasals, or one median supraethmoid, and a great "supraorbital" on each side, which passes back to the extremity of the head at a considerable height above the other cranial structures. The "parieto-frontal" and parasphenoid send processes towards one another in the side walls of the cranium, which unite in front of the exit of the hinder division of the trigeminal nerve. The suspensorium is continuous with the side wall of the cranium, and has on its anterior and outer margin

the large palatopterygoid, meeting its fellow in the middle line, and bearing dentary plates. The vomerine region bears teeth but no distinct bone at their base. Prof. Huxley says [1], "The parasphenoid, the rudimentary vomers, and the pterygopalatine plates correspond in the two genera (Lepidosiren and Ceratodus). The exoccipitals are much larger in Lepidosiren. The descending process or pre-opercular part of the squamosal is best developed in Lepidosiren, whilst its dorsal part (proper squamosal) is larger in Ceratodus. In both, there are two opercular bones, an operculum and an interoperculum; in Lepidosiren, as in Ceratodus, there are cartilaginous plates attached to the inner faces of these bones. The branchial apparatus of Lepidosiren differs from that of Ceratodus mainly in the greater number of complete branchial arches." It must be confessed that a knowledge of the development of the skull of Lepidosiren is very much needed before a satisfactory account of it can be given.

215. In *Chimæra*[2] the suspensorium is continuous with the side of the brain-case, but its principal development is far forwards, carrying the condyle for the mandible. There are two palatine and two mandibular teeth fitting into one another, and two vomerine teeth exist in front of these, as in Ceratodus and Lepidosiren. The cranial cartilage is produced into a remarkable interorbital crest *above* the anterior part of the brain. The notochord has disappeared, and the skull is articulated with the anterior coalesced vertebræ. The hyoid arch is substantially similar to the branchials, though larger. It terminates dorsally in a flat expanded triangular piece, connected with the superjacent floor of the skull by muscles and ligaments, but by no direct articulation. The dorsal pieces of the succeeding branchial arches have the same form and attachments, and unite with the ventral segments at a sharp angle. An opercular membrane covers the gill-clefts. There are large labial cartilages.

216. In Sharks and Rays there is a great diversity of form and relative size of parts, but the only members of the group that need special notice are Notidanus (Heptanchus and Hexanchus) and Cestracion. In *Notidanus* the hyomandibular is much reduced, and only partially supports the jaws; the gape is of great extent, and the articulation of the lower jaw is far back; the upper jaw is much

---

[1] *l. c.* p. 38.
[2] Huxley, *l. c.* p. 40.

enlarged vertically behind its middle, and forms a definite articulation with the strong postorbital or sphenotic process of the cranium. Further forwards there is also a small process which is tied to the basal prominence of the skull behind and below the optic foramen. There is no spiracular cartilage. Heptanchus has seven branchial arches, Hexanchus six. In *Cestracion* the fore part of the skull is very long in proportion to the posterior; the hyomandibular is small, and articulated with the cranium low down and far back, in a completely normal relation to the exit of the facial nerve; while the upper jaw has an almost semicircular superior edge, and derives its chief support from an elongated area at the base of the cranium *in front* of the orbit. There is a distinct spiracular cartilage. In Centrophorus there are three cartilaginous rays close together in the anterior wall of the spiracle.

217. The descriptions here given are considered not to go farther in identification than the facts of development at present known will warrant. An independent judgment can be formed upon them by each observer. The skulls of the Lampreys and Myxinoids are not similarly treated, owing to our ignorance of their development, and of the extent to which our present knowledge may be sufficient for their comprehension[1]. Generally speaking, it may be stated that they have a simple chondrocranium arising in the fore part in the form of trabeculæ and in some retaining that form permanently, together with more or less rudimentary visceral arches. To these elements very large labial developments are added. It does not appear that sufficiently sure grounds exist for definite conclusions respecting the relation of the structures found in the head of Amphioxus to the skulls of other vertebrates, although highly interesting and instructive speculations have been made on the subject[2].

[1] See Huxley, "On Petromyzon," *Jour. Anat.* x. 412.
[2] See Huxley, "On Amphioxus," *Proc. Roy. Soc.* 1875.

CHAPTER IV.

THE SKULL OF THE AXOLOTL.

First Stage: Embryo Axolotls, a day or two before hatching.

218. THE embryo lies coiled crescentically in its gelatinous envelope, and is only three lines in length. The tail is very short; the abdomen is protuberant, enclosing the yelk; the head constitutes fully one-third of the length of the little creature. The head is broadest over its hinder part, in the gill-region, and tapers gently forwards. The mesocephalic flexure is fully established, and the brain-vesicles are strongly protuberant. The third vesicle is well seen above, occupying much the greater portion of the upper surface of the head. In shape it is elongated oval, pointed behind, and covered by a thin dermis, beneath which there is a very watery stroma down to nearly half the depth of the chamber. The second vesicle is sub-globular, and forms the anterior termination of the head. The first vesicle is small and completely on the under surface, below the hinder half of the second vesicle.

219. The sense-capsules lie in a lateral oblique line, from just behind the middle of the third vesicle to the side of the first. The olfactory pit consists of an annular elevation surrounding a crater-like depression, lying on the supero-posterior face of the first vesicle; it is the smallest of the sense-capsules. Immediately above and behind it is the commencing eyeball, a reniform mass of cells with a stalk looking downwards and backwards. A space inter-

venes between this and the next organ, the auditory, which is also at a higher level: it is a reniform mass with its stalk above.

220. Looking on the under surface of the head, there is seen a narrow crescentic membrane convex backwards, behind the first vesicle. This is the nasofrontal process, forming a curved commissure between the two olfactory sacs. Behind this there is a slight depression, which is in front of the convex anterior edge of the mandibular fold: the boundaries of the depression being convex towards each other, it is necessarily broad at each side and narrow in the middle, so as to be somewhat hourglass-shaped. This is all that is manifest externally as a rudiment of the mouth.

221. Behind the mouth two rib-like elevations lie transversely one behind the other, ascending laterally. The anterior of these reaches nearly to the eyeball, the posterior to the auditory rudiment. There is no definite solution of continuity between these two bars, which are the first indications of the mandibular and hyoid arches. They are flat below and project somewhat forwards in the mid line. On either side of the rudimentary mouth-cavity below the eyeball, and immediately behind the nasal rim, is a small bud of tissue (maxillopalatine) which is an outgrowth of the upper part of the mandibular arch.

222. The hinder margin of the hyoid arch is crescentic in the middle ventral line, the concavity being backwardly directed. The horns of the crescent project backwards on each side, distinct from the body-wall and overlapping the structures behind. These are the rudiments of the opercular folds. The under surface of the body immediately behind the hyoid arch presents no other structures besides the continuous thin skin, through which the heart is just visible. Infero-laterally, behind each opercular fold, three oblique slits are seen ascending backwards, leading into the pharynx or fore part of the alimentary cavity. Thus there is marked out a series of visceral

arches behind the hyoid, and the fissures must be called the second, third, and fourth visceral clefts.

223. The visceral arches are slightly distinguishable by means of grooves on the lateral surface of the body, above the visceral clefts; and from the region of each arch there projects backwards a small claw-shaped bud. These buds lie very obliquely one behind the other, so that the second is higher up than the first, and the third than the second. Seen from above they are trilobate; but from below it appears that the proximal lobes form a sort of sheath to the median terminal lobe, so that the gill-buds bear some resemblance to the ungual phalanx of a cat.

224. The notochord extends forwards to the upper and posterior region of the pituitary body, tapering to a rounded extremity, after curving a little downwards. It underlies only the hinder two-thirds of the third vesicle; the anterior third of that vesicle rests on a fold of membrane which is evidently a continuation of the sheath of the notochord (middle trabecula of Rathke). This membrane, where it lies upon the infundibulum immediately behind the second vesicle, is sharply folded upon itself to form a lower layer, which is reflected backwards on the infundibulum and pituitary body, and then downwards around the latter so as to invest it entirely behind. At present there is no distinct indication of the primary skeletal elements of the cranium.

225. Within the rib-like elevations of the mandibular and hyoid arches, and in the intervals between the visceral clefts, the future visceral rods are already marked out, but have only commenced to chondrify. Thus the rudiments of five visceral arches are indicated, and a median section shows the heart lying below the two hindermost of them. Behind the heart and these arches the whole body is full of yelk; but *above* the section of the rods, a considerable visceral or pharyngeal cavity exists, in the wall of which the visceral clefts are seen. Anteriorly, above the rudimentary mouth, and behind the pituitary body, a considerable mass of cells still obstructs the alimentary

passage; but where the mandibular arch is in contact with the nasofrontal lip there is a fissure, which is all that at present represents the mouth; this slit however does not yet communicate with the visceral cavity.

226. In this embryo we have, as it were, the complete soft model of the parts to which the embryonic skeleton is to be related. The cranial cavity is bounded by unsolidified layers of cells, and a distinct membranous cranium can hardly be said to exist. Of definitely formed axial structures we have only the notochord present: while the more advanced visceral arches, five in number, represent the ventral series of skeletal parts.

Second stage; Larval Axolotls just after hatching.

227. The little creature has not undergone any very striking external change, having increased in size about one-fourth. The general shape of the head is more rounded and even, the cerebral vesicles protruding less markedly than before. Looking from above, the third vesicle occupies only about half the length of the head; the second vesicle extending over the remainder of the upper surface. The first vesicle, though still inferior, is at the anterior extremity of the head, immediately below the second. The eyes are now just intermediate in position between the nasal and auditory sacs, separated from them by a short interval. Each eye is almost a complete globe, still showing a narrow stalk below. The nasal folds are much nearer the front wall of the face, and are more complete; the cells of the crater are beginning to be absorbed to form the nasal opening. The auditory masses are globular, but a pit remains on their supero-lateral surface.

228. The oral cavity is perfect, and communicates with the pharynx or cavity enclosed by the visceral arches. It is surrounded by well-formed lips that grow from the nasofrontal process in front, the mandibular arch behind, and from the lateral buds described in the last stage (§ 221, p. 92), which are the maxillopalatine rudiments.

THE AXOLOTL: SECOND STAGE.

These lateral flaps are very considerable, overlapping the mandibular arch; they bear a strong resemblance to opercular folds. No cleft can be discovered between the mandibular and hyoid arches.

Fig. 22.

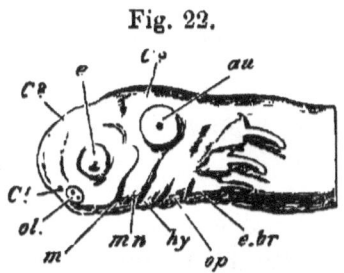

Head of Axolotl, just after hatching, side view.

C 1, 2, 3, cerebral vesicles; *ol.* nasal sacs; *e.* eyeball; *au.* auditory mass; *m.* mouth; *mn.* mandibular, *hy.* hyoid arch; *op.* opercular fold; *e.br.* points to the base of the first branchial bud. The maxillopalatine flap is discernible in front of the point to which the line from *m* extends.

229. The opercular fold (*op.*) on each lateral inferior angle of the hyoid arch has become very distinct from the solidified arch within. It has increased greatly in size and covers the roots of the trilobate external gills; moreover it is continued upwards and backwards by growths in the undefined upper regions of the first and second gill-arches. The trilobate gill of each of the three arches behind the hyoid has become twice as long as in the last stage, the median finger-like lobe retaining its preeminence; but there are no new lobes or filaments. The relative positions of the gill-rudiments are unchanged; the visceral clefts are more perfect, but they only ascend as far as the contiguous filament, without passing beyond.

230. The following description of skeletal structures is given from specimens slightly above one-third of an inch in length (four and a half lines)[1]. When the cranial roof and the whole of the brain are removed, the notochord is seen, of great relative size, ending immediately behind the pituitary body. On either side of the notochord, in the

[1] An idea of the structures at this stage may be gathered from Fig. 23, which represents a more advanced condition.

region of the third vesicle, are the rudiments of two segmental muscular masses one behind the other, indicating the future myotomes. Outside the first muscular segment lies the auditory capsule, having as yet no chondrified walls, but showing a number of shining otolithic crystals in its interior. The semicircular canals are not yet differentiated.

231. Two aggregations of nervous cells cling to the inner wall of the ear-sac. The anterior, parallel with the apex of the notochord and extending in front, is the first appearance of the Gasserian ganglion: the other is at the side of the first muscular mass and its junction with the second; it will become the ganglion of the glossopharyngeal nerve. Behind these, and parallel with the second muscular mass, is a small ball of tissue from which the thymus gland is developed. There is no indication of cartilage at the base of the cranium posteriorly.

232. The trabeculæ are distinct, though small relatively to the notochord. At present they are not cartilaginous, but consist of very dense granular tissue. They closely embrace the fore end of the notochord laterally, but do not meet beyond its apex. In front of the notochord the trabeculæ diverge outwards and forwards in a crescentic curve. Each trabecula constitutes the lateral third of a circle whose anterior (transverse) third is deficient. Within and upon this circle lie not merely the pituitary body and infundibulum which are relatively much smaller than in the preceding stage, but also the basal part of the first cerebral vesicle. Thus the primary relation of the trabeculæ is to the sides and base of the forebrain.

233. The visceral arches are already chondrified, although very delicate. They are all continuous across the middle line below, except the mandibular; but the median connecting substance is not as yet true cartilage. In this and in several succeeding stages there is no cartilage in the maxillopalatine process.

234. The mandibular arch consists of two pieces on each side; an upper, broad and three-sided, lying on the lateral aspect of the face between the eye and the ear; and a lower sigmoid bar, the axis of the mandible. The upper or suspensory portion is not cartilage in this stage, but is formed of somewhat solid indifferent tissue like the trabeculæ. The apex of this segment is growing towards the anterior third of the corresponding trabecular bar, but is at some distance from it and on a rather higher level. The anterior end of the Gasserian ganglion is wedged in between the hinder margin of the suspensorium and the antero-internal face of the auditory sac.

235. The sigmoid mandibular or meckelian cartilage is expanded at its upper end, and has a bevelled face in relation to the antero-external angle of the three-cornered suspensorium. The distal ends of the meckelian cartilages are separated by a considerable space at the chin: they lie above and behind the lower lip. In relation to the inner face of each of these cartilages at their middle is an oval mass of large globular cells. This is the stroma which in a few days will develope into the splenial bone with its teeth.

236. The succeeding arches, five in number, are not markedly distinguished from each other morphologically. They are closely connected in the middle line by granular tissue. The hyoid arch is shorter than the first branchial, and does not reach so high; the branchial arches decrease in size from before backwards, and each arch extends higher up laterally than its successor. The hyoid is crescentic, and convex forwards: the branchial arches, especially the first three, are sinuous. Each branchial arch except the last bears a trilobate gill.

237. The skeletal elements here described consist of the trabeculæ, paired mandibular arches, each segmented into two pieces, and simple hyoid and (4) branchial arches. The relations of the trabeculæ to the front end of the notochord and to the sides and floor of the forebrain, are particularly to be noted, together with the position of the

mandibular suspensorium with regard to the trabeculæ. The appearance of muscular segments in the floor of the hindbrain is of considerable interest.

Third Stage: Larval Axolotls, from five to five and a half lines in length.

238. The head has acquired its permanent perennibranchiate shape, being broad and flat; the cerebral vesicles lie nearly on the same plane, and are but little distinguishable externally. The mouth is wide and gaping, the lower jaw having increased in length very greatly, so as to project in front of the rest of the head. The sense organs are rapidly becoming perfect. From being trilobate the gills pass to a pinnate form by the growth of lateral basal filaments.

239. The hinder part of the cranial notochord is still unflanked by cartilage; and the muscular segments (Fig. 23, *m. s.*) are perfect and functional, the anterior pair being attached to the posterior pointed ends of the trabeculæ. The auditory capsules are elegantly ovoidal, having become cartilaginous in their floor and external surface. The roof is unchondrified, and the semicircular canals are not developed sufficiently to modify the external shape of the otic mass.

240. The trabeculæ (*tr.*) have become chondrified, and very greatly developed. They are distinguished into two regions, an anterior prechordal at right angles to a posterior parachordal. The parachordal portion is thick but flattened, and has grown back in a pointed form at the side of the notochord, closely applied to its sheath. The posterior extremity of this trabecular growth is defined by a line joining the middle of each auditory mass and passing through the middle of the first muscular segment and of the cranial notochord. The trabeculæ do not meet in front of the apex of the notochord.

241. The anterior portion of each trabecula, seen from below, appears as a rounded rod, passing directly forwards,

but curved a little inwards: and the right and left bars are widely separate from each other anteriorly, by a distance equal to the length of each prechordal part. Thus

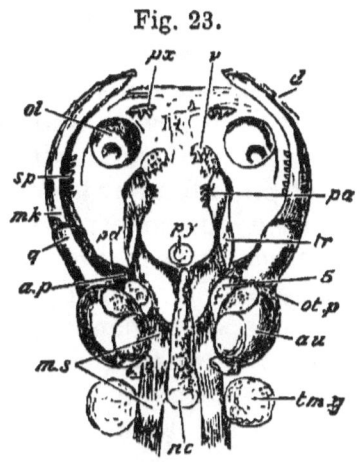

Fig. 23.

Larval Axolotl, about five lines long; upper view of skull, dissected.

nc. notochord: *py.* pituitary body; *m. s.* muscular segments; *tm. g.* thymus gland; *au.* auditory capsule; *tr.* trabecula, extending backwards by the side of the notochord (the line points to part of the trabecular crest); *ol.* nasal sac; 5. Gasserian ganglion; *mk.* meckelian cartilage; *q.* quadrate end, *u. p.* ascending process, *ot. p.* otic process, of the suspensorium; *pd.* its pedicle, which is at a lower level than *a.p.*

Bones: *pa.* palatine; *v.* vomer; *px.* premaxillary; *d.* dentary; *sp.* splenial.

the trabeculæ altogether form three sides of a somewhat square space, the fourth (anterior) side being deficient.

242. Viewing the trabeculæ from above, a thick crest is seen to be growing from the hinder three-fourths of each prechordal tract. It is highest opposite the apex of the notochord, and then suddenly diminishes behind. The height of the crest is equal to the distance between itself and the notochord. It is gently concave inwardly, and sinuously convex on the outer surface, the sinuosity arising from the fact that the temporal muscle has its origin from a depression upon it behind the eye. The muscle extends downwards, outwards, and *forwards* to be attached to the proximal part of the mandible.

243. The mandibular arch in both its divisions is now well chondrified. The suspensory cartilage, broad above, narrower and more rounded below, has receded in relation to the trabeculæ; its upper extremity grows towards the trabecular crest, parallel with the apex of the notochord; but it is still some distance from it. The postero-external part of the suspensorium is produced behind this apex; it is concave, and is applied to the convex antero-external face of the auditory cartilage. This is the otic process (*ot. p.*), afterwards of so much importance. The distal part of the suspensorium is directed forwards and a little outwards, reaching to the middle of the prechordal rods. The articular condyle is a slight simple convexity, fitting into a corresponding concavity on the proximal end of the meckelian rod (*mk*). The latter is gently arcuate, gradually tapering forwards, and ending at a little distance from its fellow in the projecting chin.

Fig. 24.

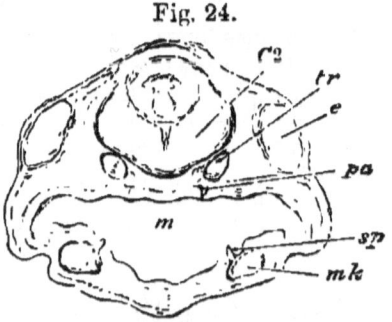

Larval Axolotl, about five lines long; transverse vertical section of head, through eyeballs.

C 2, midbrain; *tr.* trabecula; *e.* eyeball; *pa.* palatine tooth; *m.* mouth; *mk.* meckelian cartilage; *sp.* splenial tooth.

244. In the mouth several pairs of dentigerous osseous nuclei have appeared, arising parosteally in definite relation to the cartilaginous rods. The foremost of these are above and behind the upper lip, and a little in front and to the inner side of the nasal sacs; these are the rudiments of the *premaxillaries* (Fig. 23, *px.*), and each carries two denticles. The nasal sacs are becoming perforated, to open into the roof of the mouth.

245. Behind and internal to the nasal sacs, and immediately in front of the anterior termination of the trabeculæ, is another pair of bony patches; and yet further behind these, adherent to the under surface of the rounded ends of the trabeculæ, is a third pair. These little patches bear a number of prickly denticles, very much resembling the groups of teeth found on the inner edges of the branchial arches of osseous fishes, constituting their straining apparatus. The foremost pair of these patches are the rudiments of the *vomers* (*v.*); the hinder, of the *palatines* (*pa.*).

This stage shows the vomers in their interesting normal relation to the primary inferior nares. Each of these openings has one of the vomers directly internal to it.

246. The mandible has already two delicate dentigerous laminæ of bone on each ramus. The foremost of these, the *dentary* (*d.*), is on the outside of the distal half of the meckelian cartilage, and even projects somewhat beyond it towards the middle line of the chin. The middle of the inner face of the cartilage is covered by a shorter and broader patch bristling with teeth, and very similar to the vomers and palatines; this is the *splenial* (*sp.*).

The dentigerous splenial, appearing very early in development, is characteristic of the Urodela: the Anura, as a rule at any rate, have no splenial. In this minute Axolotl, five lines and a half in length, hatched about a week, there are no fewer than ten bones distinguishable in relation to the mouth.

In Axolotls two or three days younger and half a line shorter, the meckelian cartilage is not nearly so long, and is curved downwards in front. It then has only one dentigerous plate, the splenial, bearing a few teeth; but this splenial is on the inside of the fore part of the cartilage. The region in relation to which the dentary is soon developed is formed by the elongation of the cartilage in front of the splenial.

247. The succeeding visceral arches are becoming segmented. No part of the hyoid arch is developed into a counterpart of the suspensory portion of the mandibular: but the distal or ventral end has a short segment cut off

from it; this is the *hypohyal*, and the main bar is the *ceratohyal*. The rounded upper end of the ceratohyal articulates with a scooped surface on the hinder margin of the lower part of the suspensorium; and it extends no further upwards towards the auditory capsule.

This is a constant relation of the two arches in the larvæ of the Amphibia, which also have no upper hyoidean element corresponding with the hyomandibular of Fishes in their larval condition.

248. The two large anterior branchial arches segment into two pieces each, the upper being twice the size of the lower. The superior is the *epibranchial* element (with no pharyngobranchial above it); the inferior is the *ceratobranchial* (without a hypobranchial below it). The third and fourth branchial arches are simple; and the median azygous branchial elements are not yet developed.

249. In this stage the trabeculæ, with the partially chondrified auditory capsules, remain the only cartilaginous elements of the cranium. The trabeculæ possess a definite parachordal part, and send up lateral crests in front of the ear masses, constituting a rudimentary sphenoidal wall. The mandibular suspensorium is related to the hinder part of this wall and to the otic cartilage. The hyoid and branchial arches are segmented. Bony parosteal centres have arisen in the mouth, all of them dentigerous: three pairs, the premaxillaries, vomers, and palatines, belong to the palate: two pairs, the dentaries and splenials, to the lower jaw. The denticles appear almost before any true osseous deposit begins in the membrane.

Fourth Stage: Young Axolotls three-quarters of an inch in length.

250. With very little change in outward appearance, there has been great progress in skull-building. The posterior half of the cranial notochord has now a distinct tract of cartilage investing it on either side (Fig. 25,

pa. ch.). These are arcuate bands, pointed in front where they lie between the notochord and the auditory capsules; and broader posteriorly where they curve outwards behind the ampullæ of the posterior semicircular canals. These cartilages form the base of the occipital region of the skull, but the occipital roof and side walls are still membranous.

251. The anterior half of the cranial notochord is closely invested by the parachordal portions of the trabeculæ, which extend backwards to and even overlap the occipital parachordals. The trabeculæ have met just in front of the notochord, and the apex of the latter becomes more and more perfectly ensheathed by trabecular cartilage. Furthermore, the apex of the notochordal sheath becomes distinctly ossified.

252. The anterior portions of the trabeculæ enclose an elegant ovoid space in the cranial floor, the broad end of the ovoid being behind. They are connected in front by a considerable transverse tract of cartilage. This anterior plate presents three regions: a median internasal (*i. n. c.*), and a pair of lateral horns (*c. tr.*), extending outwards and curving somewhat backwards. The median part projects farthest forwards, and has a convex margin anteriorly. The lateral horns half surround the anterior and inner face of the nasal sacs below. At present there is no chondrification of the nasal walls. With the elongation of the trabeculæ, their longitudinal crests have increased in height and length, forming considerable lateral cranial walls. Close to their hinder termination the ascending region of each suspensorial cartilage (*a. p.*) has coalesced with the corresponding crest near its upper edge.

253. The auditory masses are considerably modified by the greatly altered condition of the enclosed labyrinths, and the bulgings of the semicircular canals. The whole of the periotic wall is cartilaginous, but on the lower surface there is a large crescentic cleft towards the

outer side. Mesiad of the fissure itself, the cartilage becomes extremely irregular and partially deficient, and

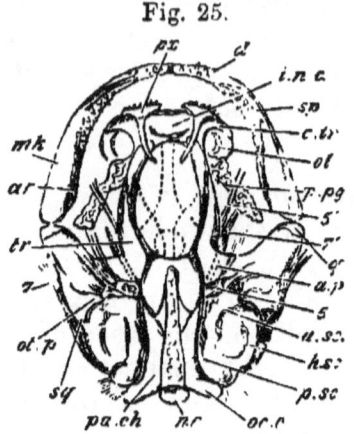

Fig. 25.

Larval Axolotl, three-quarters of an inch long; upper view of basis cranii and lower jaw, dissected.

nc. notochord; *pa.ch.* parachordal cartilage; *oc.c.* occipital condyle; *tr.* trabecula, pointing to its crest; *i.n.c.* internasal cartilage; *c.tr.* cornu trabeculæ; *ol.* olfactory sac; *a.sc.*, *h.sc.*, *p.sc.*, anterior, horizontal, and posterior semicircular canals; *mk.* meckelian cartilage; *q.* quadrate condyle; *ot.p.* otic process; *a.p.* ascending process united with trabecular crest; under it the dotted lines represent the anterior branch of the trigeminal nerve, 5'; 7', anterior branch of facial nerve; 5, Gasserian ganglion; 7, main facial nerve.

Bones: *px.* premaxillary; *p.pg.* palatopterygoid; *sq.* squamosal; *ar.* articular; *sp.* splenial; *d.* dentary.

through its hinder part a massive otolith is visible. This cleft is the first appearance of the fenestra ovalis.

254. The articular extremity of the suspensorium projects further outwards than in the last stage. The ascending process, as we have seen, has coalesced with the trabecular crest. The postero-external or otic region has developed a pedate otic process (*ot. p.*), which adheres to the auditory wall outside the ampullæ of the anterior and horizontal semicircular canals. On the outer surface of the suspensorium a thin rudiment of the *squamosal* parostosis has appeared (*sq.*).

255. The meckelian rod (*mk.*) is twice as thick behind as in front; but its posterior extremity is narrowed

to a rounded angle receiving the articular condyle. The two halves of the mandible are closely approximated in front; the dentary and splenial bones are much more extensive than they were, and very largely embrace the anterior part of the cartilage. On the inner face of the posterior part of the mandible a thin non-dentigerous splint of bone has appeared; this is the rudiment of the articular (*ar.*).

256. There is one new bone in the roof of the mouth; but those which existed before are greatly modified. The premaxillaries are pedate in front, with many teeth, and have long sigmoid nasal processes. These grow backwards, over the junction of the internasal plate with each cornu, and extend for some distance over the frontal region of the cranium[1]. They overlap the anterior ends of a pair of filmy bones, the *frontals* (*f.*), in the fore part of the cranial roof, and these in their turn overlap a similar pair behind, the *parietals* (Fig. 27, *p.*).

257. The vomers have increased in size, and are now separated by a little distance from the palatines. The latter have broad outwardly-extended dentigerous plates; and from each of their postero-external angles there projects a long flattened jagged handle, directed somewhat backwards and outwards beneath the eyeball, to reach the suspensorium a little behind the articular condyle. Thus the bone has become a palatopterygoid (*p. pg.*), the new portion being the pterygoid. Beneath the intertrabecular space and the hinder parachordal part of the trabeculæ is a median lanceolate parostosis, the *parasphenoid*, forming a support for a considerable portion of the cranium, and lying in the roof of the mouth.

258. The primary relations of the chief cranial nerves are distinctly manifest in this stage. Instead of the large anterior ganglionic mass formerly described (§ 231, p. 96), there are now two ganglia at the fore end of the auditory capsule, in contact with the posterior margin of the tra-

---

[1] This relation is precisely similar to that of the same processes in Birds.

becular crest. The first or *Gasserian* (5) gives off three branches, constituting the trigeminal nerve. The *orbitonasal* branch passes forwards beneath the ascending process of the suspensorium. The other two pass *over* the same process lower down, gently diverging as they pass outwards and forwards.

259. Beneath the Gasserian ganglion, and well seen on the under surface of the skull, is the ganglion of the facial (7); the two principal nerves given off from it are (1) the anterior or *Vidian*, passing forwards *beneath* the ascending process, below and a little external to the orbitonasal; (2) the *facial*, directed outwards and backwards, and then forwards, skirting the auditory capsule, and crossing beneath the otic process of the suspensorium. On the inner side of the auditory capsule behind, wedged between the occipital parachordal and the periotic cartilage, is the ganglion of the glossopharyngeal nerve, and the vagus passes out close behind it. The auditory nerve enters the capsule behind the facial ganglion. The optic nerve pierces the trabecular crest near its middle. The olfactory crus passes over the anterior part of the trabecula, where it gives off the lateral horn.

260. This stage is marked in the chondrocranium by the development of the proper occipital parachordals, by the complete chondrification of the ear-capsules and the formation of the fenestra ovalis, and by the union of the trabeculæ anteriorly and posteriorly. The internasal plate and trabecular cornua become distinct. The mandibular suspensorium has coalesced by its ascending process with the sphenoidal wall of the cranium, and possesses a distinct otic process. The ossification of the apex of the notochordal sheath is very interesting; other new bony developments (parostoses) are the parasphenoid, parietals, frontals, squamosals, and articulars; the nasal processes of the premaxillaries, and the pterygoid portions of the palatopterygoids.

Fifth Stage: Young Axolotls of 1¼, 2¼, *and* 3¼ *inches long*[1].

261. In the following account, observations made at three different periods of growth are combined for the sake of brevity. The changes described occupy several months, and are of great importance. The most notable phenomena are the growth and straightening of the trabeculæ, the formation of the stapes, the appearance of a palatopterygoid series of cartilages, and the subdivision of the palatopterygoid bone.

262. The cartilaginous bridge formed by the trabeculæ in front of the apex of the notochord increases considerably; and the cartilage gradually creeps over the notochord above and below, so as nearly to enclose it. The anterior half of the cranial notochord becomes ensheathed by the growth of the bony matter at its apex (Fig. 26, *c. st.*), described in the last stage (§ 251, p. 103).

263. The parachordal portions of the trabeculæ unite completely with the occipital parachordals; and the whole basilar cartilage thus constituted coalesces with the periotic capsules. The occipital condyles are formed, immediately external to the notochord, in *A*. Lateral occipital walls arise from the hinder part of the basilar cartilage, in continuity, however, with the ear-capsules, but extending some distance behind them. These curve inwards above, and have coalesced to form an occipital roof in *B* (Fig. 27, *s. o.*). In the same stage a small patch of bone is developed around the foramen of exit of the glossopharyngeal and vagus nerves (Fig. 26, *e. o.*). Later on, this exoccipital bone grows into the base of the condyle, which has increased in size and appears shortly pedunculated.

264. The trabeculæ lengthen considerably, so as to outstrip the other cranial structures in growth. Thus the intertrabecular space becomes a parallelogram with rounded angles. The trabecular crest, or sphenoidal wall, is con-

[1] These will be referred to as *A*, *B*, and *C*.

tinuous behind with the periotic cartilage over the Gasserian ganglion; in front it extends to the nasal region. A transverse section in any part where the crest is developed presents a crescent standing erect with the convexity outwards; *there is no distinct basal part:* the base of the trabecula does not develope beyond its first condition, and consequently it becomes insignificant by the greater growth of other parts. Thus the cranium has its roof and base very largely membranous, while its lateral walls are completely cartilaginous.

265. Instead of the internasal plate projecting a little beyond the lateral horns of the trabeculæ, these latter grow into considerable fan-shaped lobes (*c. tr.*), which by their inward approximation bound three-fourths of a circular space in front of the internasal plate. This circle is completed in *B* by the premaxillaries (Fig. 26, *px.*) growing transversely to meet each other.

266. A new structure appears in the nasal region. The nasal sac is floored in front by the trabecular cornua, but its roof is formed above and behind by a distinct crescentic cartilage (Fig. 27, *ol.*). The superior nasal opening lies in the concavity of the nasal cartilage, but is narrowed by membranous folds. In *C* the trabecular horns have grown very much, and the capsular cartilages have united with them by their anterior and inferior edges, the hinder part remaining free.

267. In *B*, the cleft which is to form the fenestra ovalis in the periotic capsule, is more lateral than in the earlier stages; and the irregular cartilage previously described on its antero-internal margin has developed into a broad thin lamina which forms a lid over the fenestra. This lamina is connected by a broad isthmus antero-externally with the general periotic cartilage, but it is commencing to be segmented off (Fig. 26, *st.*). In *C* this is completely separated as the *stapes*, lanceolate in shape, with the point behind. From its outer surface a fibrous ligament passes to be inserted into the hinder and under

surface of the otic process of the suspensorium: this is the suspensorio-stapedial ligament.

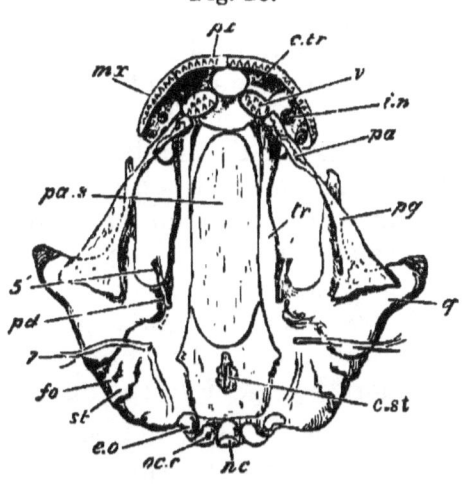

Fig. 26.

Young Axolotl, 2¼ inches long; under view of skull, dissected, the lower jaw and gill arches having been removed.

nc. notochord; *oc.c.* occipital condyle; *f.o.* fenestra ovalis; *st.* stapes; *tr.* trabecular cartilage; *i.n.* internal nares; *c.tr.* cornu trabeculæ; *pd.* pedicle of suspensorium; *q.* quadrate region; under *pg.* outline of pterygoid cartilage; 5′, orbitonasal nerve; 7, facial nerve.

Bones: *pa.s.* parasphenoid; *c.st.* cephalostyle, part of the parasphenoid having been removed to shew it; *e.o.* exoccipital; *v.* vomer; *px.* premaxillary; *mx.* maxillary; *pa.* palatine; *pg.* pterygoid.

268. The quadrate extremity of the suspensorium (*q.*) gets progressively turned outwards, making the mouth broader. The ascending process continues coalesced with the cranial wall; and a lower process in the same region, the *pedicle* (Fig. 26, *pd.*), becomes distinct as a rounded bud which articulates with the antero-inferior edge of the auditory capsule. Above the condylar facet for the mandible a considerable mass of bone arises in the suspensorium, forming the *quadrate* bone.

It is worth noting that this bone and the exoccipital, while arising at first as ectostoses, rapidly metamorphose the deep cartilage and appear very like true endostoses.

269. In *A* there is no pterygoid process from the anterior edge of the suspensorial cartilage: but in the

antorbital region a small transverse rod of cartilage has appeared, attached by its inner end to the trabecula somewhat behind the internal nares, above the broad tooth-bearing part of the palatine bone. In *B* the suspensorium sends forward from the middle of its anterior border a large (pterygoid) tongue of cartilage, gradually narrowing to its termination half-way between its origin and the nasal capsule. In front of this is a distinct but much smaller (post-palatine) cartilage, tending to attach itself to the antorbital rod. The latter has now a broad base towards the trabecula, and narrows as it proceeds outwards and forwards, nearly reaching the posterior angle of the maxillary bone. In *C* the small post-palatine cartilage has completely coalesced with the fore end of the pterygoid process, and is attached by a short ligament to the antorbital.

270. The parasphenoid in *A* extends from the hinder end of the internasal plate to the basioccipital region. Opposite the apex of the notochord it gives off a small angular basitemporal process directed outwards. Later, the parasphenoid (Fig. 26, *pa. s.*) has become broader and longer, underlying the internasal plate to a considerable extent, and behind reaching nearly to the extremity of the skull. Still later, the apex of the shrunken cranial notochord is found lying in a groove on the upper surface of the parasphenoid, and the bony matter which surrounded the notochordal apex cannot be distinguished from that bone.

271. The vomers in *A* and *B* are oval plates directed forwards and inwards, close to the fore end of the parasphenoid. In *C*, harmonising with the greatly increased breadth of the front parts of the skull, the vomers have become elongated, narrow, and subcrescentic, the concavity looking backwards. They extend from the emargination of the front of the internasal plate to the inner side of the internal nares.

272. In *A* the pterygoid (hinder) part of the palato-pterygoid bone is much elongated, being proportionately broader behind, where it partly underlies the quadrate

THE AXOLOTL: FIFTH STAGE.

region of the suspensorium. In *B* the dentigerous palatine (*pa.*) is nearly segmented from the edentulous pterygoid (*pg.*). They tend to separate obliquely, so as to overlap one another by a blunt point. In *C* the separation is complete, and on the right side a space intervenes between them. The axes of the two bones are coincident, and the vomer continues the line, but with a curve inwards. In *B* the hinder part of the pterygoid is still broader, and underlaps the quadrate more and more: but in *C* it has become extended almost to the hinder margin of the suspensorium.

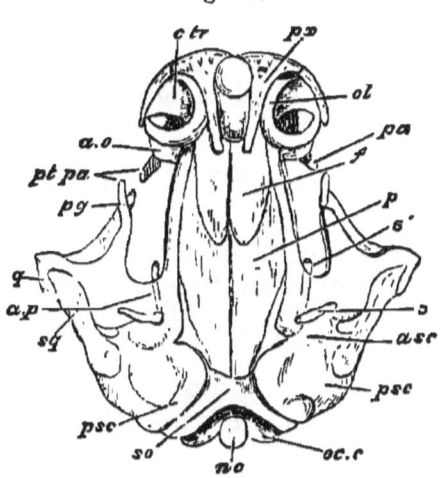

Fig. 27.

Young Axolotl, 2¼ inches long; upper view of skull; lower jaw removed.

nc. notochord; *oc.c.* occipital condyle; *s.o.* supraoccipital cartilage; *a.s.c.*, *p.s.c.* anterior and posterior semicircular canals; *c.tr.* cornu trabeculæ; *ol.* nasal capsular cartilage; *a.o.* antorbital cartilage; *pt.pa.* post-palatine cartilage; *pg.* pterygoid process; *a.p.* ascending process of suspensorium; beneath it, 5', the orbitonasal nerve; 5, hinder part of trigeminal nerve.

Bones: *p.* parietal; *f.* frontal; *px.* premaxillary; *pa.* palatine; the pterygoid is partially seen under *pg.*; *q.* quadrate; *sq.* squamosal.

273. In *A* the premaxillaries do not meet at the middle line; in *B* they meet; in *C* their nasal processes have become broad. A slight maxillary style, bearing teeth, appears in *A*, continuing the arcuate line of the

premaxillaries, so that the four bones (two on each side) form nearly a semicircle. In *C* a small ascending plate is found on the upper margin of the maxillaries, lying outside the capsular cartilage of the nose. Attached to this process is a small bone lying on the nasal roof, the *ectethmoid*; and between this and the nasal process of the premaxillary, is another small bony plate, the *nasal*. The external nostril is now bounded, in front by the premaxillary, externally and below by the maxillary, behind by the ectethmoid and nasal, and internally by the nasal and nasal process of premaxillary. This relation is substantially maintained in subsequent stages.

A bone in a precisely similar position to the ectethmoid here described arises as an ectostosis in the Salmon.

In the Axolotl and Urodeles generally it appears to be always a parostosis.

274. The frontals (Fig. 27, *f.*) and parietals (*p.*), about equal in size, continually enlarge, gradually covering the cranial fontanelle; and, in *B* and *C* the parietals reach back to the supraoccipital cartilage. They also send out an external process which strongly clamps the auditory capsule close behind the anterior ampulla and the exit of the two hinder divisions of the trigeminal nerve. The squamosal (Fig. 27, *sq.*) becomes much broader above, where it spreads over the otic process and the antero-external part of the auditory capsule. Below, it extends in a pointed shape nearly to the quadrate hinge. The bones of the mandible thoroughly invest the meckelian cartilage. There is no change of importance to notice at present in the remaining arches.

275. The conclusion of this series of changes leaves the chondrocranium fairly complete. We have the basilar plate formed out of trabecular and occipital elements, and the occipital investment united with the periotic cartilages. The cranial cavity is bounded by cartilaginous side walls derived from the trabeculæ, while the parasphenoid and the frontals and parietals protect the brain,

where it lacks a chondrified floor and roof. In front of the cranial cavity is a thick internasal plate, with large trabecular horns, and a cartilaginous nasal roof. The formation of the stapes, at first in continuity with the periotic cartilage, and afterwards separating from it; the development of a considerable pterygoid process from the suspensorial cartilage; and the appearance of antorbital and post-palatine cartilages, are the remaining principal events of chondrification. The exoccipitals and quadrates are the only bones which are at present transforming cartilage. The maxillaries, ectethmoids, and nasals are new parosteal centres. The transformations which the vomers, squamosals, and palatopterygoids pass through are of very great interest.

The Skull of the adult Axolotl.

276. Comparing this form with C in the last stage, the changes that have occurred are seen to be on the whole simple, principally consisting in the addition of cartilage-bones, and alterations in the relative size of parts during the increase (about threefold) of actual dimensions. Thus the precranial region, including the internasal plate, nasal capsules, and trabecular cornua, has increased to twice its former relative size; whilst the auditory capsules are not much more than half as large proportionally; and the same is the case with the basilar cartilage. The latter is now only about one-sixth of the length of the skull: while in the fourth stage (§ 250) the parachordal cartilages are one-half its length. The new bones developed are the sphenethmoid, the prootic, the pterotic, and the stapes: all paired bones.

277. The cranial cartilage and its bones will now be described in detail. The chondrocranium as a whole may be characterised as an elongated flattened box, deficient in the middle below, and in a large portion of its length above. In front it bears a much broader flat nasal expansion; while behind, its breadth is extended by the rather small ear-sacs, to which the outspread forwardly-

directed suspensoria of the mandible are firmly fixed. A bar on each side, parallel to the axial box, extends from the front part of the suspensorium to the hinder region of the nasal expansion.

278. The notochord is not visible on an external examination of the cartilaginous basilar plate. The groove for its reception on the upper surface of the parasphenoid has disappeared, and the latter comes clean away from the cartilage. The basilar plate as it passes into the prechordal region consists of more solid and massive cartilage, and is continuous with the cranial walls on either side (Fig. 28, *tr.*). In their anterior third the cranial walls are converted into solid bones, which widen anteriorly, and in front of the cranial cavity are thrust like thick wedges into the sides of the internasal plate, to one-third of its extent; thus we have a pair of *sphenethmoid* bones (*G*).

279. The projecting occipital condyles are somewhat reniform, and their cartilaginous articular face is principally inferior. They are largely ossified by the exoccipitals (*e. o.*); but a basioccipital region, equal in width to the exoccipitals, remains unossified. These bones occupy the side walls of the hindmost region of the skull, and come nearer together above than below. The obliquity of the foramen magnum causes the supraoccipital plate to have a scooped margin behind; its front outline is also concave forwards. The whole of the broad supraoccipital region except the small part occupied by the exoccipitals behind, is entirely unossified: as is also a projecting angle in the auditory cartilage on each side, over the junction of the anterior and posterior semicircular canals (the epiotic region).

280. On the inferior surface of the ear-capsule, mesiad of the fenestra ovalis and not reaching to it, is a shell-like *opisthotic* plate of bone, semidistinct from the exoccipital, underlying the main portion of the vestibule, and extending partially behind it. On the front region of the vestibule below, a narrow tract of bone lies transversely, parallel with the scooped surface in which the pedicle of the sus-

pensorium fits. This is part of the *prootic* ossification (Fig. 29, *pr. o.*), which is more largely seen from above, fitted, shell-like, over the neck of the anterior ampulla, and running down on the antero-internal face of the ear-capsule. Above, it grows forwards along the upper edge of the cranial side-wall, nearly reaching to the sphenethmoid. This may be called the alisphenoidal spur of the prootic.

Fig. 23.

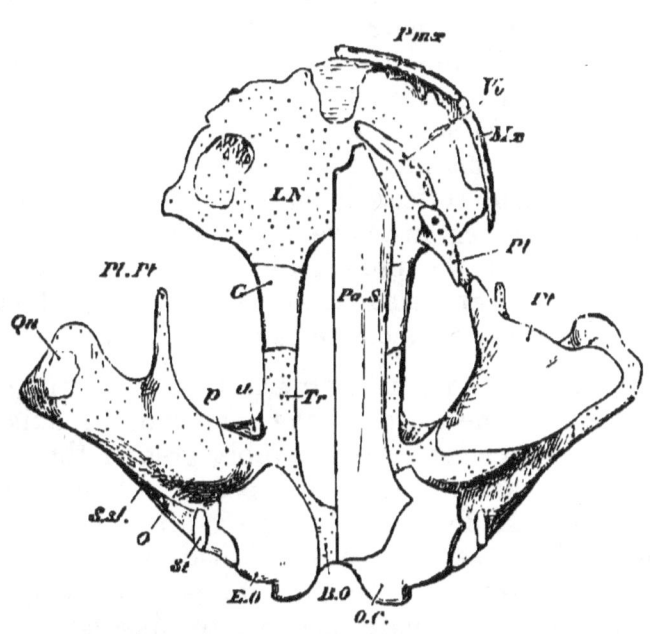

Adult Axolotl; under view of skull, the lower jaw and arches being removed, and also the investing bones on the right side.

b.o. basilar cartilage; *tr.* trabecular cartilage; *i.n.* internasal plate; *s.s.l.* suspensorio-stapedial ligament; *p.* pedicle, *a.* ascending process, *o.* otic process, *pt.* pterygoid process of suspensorium; the antorbital cartilage is not shown.

Bones: *e.o.* exoccipital; *o.c.* occipital condyle; *pa.s.* half of parasphenoid; *g.* sphenethmoid; *pmx.* premaxillary; *mx.* maxillary; *vo.* vomer; *pl.* palatine; *pt.* pterygoid; *qu.* quadrate; *st.* stapes.

281. The contiguous ampullæ of the anterior and horizontal canals are partially uncovered by bone; but behind this bare space, covering the horizontal canal and most of the roof of the capsule, is a massive bone in the

cartilage, corresponding to the *pterotic* of the Salmon. Above, it articulates by its antero-internal angle with the prootic, and behind with the outer half of the front margin of the exoccipital. It extends through the substance of the external longitudinal ledge (pterotic ridge) of the capsule. The cartilage is bevelled inwards below this ridge as far as the *stapes* (*st.*), which is now a somewhat curved plug of bone with its broad end inwards, fitting into the fenestra ovalis, and its narrow end looking outwards and forwards. From its narrowed extremity there extends a strong fan-like ligament, which spreads to be attached anteriorly to the posterior margin of the suspensorium (suspensorio-stapedial ligament, *s. s. l.*). Thus we have, almost perfectly distinct, four bony elements in the ear-capsule, prootic, pterotic, opisthotic, and stapes. The sphenotic and epiotic regions have no ossification; and below, the lower part of the prootic and the outer part of the opisthotic tracts are also unossified. In all, about two-thirds of the superficies of the capsule is bony.

282. The broad flat internasal tract of cartilage (*i. n.*) occupies about one-third of the longitudinal axis of the skull, and its breadth is considerably greater than this. It has a concave margin before and behind. The hinder concavity lodges the fore part of the brain; the anterior is the emargination separating the broad trabecular cornua. The latter are subquadrate, and project further forwards from the internasal plate than in the last stage. They are thinned away from below forwards, terminating by a sharp edge.

283. The height of the internasal plate behind is equal to that of the fore part of the cranial cavity; it is scooped in such a manner that the superior cranial fontanelle extends further forwards than the inferior fontanelle. Each sphenethmoid bone sends down a process which covers the upper surface of the cartilage bounding this inferior fontanelle in front, and these processes nearly meet. The olfactory crura pierce the shelving front wall of the cranium (in the internasal plate) just on the inner

surface of the sphenethmoids near their upper edge. The nasal capsules are relatively twice their size in the last stage. The cartilage is more than crescentic, the two horns of the crescent nearly meeting externally; and the internal part of the cartilage is broadly confluent with the internasal plate and trabecular cornu. The external nares (Fig. 29, *a. n.*) are bounded in front by the incurved anterior cornu, and posteriorly by a valvular fold of fibrous tissue.

284. The broad parasphenoid (*pa. s.*) underlies almost the whole of the skull. Its sides are nearly parallel in the greater part of its extent; but it sends out a small process on either side in the basitemporal region; and each antero-external angle projects prominently so as to touch the inner extremity of the corresponding vomer. The latter crescentic bone (*vo.*) is just behind the median notch terminating the basicranial axis: it extends outwards and backwards to the fore part of the palatine. The anterior bones (premaxillaries and maxillaries, *pmx. mx.*) being apposed, form a very regular semicircular margin to the mouth, having no palatal plates; the arcs of the two vomers are parts of another semicircle behind this, and almost concentric with it.

285. On the upper surface of the skull, the premaxillaries send a long nasal process backwards. overlapping the frontals. The two processes converge, but do not meet, part of the internasal cartilage being visible between them. In front of the cartilage, the emargination between the lateral cornua is left open as a round membranous space, bounded anteriorly by the apposed premaxillaries. The maxillary bones have also a short nasal process extending upwards and backwards, attached to the anterior and outer end of the ectethmoid. Thus the superior narial opening is bounded on one side by the nasal process of the premaxillary, on the other by the corresponding process of the maxillary. Above and behind, the ectethmoid (*l. eth.*) and the nasal bound the nostril, the latter bone being wedged in between the former and the nasal process of the premaxillary.

286. The frontals (Fig. 29, *fr.*) appear to be larger than the parietals (*pa.*), but the latter are considerably overlapped by the frontals: they really reach nearly to the antorbital region. The frontals are partly covered in front by the nasal processes of the premaxillaries, by the nasals, and by the lateral ethmoids. The frontals and parietals together form a nearly regular oblong, much resembling the parasphenoid, but broader.

The parietals just in front of the auditory capsule are curved downwards on the side walls of the cranium, and their lateral boundary lies alongside of the alisphenoidal spur of the prootic bone. The fore part of each auditory capsule is covered by a distinct spur-like process of the parietal, which, in the angle between this process and the decurved part of the bone, is grooved downwards, forwards, and outwards, after the manner of a temporal fossa, leading downwards to the trigeminal foramen. The hinder margin of the two parietals is notched in the mid line in correspondence with the notching of the supraoccipital cartilage, which terminates a short distance behind the bones.

287. The suspensorial cartilages are very large, turning almost directly outwards, but also a little forwards; they are usually somewhat unsymmetrical. The main part of the suspensorium is ossified by the spatulate quadrate bone (Fig. 28, *qu.*), which has a broad end downwards, and a long handle passing upwards towards the auditory capsule, terminating in the otic process. The latter is a broad process articulating with the front of the external and upper surface of the auditory capsule. To it partly, but most largely to the hinder edge of the main part of the suspensorium, are attached the fibres of the suspensorio-stapedial ligament. The ascending process of the suspensorium remains coalesced by cartilage with the upper edge of the cranial wall. Below this, the *pedicle* (§ 268, p. 109) has become a large condylar head, fitting in a scooped socket on the anterior and inferior surface of the auditory capsule, which socket is overarched

by the prominence of the anterior ampulla above. Thus the suspensorium has four regions of attachment to the cranium: to the sphenoidal wall by the ascending process antero-superiorly: and to the ear-capsule by the pedicle in front, by the otic process externally, and postero-inferiorly, by the suspensorio-stapedial ligament proceeding from the stapes. The quadrate is covered distally and inferiorly by a deeply-scooped surface of cartilage, in which the large rounded mandibular condyle plays. The squamosal (*sq.*) is a strong splint on the upper and hinder surface of the suspensorium, lying in digitate lobes on the outer edge of the pterotic, and extending by a pointed extremity to the quadrate articulation.

288. The cartilaginous pterygoid process (*pt.*) arises by a broad base from most of the anterior edge of the suspensorium, becoming much narrowed as it passes forwards parallel with the cranium to reach the antorbital cartilage. The latter is now very small relatively, but is quite distinct, and attached by its inner face to the lateral ethmoidal region of the skull. The tooth-bearing palatine bone (*pl.*) extends farther in front of this cartilage than behind it, and is attached to the outer extremity of the vomer. The inferior narial aperture is just outside the meeting-place of the vomer and the palatine. The latter becomes pointed behind, where it lies external to the cartilaginous pterygoid process, and touches the pointed anterior end of the pterygoid. This latter bone (Fig. 28 *pt.*) is five or six times larger than the palatine, very broad behind, underlying the greater part of the suspensorium, and nearly reaching to the cranial wall in the region of the pedicle; it is narrowed to half the width at the base of the cartilaginous pterygoid process, and in front becomes suddenly reduced to a point.

289. The mandibular arc is semicircular, with sides produced behind. The meckelian cartilage persists as the axis of each ramus and is continuous with the condylar surface, which is at the back of an elevated coronoid boss,

The articular bone lies like a thin splint on the inner and under surface of about three-fourths of the arc on each side; it extends almost to the extremity of the angle of

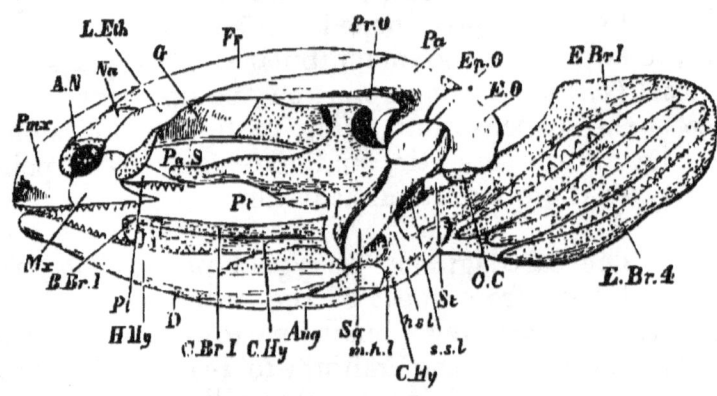

Axolotl, nearly adult; side view of skull.

e.o. exoccipital; *pt.o.* pterotic; *pa.* parietal; *pr.o.* prootic, the ascending process of the suspensorium reaching its alisphenoidal spur; *fr.* frontal; *g.* sphenethmoid; *l.eth.* ectethmoid; *na.* nasal; *a.n.* anterior nostril; *pmx.* premaxillary; *mx.* maxillary; *pa.s.* parasphenoid; *pt.* pterygoid seen at the edge of the pterygoid cartilage and the suspensorium; *pl.* palatine; *d.* dentary; *art.* articular; *sq.* squamosal; *st.* stapes; *s.s.l.* suspensorio-stapedial ligament; *h.s.l.* hyo-suspensorial ligament; *m.h.l.* mandibulo-hyoid ligament; *c.hy.* ceratohyal; *h.hy.* hypohyal; *b.br.* 1, first basibranchial; *c.br.*1, first ceratobranchial; *e.br.* 1, 4, epibranchial bars.

the jaw. The dentary (Fig. 29, *d*) occupies most of the outer surface, extending back as far as the condyle, and forward to the symphysis; the splenial lies on the anterior two-thirds of the inner surface. Each of these bones bears teeth, so that there are two series in the lower jaw as well as in the upper.

290. The ventral ends of the hinder arches and their azygous pieces all lie within the intermandibular space. No more of the hyoid arch is chondrified than corresponds with the mandible. Each moiety (*c. hy.*) is falcate, turning inwards above to be attached to the arch in front. An oblong ligament, the mandibulo-hyal, passes from the angle

of the mandible to the outer face of the upper end of the hyoid bar. Another ligament, the hyo-suspensorial, arises from the hinder margin of the suspensorium, behind the broad part of the quadrate bone, and passes outside the last ligament to a broad insertion on the hinder edge of the hyoid. The thicker ventral end of the hyoid arch has a short anterior segment detached, as in the Salmon: this is the hypohyal, and it is attached by its apex to the fore end of the first basibranchial. The main portion of the bar is the ceratohyal.

291. Behind and between these bars there are four pairs of branchial arches. The anterior thirds of the first two are segmented off, forming ceratobranchials, with no hypobranchials. The upper and posterior pieces of these arches (epibranchials) are rounded arcuate bars with the convexity outwards, extending upwards and backwards into handle-shaped extremities. The two hinder branchials are unsegmented, and are much smaller than the rest. They are entirely parallel with the other epibranchials, and correspond with them. The ceratobranchial pieces are suppressed, and these arches do not attach themselves to the azygous basibranchials. The gill-clefts are narrow and long, and are partly closed by very regular series of interdigitating tooth-like processes on the corresponding surfaces of each arch. There are two basibranchial cartilages; the first of these bears on its hinder third the distal extremities of the first two branchial arches: the second basibranchial is half as long again; and it lies on a lower plane. It is a thin spatular cartilage, with the broader end behind; it is quite free from the arches, though attached to the first basibranchial. Each of the three first branchial arches carries near its upper part a long pinnate external gill. All the postmandibular arches are wholly unossified.

292. In Axolotls in which the gills are almost absorbed, and which may be considered practically cryptobranch forms, a certain amount of morphological change has taken place, tending in a salamandrine direction. All the bones

are more solid and dense, and are enlarged. The parasphenoid especially shows progress towards the caducibranch types, being broadened in the ethmoidal region, and its basitemporal processes being more marked. In the lower jaw the articular bone has developed in front of the condyle a large coronoid region embracing the cartilage both without and within. A bone has appeared in the branchial system, occupying the posterior two-thirds of the second basibranchial, the hinder end of which has grown into two sharp horns, whose apices alone remain cartilaginous. The branchial arches retain their massiveness, notwithstanding the absorption of the gills. They are so large that the first epibranchial equals the mandibular ramus in size.

The skull in Amblystoma opacum, a North American species developed from an Axolotl form (not the same species as the last described).

293. This skull is much more perfect in many ways than that of the Axolotl, having the aspect of a higher form. The front of the face instead of being semicircular is semi-elliptical; and the suspensorium is directed further backwards, making rather more than a right angle with the fore part of the skull. A detailed comparison with the Axolotl shows many points of interest.

294. The exoccipitals have completely coalesced with the pterotics, opisthotics, and prootics: but a considerable tract of cartilage persists between each occipito-otic mass both above and below. The sphenethmoid ossification (Fig. 30, $o.\,s.$) reaches from the antorbital region to the hinder margin of the optic nerve, which it nearly encloses. Behind this there is a tract of cartilage as far as the ascending process of the suspensorium. Here the alisphenoidal spur of the prootic has more largely ossified the cranial wall above, and the bony matter extends into the ascending process. No additional bony centre arises in the substance of the chondrocranium.

295. The general form and position of the parasphenoid remain unchanged; the basitemporal processes are much larger, extending nearly to the pedicle of the suspensorium. But the palatal surface has undergone a very striking metamorphosis due to the growth of large flat palatal plates of the vomers, anterior to their dentary tract. The parasphenoid terminates emarginately just behind the primary notch between the trabecular cornua; but its anterior region is underlaid by the vomers, whose dentary portion now lies further back than the antorbital region, with a slight curve backwards, strikingly contrasting with the forward curve they previously had. This curve is continued outwards by the small dentigerous palatines, which have become transversely placed. In front of its dentary region each vomer sends out a broad lamina, fan-shaped in front, having its outer edge parallel with the premaxillary and maxillary bones. The external edge of the vomerine plate is notched behind so as to surround the greater part of the inferior nostril, which is also slightly bounded by the palatine: a portion of the outer margin of the orifice is, however, unenclosed by bone.

296. The premaxillaries do not develope a palatal plate, but the maxillaries, which are very much lengthened backwards towards the mandibular articulation, have an inturned edge which is a rudimentary palatal plate. The transversely-placed palatine bone reaches this palatal edge of the maxillary. In contrast to former stages, scarcely any cartilage is visible on the under surface of the palate, and there is much less of it actually, for the cartilage has become largely absorbed concurrently with the growth of the vomers.

297. On the upper surface of the skull one of the most notable changes affects the nasal processes of the premaxillaries, which are now longer and broader, and meet in the middle line through almost their whole extent: so that the internasal cartilage is no longer visible, and the membranous space that formerly existed in

front of the cartilage is reduced to a fissure. The small ascending process of the maxillary has become much larger and ascends higher behind the narial opening; and on the anterior margin of this process on the left side there is a separate granule of bone (parostosis) immediately external to the nostril.

This is the first appearance of a parosteal bone which occurs in a great variety of forms in the outer, under, and inner or septal wall of the nasal capsule. In its most developed and specialised form it is the "septo-maxillary" of lizards and snakes.

298. The nasals are denser, shorter, and broader; they overlap the ectethmoids by their outer edge, as do the maxillaries by their ascending plate. The ectethmoid extends much further back than it did; and it overlaps the frontal in the outer two-thirds of its length, and even slightly overlaps the pointed outer end of the parietal. The anterior part of the ectethmoid, like the nasal, is closely applied to the olfactory capsule, but does not graft

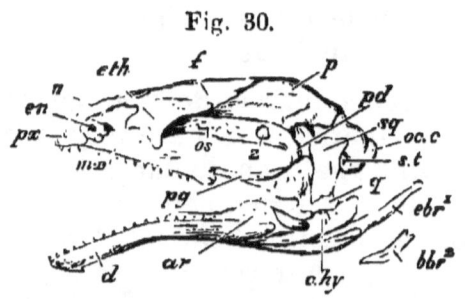

Fig. 30.

Skull of Amblystoma, side view.

Oc.c. occipital condyle; *p.* parietal; *f.* frontal; *eth.* ectethmoid; *n.* nasal; *e.n.* external nostril; *p.c.* premaxillary; *mx.* maxillary; *o.s.* orbitosphenoid region of bony cranial wall; 2, optic foramen; *sq.* squamosal; *pd.* pedicle; *pg.* pterygoid; *st.* stapes; *q.* quadrate; *ar.* articular; *d.* dentary; *c.hy.* ceratohyal; *e.br.* 1, first epibranchial; *b.br.* 2, second basibranchial.

itself upon the cartilage so as to ossify it. Behind the lateral ethmoidal or antorbital region the ascending lamina of the maxillary is continued backwards, though it does not reach so high as in front. This facial lamina is coextensive with the dentary portion of the bone, and makes an acute angle with the slight horizontal (palatal) plate of the same bone. Lying on its hinder extremity is a small graniform jugal bone, and the same line is continued throughout the posterior third of the gape by a dense band of fibrous tissue attached to the hinge of the lower jaw; this may be termed the jugal·ligament.

299. Each frontal (*f.*) is overlapped by the nasal porcess of the premaxillary, by the nasal, and by the ectethmoid. The bone is much larger than it was, reaching further forwards under the anterior bones, and itself overlapping the parietal more considerably. The parietals (*p.*) are larger, more solid, and more accurately adapted to the auditory capsules and to the occipito-otic bony masses. A transverse ridge is developed on each parietal behind the temporal fossa, for muscular attachment. The outer portion of the parietal exactly covers the epiotic region.

300. The quadrate has ossified the greater part of the suspensorium, there being only a little cartilage at the mandibular joint, and in the proximal part. The pterygoid bone (*pg.*) is only two-thirds its former size, and in correspondence with the more directly transverse position of the suspensorium, the anterior end of the pterygoid is carried to a distance from the cranium, so as to be nearly in contact with the posterior end of the maxillary. On the under surface the hinder part of the pterygoid is applied accurately like a flexible splint to the quadrate, so as to hide it very largely; and this is specially the case in relation to the pedicle. In its anterior portion the pterygoid ossification has grafted itself upon the pterygoid process of cartilage so as to ossify the greater portion of its substance; but a continuous epipterygoid rod of cartilage is left on the upper and inner grooved edge of the bone, extending also somewhat in front of it.

301. The squamosal (*sq.*) has become a very dense wedge-like bone, not reaching so high as it did, but ter-

minating upwards in a prominent longitudinal tuberculated ridge, and at the same time impressing the pterotic region of the auditory capsule beneath it. It narrows almost to a point below, being closely bound on the outer and hinder edge of the quadrate. On the right side the postero-superior angle of the squamosal is converted into a distinct small supratemporal bone.

302. The lower jaw is intensely ossified, but the elements can be distinguished: the splenial carrying its teeth at a lower level than those of the dentary. The articular (*ar.*) has invested and aborted all the cartilage of the mandible except that which forms the condylar facet. In front of the latter the outer and inner laminæ of the articular have met together, coalesced, and formed a high sharp coronoid process.

303. The most remarkable changes have taken place in the hyobranchial apparatus. The two hinder arches have been absorbed, as well as the anterior part of the second basibranchial. The ceratohyals (*c. hy.*) are of about the same length as before, but flattened into thin tapelike bands of cartilage. The hypohyal segment is very distinct from the ceratohyal, and turns suddenly backwards and inwards at an acute angle with that piece; it is a very slender terete rod, and has coalesced by its hinder extremity with the corresponding angle of the first basibranchial.

304. The first epibranchial is completely separated from its ceratobranchial, being articulated immediately by its ventral extremity with the middle of the first basibranchial. The first ceratobranchial is shorter than the hypohyal, and altogether much reduced in size. It retains its original attachment to the first basibranchial immediately in front of its own epibranchial, and is nearly parallel with the hypohyal, its former backward direction having been changed into a forward one; its outer end is free. The second branchial arch has lost all distinction of epi- and ceratobranchial elements, consisting of a very small rounded arcuate rod on each side, with its

convex margin looking inwards and backwards, attached ventrally to the hinder extremity of the first basibranchial, and at its outer and upper end to the first epibranchial at its upper third.

305. The first basibranchial is considerably enlarged; the fore part broader and ossified; the hinder part narrower and cartilaginous. The anterior part of the second basibranchial (*b. br.* 2) is absorbed up to its forks (§ 292, p.122), both cartilage and bone. The terminal cornua now form a Λ-shaped bone with cartilaginous ends behind, brought into relation with the rudimentary larynx as a laryngobranchial.

306. SUMMARY. In the skull of the Axolotl, while there is not such an extreme diversity and number of parts as in the Salmon, we find a great extent of metamorphism of the most highly interesting description. The visceral arches begin to acquire skeletal elements before the trabeculæ are manifest, and the mandibular arch is very early divided into suspensorium and free mandible. The trabeculæ and the notochord constitute the first basicranial development; and the former have a very notable extension by the side of the notochord before the occipital parachordals arise. The side walls of the fore part of the cranium are formed by trabecular crests, confluent behind with the auditory capsules, while an occipital ring grows up behind them; much of the cranial roof as well as a considerable portion of its floor in front of the notochord remains unchondrified. An internasal plate becomes continuous with the trabeculæ, and developes broad lateral cornua, which at a later stage are produced in front of the median cartilage. This precranial mass is massive, and has lateral ethmoidal or antorbital processes; it coalesces with separate nasal capsular cartilages.

307. The mandibular suspensorium is at first apposed to the anterior third of the trabecula, but ultimately has a special relation to the hinder part of its crest, with which it coalesces, developing also an otic process, and later an inferior auditory attachment or "pedicle." The suspensorium,

from being directed forwards, is gradually retracted, and at the same time carried outwards. Palatine and pterygoid cartilages do not arise till after the ossifications which bear the same name. The pterygoid process from the suspensorium is triangular, but becomes absorbed or aborted until only a narrow inner and upper epipterygoid tract remains. The hyoid arch is chondrified only by two segments in its ventral region: of the four branchial arches, the first two are segmented into two pieces; and there are two basibranchials. In Amblystoma a great abortion and transformation of the gill-arches has taken place.

308. The ossifications of the cranial cartilage are few, namely, exoccipitals, prootics, pterotics, opisthotics, and sphenethmoids, all arising late. In Amblystoma one occipito-otic mass is seen on each side. The mode of formation of the stapes is to be noted (§ 267, p. 108). Parostoses arise remarkably early in the mouth, and a little later in the roof of the skull. In the latter region they are parietals, frontals, and nasals; in the upper margin of the gape, premaxillaries and maxillaries, with minute jugals; in the palatal roof, parasphenoid, vomers, palatines, pterygoids; in the lower jaw, articulars, dentaries, and splenials. The suspensorium is ossified by the quadrate, the squamosal lying as a splint upon it above, and the hinder part of the pterygoid beneath. The ectethmoid is a similar splint on the lateral ethmoidal region. The origin of the palatine and pterygoid as one bone, subsequently cut into two and variously related, is of great interest; the transformations of the vomer, especially in reaching the Amblystoma form, are similarly noteworthy. The ossification of the notochordal apex becomes confounded with that of the region of the parasphenoid on which it lies. The very late ossification of the second basibranchial is the only exception to the persistent cartilaginous condition of the hyoid and branchial rods. The differences between the Axolotl and Amblystoma become the more significant as the student's knowledge advances, and may be left to convey their own teaching.

APPENDIX ON THE SKULLS OF AMPHIBIA URODELA.

309. In *Proteus* the cranial notochord is slightly persistent, but does not reach half-way to the pituitary body. The parachordal moieties are extremely small, and but slightly connected with one another; there is a small narrow occipital arch. If any trabecular cartilage surrounded the apex of the notochord in the embryo, it has been again absorbed; a process from the prootic bone ensheaths the hinder extremity of each trabecula. The trabeculæ run forwards as narrow rounded rods, straight and parallel, to the ethmoidal region, where they bend gently inwards, but do not unite to form an internasal plate till they reach the middle of the internasal region. At their anterior termination they again separate as cornua. The large basal fontanelle extends from very near the foramen magnum to the middle of the internasal region; the superior fontanelle is of similar dimensions. The ethmoidal region is marked externally by the attachment of a curved antorbital cartilage passing outwards and backwards. From this region to the internasal junction the cranial (trabecular) cartilage is ossified by a lateral ethmoid on either side. The nasal capsule is wholly membranous. The auditory cartilages retain an early embryonic form, a long ovoid. There is a considerable prootic anteriorly, then a median cartilaginous mass; and posteriorly a bone including exoccipital, epiotic, and opisthotic elements. The right and left bones are somewhat united. There is a long well-ossified stapes.

310. The suspensorium, like that of the larval Axolotl, is directed forwards at an acute angle with the axis. Above and behind, it passes under the trigeminal nerve and coalesces with the trabecula at its extremity. There is no ascending process, and the otic process is free. The lower half of the suspensorium is well ossified by the quadrate. There is a large cartilaginous hyomandibular attached to the back of the suspensorium, the otic mass, and the stapes, by a strong band of fibres. It is scooped on its inner face for the ceratohyal, which is ossified. There is no hypohyal nor basihyal. The first branchial arch is large and long, with a well-ossified epibranchial and a short ceratobranchial; behind this are two slenderer ossified arches, (epibranchials), with a small basal cartilage bearing both; below this

again is a single slender ceratohyal attached to the first basibranchial. There are two ossified basibranchials; the second is attached to the first, but has no other elements springing from it.

311. The bones on the roof of the skull are premaxillaries (extending to the ethmoidal region), frontals (to auditory capsules), and parietals. The squamosals are long, lying on the side of the ear-capsules and running down on the suspensorium, giving off a long process over the facial nerve to reach the stapes. There are no nasals or maxillaries; the dentary part of the premaxillaries is very short. Behind these are the two large and long vomers, with dentigerous outer margins. The palatopterygoids resemble those of the larval Axolotl three quarters of an inch long; they are broad spatulas in front, with a dentigerous tract carrying on the line of the vomers; they end at the middle of the inside of the suspensorium by a blunt point: there is no pterygoid cartilage. In the mandible cartilage is largely persistent; the ossifications are dentary and articular; there are no splenials, but the dentary ossifies a tract of cartilage in front after the manner of the mento-meckelian of the Frog.

The Skull of Siren.

312. In *Siren* there is a postpituitary trabecular bridge of cartilage, but behind it the parachordals are widely separate, forming a triangular posterior fontanelle; they are scarcely united even in the condylar region. The supraoccipital arch is of considerable extent; the exoccipital bones (which include an opisthotic region) nearly meet above, and pass partially into the cranial floor below. The auditory capsules are largely ossified, especially by the prootics, which extend into the cranial roof, and to a small extent into its floor. In the top of the capsule is a large permanent *aqueductus vestibuli;* the stapes is considerable, but unossified. In front of the ear-masses the cranial walls are steep, with cartilage behind continuous with the auditory capsules and the postpituitary bridge. From near the trigeminal foramen to the middle of the internasal region the cranial walls are ossified by sphenethmoids, solid and broad in front, high behind; they do not unite with one another, but the olfactory crura perforate them. The internasal plate is short and broad, ending in three triangular processes, prenasal and cornual. There is a pair of supraethmoid cartilage-bones. The nasal capsule has a small cartilage in its posterior boundary; and the ethmoidal region has a

considerable antorbital attached to it, and connected by ligament with a large sphenotic process from the front of the auditory capsule: the small eyeball is also fixed to the antorbital.

313. The premaxillaries extend on to the upper surface of the skull outside the supraethmoids: there are frontals and parietals, but neither nasals nor ectethmoids. The maxillaries are minute; the vomers are long and arcuate, with many teeth in several rows, which are continued on the short palatine; there is no pterygoid ossification or cartilage. The parasphenoid is much like that of Proteus. The short broad suspensorium is wholly unossified; it is directed nearly outwards, the condyle looking somewhat forwards; the pedicle passes into the trabecular cartilage beneath the trigeminal nerve, and there is an ascending process above its orbitonasal division. The squamosal lies upon the suspensorium, and has no backward process over the facial nerve as in Proteus. The lower jaw is of great depth, with a solid dentary and articular, and a considerable meckelian cartilage extending nearly to the middle of the ramus; the splenial is a very delicate long dentigerous spicule. The hyomandibular is a small oval cartilage attached by ligament to the stapes, the suspensorium, and the angle of the jaw. The large hyoid bar below is covered with a bony sheath in its lower three-fifths: there is no hypohyal. The long first basibranchial is largely ossified; it carries, in its hinder part, two pairs of branchial arches, the first large, with nearly the lower half ossified; the second with a ceratobranchial piece bearing three short cartilaginous epibranchials. The second basibranchial is a remarkable bony mass, terminating behind in three flat wings of bone.

The Skull of Menopoma.

314. This is a very flat broad skull, with comparatively little cartilage persisting in it. The exoccipitals pass far forwards, ossifying the hinder part of the ear-capsule. The prootics are distinct and large in front: the region surrounding the vestibule where the stapes fits is unossified. The hinder part of the cranial wall near the auditory mass is cartilaginous, but there is a strong sphenoid bone from the ethmoidal region to near the trigeminal foramen. The internasal region is flattened, and there are alinasal cartilages over the large nasal sacs; the external nares are small, and far forwards. The dentary margin of the upper jaw is very extensive, and semicircular

in shape. In the middle line are two triangular premaxillaries with a small round supraethmoid between them in front. Two wedge-shaped nasals are related to the external nostrils, while over the upper edge of each nasal sac lies an irregular lateral ethmoid. Between this bone and the nasal the frontal curves forwards to the maxillary close to the nostril; it extends far backwards in contact with its fellow between the parietals; which stretch from the ethmoidal region nearly to the foramen magnum.

315. The maxillaries occupy a much greater region of the gape than the premaxillaries. Both these bones have palatal plates, and within them are the broad flat vomers, having dentary regions parallel with the external series of teeth. The parasphenoid extends partially between the vomers, and passes back to the end of the skull as a broad flat solid bone with basitemporal wings. The suspensorium has a large oblong quadrate ossification with a condyle looking forwards. Internally and above is a long tract of cartilage passing beneath the trigeminal nerve and coalescing with the unossified trabecular cartilage close behind the sphenoid bone. Another tract of cartilage (ascending process) passes upwards over the orbitonasal and vidian nerves, runs into the front of the prootic, and is even ossified from it. The squamosal is oblong, directed outwards, with a short posterior process. The palatine and pterygoid are represented by one big flat non-dentigerous bone, stretching far forwards, touching by its forward angle the ethmoidal cartilage, and by an outer fork nearly reaching the maxillary. The palatopterygoid narrows behind, passing sharp and wedge-like to the inner and hinder part of the quadrate. The lower jaw is composed of very solid bones, the dentary being predominant; its teeth fit between the maxillary and vomerine series above. The dentary is grooved for the little persistent meckelian cartilage, which is covered internally by the thick and large splenial. The articular is small; and the condyloid facet oval.

316. The large stapes is well ossified, and attached by its anterior third to a thick triangular cartilage fitting against the suspensorium and lying *over* the facial nerve; by this relation it is shown to belong to the same category as the cartilage which forms the tympanic annulus of the Frog. Behind the quadrate, attached to it at a deeper level, is the pyriform hyomandibular, tied by its point to the main hyoid cartilage which is massive, ossified at its upper third, bent

outwards, and unossified in its lower portion. Mid-ventrally there are three pairs of basal cartilages, one pair only in contact, with a small median piece projecting backwards. The first basibranchial is a broad leaf of cartilage in front, and a thick stalk behind, where it bears two branchial arches, of which the first, unsegmented, is ossified in its middle third. Nearly the whole of the second branchial bar (attached to the extremity of the basibranchial) is ossified, and it bears at its lateral termination three more branchial bars, all more or less ossified. All that remains to indicate a second basibranchial is a pair of small cartilages applied to the larynx.

The Skull of Menobranchus[1].

317. The adult skull, which is much depressed, has no proper cartilaginous floor: the low side walls are cartilaginous in front of the auditory capsules: an antorbital cartilage is united by fibrous tissue to the cranial wall, which is continuous with an ethmoidal, internasal, and prenasal tract, extending to the end of the snout. The conical extremity of the notochord is embedded in fibrous tissue in the basioccipital region. The suspensorium is inclined downwards and forwards at an acute angle with the cranial axis; it has a pedicle continuous with the cranial wall in front of and below the level of the exit of the trigeminal nerve; an ascending process above the orbitonasal branch of that nerve; and an otic process lying above the main facial nerve. There is a short triangular pterygoid process on the middle of the front edge of the suspensorium. The meckelian cartilage is largely persistent, stout behind, tapering forwards. The hyoid arch includes a small hypohyal and a long straight ceratohyal, united by ligament to the posterior edge of the suspensorium, to the angle of the mandible, and to the stapes. The first basibranchial bears posteriorly at an acute angle with its axis a pair of ceratobranchials, which themselves bear three other branchial pieces from their hinder ends, an outer large and much curved, a middle bar less curved, and an inner nearly straight, attached to the last. These are called epibranchials by Prof. Huxley, a little nodule at the base of the middle piece being denominated ceratobranchial. An osseous second basibranchial, rod-like, projects from behind the connection of the two principal ceratobranchials with the first basibranchial.

[1] Huxley, *Proc. Zool. Soc.* 1874, p. 186.

318. The cartilage bones in the skull of Menobranchus are (1) the exoccipitals, (2) the epiotics, ossifying also the opisthotic region; they are conical caps of bone distinct from the exoccipitals, separated by cartilage from (3) the prootics, covered above by the parietals; (4) the quadrates, in the distal ends of the suspensoria; (5) the second basibranchial. The hyoid and branchial arches are unossified. The parostoses are (1) the parietals, of remarkable shape; broad behind, covering almost the extremity of the exoccipitals and epiotics; meeting in the middle line by suture, passing forwards between the frontals; extending forwards also outside the frontals as far as to the antorbital region by a rod-shaped process which supports the side wall of the cranium; and having another short process lying on the ascending process of the suspensorium; (2) the frontals; they project posteriorly in the angle between the median and lateral regions of each parietal (united by suture in the greater part of their extent); they send a small spur between the nasal processes of the premaxillaries, and are also extended far forwards outside those processes; further, they surround the olfactory foramen and reach behind it to the antorbital cartilage; (3) the premaxillaries, bounding the gape, having a narrow dentigerous part not reaching the internal nares, and a nasal process ascending at an acute angle with the rest of the bone on the surface of the skull; (4) the vomers, with a dental series parallel to the premaxillary teeth, broad behind and within the teeth; separated mesially by the anterior region of the parasphenoid, and abutting externally against the palatopterygoids; (5) the parasphenoid, partly underlying the exoccipitals and the ear-capsules behind, and in front of them wider than the cranial cavity; (6) the palatopterygoids, flattened plates, dentigerous in front, not widening behind where they underlie the suspensorium; (7) the squamosals, long slender curved bones reaching from the epiotic region along the hinder face of the suspensorium to the condyle, with a backward process from the middle towards the stapes; (8) the dentaries, as long as the jaw, having the meckelian cartilages lying within them in a grooved shelf; (9) the splenials, much smaller, but also dentigerous.

The Salamandrine Skull.

319. In high Salamandrine forms, such as *Salamandra*, *Triton*, and *Seironota*, nearly the whole cranium consists of solid bone. There is a tract of cartilage between the prootic and the sphenoidal

wall; with a prootic postpituitary bridge in Salamandra but not in Triton. The exoccipitals are large, and nearly meet above and below. The antorbital cartilage has coalesced with the large olfactory capsules. The suspensorium has its pedicle attached by fibrous tissue to the cranium, while the large ascending process has coalesced with it. There is a small pterygoid process of cartilage: the pterygoid bone, which is small and outwardly directed, fits closely to the pedicle, forming a joint with the basis cranii and the auditory capsule. Usually the quadrate is ossified, the otic process unossified. The palatines are long and dentigerous, coalesced with the vomers; they reach nearly to the auditory capsule, and undergird the edges of the parasphenoid. The vomers, maxillaries, and premaxillaries are broad, forming the broad anterior part of the palate; the premaxillaries may be anchylosed. The ectethmoids and nasals are of considerable size, and there are frequently septo-maxillaries. The frontals in some (e.g. Seironota) have strong supraorbital ridges. The lower jaw has large dentaries and articulars, with narrow dentigerous splenials. An ossified (spiracular) leaf behind the suspensorium occurs in Desmognathus and Spelerpes. The stapes is ossified: there is no hyomandibular; a considerable upper tract of the hyoid arch is bony, and the two remaining branchial arches are more or less ossified. The second arch loses its epibranchial piece entirely. Mostly the remnants of the primitive long second basibranchial are found attached to the larynx.

CHAPTER V.

THE SKULL OF THE COMMON FROG.

First Stage: Embryos from two to three lines long, about the time of hatching.

320. IN these embryos the cerebral vesicles are neither so large relatively, nor so prominent as in forms previously described. The upper surface of the head is not especially broad; it is gently convex, and curves downwards in front. The first vesicle occupies the fore part of the head; but is partly surmounted by the second; thus a mesocephalic flexure is established. The frontal cerebral prominence is separated from the face by a distinct transverse elevation of the skin, extending from the upper border of one eyeball round to that of the other. Below this frontal ridge, and above the oral opening, is the broad nasofrontal process, emarginate in the middle line Laterally the outer margins of the nasofrontal process bound the olfactory rudiments. Immediately above and behind these, and projecting more externally, the primitive eyeballs are seen as similar involutions, with a rim deficient in front and below. In the same lateral line, at some distance behind the rudiments of the eyes, are the ear-masses, elongated oval in shape, and covered with a continuous integument.

321. The mouth is a small lozenge-shaped depression behind the nasofrontal process; bounded in front by its right and left horns and median emargination, and behind by the rounded prominences of the mandibular tracts. The swelling of the cheeks is due to the presence

of a mass of cells surrounding the fore part of the primary embryonic cavity. This aggregation of cells gives rise to most of the mucous membrane of the mouth and pharynx, and passes into the cutaneous system a little way

Fig. 31.

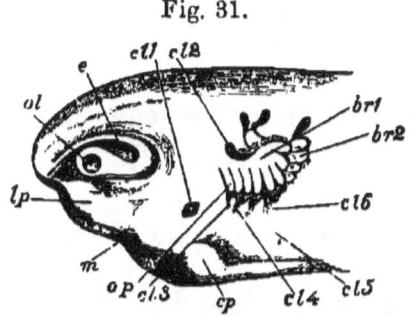

Tadpole four lines long, four or five days after hatching; side view of head.

ol. olfactory sac; *e.* eyeball; *lp.* maxillopalatine process; *m.* mouth; *cl.* 1 to 6, visceral clefts; *op.* operculum (the line is carried a little too far); *br.* 1, 2, external branchiæ; *cp.* clasper.

within the oral cavity. Below and behind the cheeks are the sucker-like organs called claspers (*cp.*), which together form a horse-shoe shaped prominence by which the little tadpole adheres to water-weeds.

322. The lateral facial wall presents four vertically-placed elevations, one behind another, occupying the tract between the mouth and the yelk-mass of the body-cavity: but there is no complete fissure between any of them. The first or mandibular is the largest, being swollen by the cell-mass in the pharynx; it is very strongly turned inwards below to the median ventral line. The succeeding arches recede more and more from the median line beneath, and are only well marked laterally. The hyoid arch has no distinctive feature; but each of the two branchial arches behind bears a small papilla, the rudiment of the external gills. Behind the branchial arches on either side there is a cleft opening into the body-cavity.

323. On dissection, the gelatinous notochord, enclosed in a distinct sheath, is found underlying the hinder part of the cranium, hardly reaching to the fore part of the third vesicle. Immediately in front of the apex of the notochord is the relatively large pituitary body; and a considerable part of the frontal region of the cranial cavity is below the line of the notochord. The mesocephalic flexure, however, very early begins to be obliterated; in embryos only two days after emergence from the egg, much progress has been made in the straightening of the cranial floor.

324. Small patches of mesoblast, approaching a cartilaginous consistency, are found by the sides of the apex of the notochord; these soon become identified with the trabeculæ. The parachordal patches are much smaller and less definite than the series of rapidly chondrifying paired rods which begins in the fore part of the cranial floor and

Fig. 32.

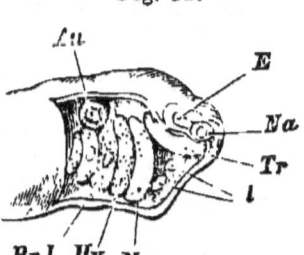

Embryo Frog, just before hatching; side view of head, with skin removed.

Na. olfactory sac; *e.* involution for eyeball; *au.* auditory investment; *tr.* trabecula; *mn.* mandibular, *hy.* hyoid, *br.* 1, first branchial, arches; the gill-buds are seen on the first two branchial arches; *l.* labial cartilages.

is continued in the ridges in the side walls of the mouth and throat. But the first pair, the trabeculæ (*tr.*), are distinguished from the others by their forward direction. The visceral arches, although the first two are parallel at their proximal ends with the trabeculæ, descend vertically in the wall of the throat, and turn somewhat backwards below. The trabeculæ and the arches are all absolutely distinct, ending proximally by blunt points. The former are shorter and thicker than the latter, and lie in the base of the prechordal part of the cranium. The oral opening is included between the anterior extremities of the trabeculæ and the mandibular arches.

325. The visceral rods resemble one another very considerably, the two foremost, however, being stouter and extending farther both proximally and distally than the branchials; nevertheless, the mandibular and hyoid bars of opposite sides are separated by some distance in the middle line below. The third branchial is very slender, and the fourth is but indistinct. The heart in its pericardium occupies the ventral line of the throat between and below the distal ends of the branchial arches. Proximally all three branchial rods approach closely to the apex of the notochord.

326. Only one of the sense-capsules, the auditory, has a wall of condensed tissue. This is continuous except on the supero-external face, where there is an area of soft cells; but there is no channel between the interior and the exterior. Two granular elements are found in the upper and lower lips. One of these rudiments of labial cartilages is found beneath the fore end of each trabecula; and another below and internally to the distal end of each mandibular rod (fig. 32, l.).

327. In the oldest tadpoles of this stage the mesocephalic flexure is almost unobservable; the distal clubbed ends of the mandibular bars are preparing for segmentation; the rudimentary tongue lies between and upon the ventral ends of the right and left hyoid arches; and the gill-buds have enlarged.

328. This stage presents us with an exceedingly simple view of the primary elements of the skull. Parachordals, trabeculæ, mandibular, hyoid, and branchial arches are all separate from one another; the auditory investment is solidifying; and labial rudiments exist. No palatopterygoid region or arch is yet discernible.

Second Stage: Tadpoles from 5 to 6 lines long.

329. At this period the external gills are at their fullest development, having tertiary papillæ; the operculum has not yet covered and enclosed them. The brain-case has greatly diminished relatively to the face, and the component parts of the brain lie nearly in a straight line. The sense-organs are subequal in size, the ear-sacs and eyeballs not acquiring such a preponderance over the olfactory organs as in the embryos described previously. The fore part of the face has grown very much faster than the branchial region; and the mouth is now a rounded cavity of considerable size.

330. Parachordal cartilages extend from the apex of the notochord to the first cervical vertebra; they are very early found perfectly continuous with the trabeculæ, which become nearly horizontal instead of being placed at an angle of 45° with the notochord. The trabeculæ form a commissural union in front of the notochord and behind the pituitary body; and a second immediately in front of the same organ; the intertrabecular space is rounded. The anterior confluence extends to the internasal region, where the fore ends of the trabeculæ diverge in the form of broad cornua, lying within and in front of the inferior nares (already distinct at this early stage) and their connection with the olfactory sacs (Fig. 33, c.tr.).

331. The auditory capsules are well-chondrified ovoid masses lying in the hinder cranial region, having the mandibular arch directly in front of them, and the branchial arches immediately behind and beneath. They are as yet distinct from other cartilaginous elements, but the mandibular arch is clinging very close to them. When the tadpoles have attained a length of about seven lines, a slit appears on the infero-lateral face of the auditory capsule, just mesiad of the projection of the external (horizontal) canal. This slit widens into an irregularly lanceolate space, and is filled with delicate cells of indifferent tissue; by the time the tadpole is eight lines

THE FROG: SECOND STAGE.

in length this has chondrified to form the simple stapes, lying in the fenestra ovalis which has thus arisen. No further development in connection with the stapes takes place until a month or more after metamorphosis is complete (see pp. 160, 161).

Fig. 33.

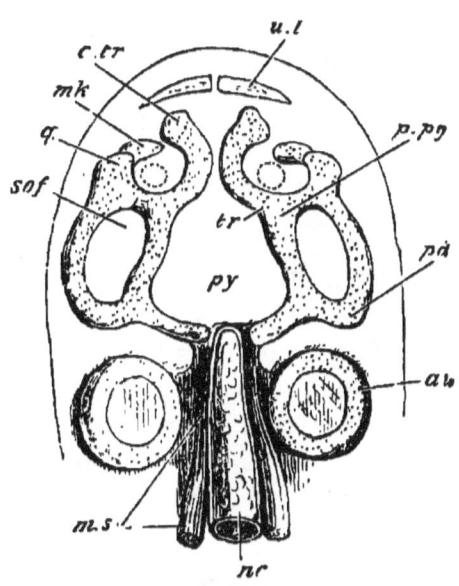

Tadpole of Common Toad, one-third of an inch long; cranial and mandibular cartilages seen from above; the parachordal cartilages are not yet definite.

nc. notochord; *m.s.* muscular segments; *au.* auditory capsule; *py.* in region of pituitary body; *tr.* trabecula; *c.tr.* cornu trabeculæ; *p.pg.* palatopterygoid conjunction; *pd.* pedicle, *q.* quadrate condyle, *mk.* meckelian piece of mandibular arch; *s.o.f.* subocular fenestra; *u.l.* upper labial cartilage. The dotted circle within the quadrate region indicates the position of the internal nostril.

332. The mandibular arch has become greatly enlarged, and much broader; and its distal end is segmented off as a small free meckelian bar (*mk.*), which is directed forwards and upwards in the lower lip at an acute angle with the main part, which may now be called the mandibular *suspensorium*. In its proximal fourth, the suspensorium is almost at a right angle with the trabecula, and is confluent

with it. It then turns forwards at a somewhat lower level, and is roughly parallel with the trabecula, a membranous space, the subocular fenestra, intervening. More anteriorly, just behind the nasal sacs, the suspensorium is united with the trabecula by a bar similar to the hinder connective. This palatine bar lies below and within the developing temporal muscle, which is also covered by the *orbitar process*, another lamina of cartilage growing from the suspensorium in the same position (see Fig. 34). The anterior or quadrate region of the suspensorium is very short; and it bears a rounded head or condyle for the meckelian rod.

333. Only the lower region of the second or hyoid arch has chondrified; it has become enlarged and thickened, and diverges backwards at a considerable angle from its articulation with the middle of the hinder edge of the suspensorium. It becomes incurved towards the middle line below, but does not yet meet its fellow. Mesially, in the hinder part of the tongue, there is a small piece of nascent cartilage, the basihyal. The four branchial arches are simple, being curved inwards above and below, and flattened laterally. Mesiad of the first and second branchials there is pyriform basibranchial cartilage.

334. The upper and lower labials have now become a pair of cartilages above and below: the upper have considerable vertical as well as lateral extension; the lower are rounded rods vertically placed in the lower lip. The internal nares open in the roof of the mouth external to the internasal plate and behind the trabecular cornua; the external nares open on the upper surface of the head in the same region.

335. The amount of development which has occurred since the first stage constitutes a notable morphological advance. The confluence of the trabeculæ to form an internasal plate, their union with the parachordals behind and with the mandibular bars laterally are events having a very high interest. Labial cartilages are well developed,

although at the same time the rudiment of the lower jaw has been cut off from the at present much larger suspensorium. The orbitar process is a remarkable but little understood structure. The hyoid arch has acquired a definite articulation with the suspensorium, and basihyal and basibranchial elements are discernible.

Third Stage: Perfect Tadpoles one inch long, with hind legs appearing.

336. The digestive canal having been completed in the last stage, growth and development have been very rapid, and great morphological advance has taken place. The external gills have become covered by the opercular skin-fold, and at the same time there have been developed copious tufts of internal gills which arise from both the surfaces of the gill-arches. On the right side the opercular fold growing backwards from the hyoid arch unites with the skin of the thorax; but on the left, some way behind the last branchial arch, a vertical oval passage for the outflowing current of water remains (Fig. 35, *op.*). The coalescence of primarily distinct skeletal elements which had commenced in the last stage, has now, after the lapse of two or three weeks, attained its maximum. The differing relations of the sense-capsules, each of which has a chondrified wall, are manifest. The auditory sacs and the eyeballs acquire their cartilaginous coat independently, but the former coalesce with other elements, while the latter remain distinct; and the nasal cavity has its floors, roof, and walls, so far as they are yet formed, entirely developed as outgrowths of the trabecular plate.

337. The parachordal cartilages have united with one another, over and under most part of the cranial notochord, forming a basilar plate which coalesces on each side with the auditory capsule (Fig. 34). Posteriorly the parachordals curve outwards behind the ear-capsules as far as their most prominent region, to form the two rounded occipital condyles (*oc. c.*); and here also lateral walls and a roof of slight longitudinal extent are developed, forming

an occipital cincture of continuous cartilage. The roof has coalesced with and extends between the hinder half of the auditory masses. More anteriorly, the parachordals have not united; but the apex of the notochord no longer occupies its former position. It appears to have relatively receded, leaving a fissure occupied by membrane, the posterior basicranial fontanelle. In front of this fissure the parachordals are perfectly continuous with the trabeculæ.

Fig. 34.

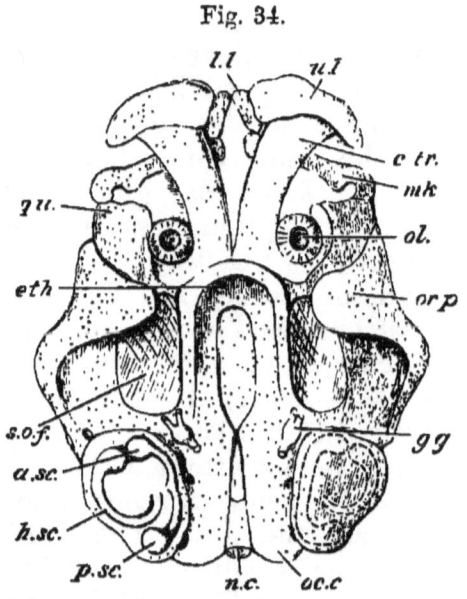

Tadpole about one inch long; view of face and cranial floor from above, the brain having been removed; the upper region of the auditory capsule has been sliced off.

n.c. notochord, covered by basilar cartilage; the pointed space formerly occupied by the apex of the notochord is seen as the posterior basicranial fontanelle; *oc. c.* occipital condyle; *u. sc., h. sc., p. sc.* anterior, horizontal, and posterior semicircular canals; *g.g.* Gasserian ganglion; in front of it the cranial side wall is seen rising from the trabecular floor; the thinly chondrified median tract of the cranial floor is distinctively indicated; *eth.* anterior concave ethmoidal wall of the cranium; *c.tr.* cornu trabeculæ; *ol.* nasal sac; *u.l.* upper labial; *l.l.* lower labial; *s.o.f.* subocular fenestra, flanked outside by the mandibular arch, which is united behind with the cranial floor, and applied round the auditory capsule; giving off anteriorly the orbital lamina *or. p.* rising over the temporal muscle; *qu.* quadrate region; *mk.* meckelian cartilage, extending to the lower labial.

338. The original intertrabecular space, at first only enclosing the pituitary body, has become much elongated; moreover it has acquired a thin but complete cartilaginous floor continuous with the trabeculæ. All along the cranial trabecular region, low side walls have arisen, which become more prominent anteriorly, and join in front of the olfactory lobes of the brain to form an anterior concave ethmoidal wall (*eth.*). The first bone has appeared as a parostosis beneath the intertrabecular space just described, and of equal extent with it; this is the *parasphenoid*.

339. Instead of small trabecular horns springing from a commissural region of no great extent, we find now a pair of gently diverging rods, lying above the fore palate (*c. tr.*). They broaden anteriorly and are strongly decurved, each bearing a transverse horizontal ridge articulating with the corresponding upper labial (Fig. 34, *u. l.*). The ethmoidal region remains restricted, forming the front wall of the brain-case, and uniting with the palatine bars laterally. The nasal sacs are relatively small, lying in the angle between each palatine bar and the corresponding cornu, and flanked externally by the quadrate end of the suspensorium.

340. The auditory capsule has lost its simple external appearance and pyriform shape, the development of the ampullæ and canals having distended it in various directions. The capsule is hollowed in the middle above, between the elevated portions; and it projects outwards in the region of the horizontal canal, forming a pterotic ridge, the rudiment of the *tegmen tympani*. The fenestra ovalis, situated infero-laterally, is elongated-oval in shape, and filled with membrane, in which the stapes has already chondrified.

341. There is a complete coalescence between the auditory capsule and the cranial investment. The glossopharyngeal nerve pierces the hinder end of the capsule below; and the vagus escapes between it and the occipital cartilage. The Gasserian ganglion (*gg.*) lies close to the

prootic region, opposite the end of the notochord, and the branches of the trigeminal nerve pass out of the cranium over the pedicle of the suspensorium. The facial nerve comes off from the same ganglion, curves round the fore face of the capsule, and passes to its distribution behind the suspensorium. The auditory nerve enters the otic mass below the junction of the anterior and posterior canals.

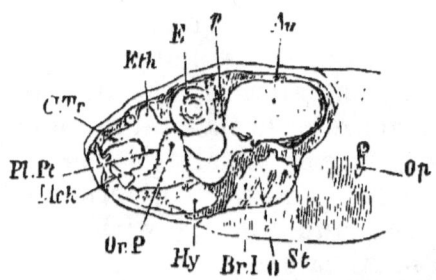

Fig. 35.

Tadpole one inch long; side view of skull dissected.

Au. auditory capsule; *e.* eyeball; *eth.* ethmoidal region; *c.tr.* cornu trabeculæ; *p.* pedicle; *o.* primary otic process of suspensorium; *or.p.* orbitar process; *pl.pt.* palatopterygoid bar; *mck.* meckelian cartilage; labial cartilages are indistinctly visible; *hy.* hyoid bar; *br. 1*, first branchial arch; the epibranchial connective of the arches is seen; *st.* stapes; *op.* opercular opening.

342. The mandibular arch has become highly complex. The suspensorium has increased in length nearly as much as the trabeculæ, and is roughly parallel with the axis of the skull. But its hinder extremity is curved inwards and upwards, giving off two forks. The anterior and longer, the *pedicle* (Fig. 35, *p.*), is confluent with the cranial base and side wall just in front of the auditory capsule; the posterior process (*o*) is closely applied to the antero-external and upper face of the same capsule, arching over the facial nerve. Between the capsule and the two forks of the suspensorium there is a hourglass-shaped membranous space.

343. The paracranial part of the suspensorium is more than half its length, extending from the pedicle to

the palatine connective. The suspensorium is at a lower level than the cranial floor, and consequently the membranous *subocular fenestra* (*s.o.f.*) is rather obliquely placed. The eyeball lies above it, and is well seen on the upper surface of the head: the temporal muscle is attached to much of the surrounding cartilage, and covered antero-externally by the orbitar process (*or. p.*), which has become large, elevated, and incurved over the side of the face. The palatine bar remains a small narrow pedicle, but it also developes on its posterior edge a small backwardly-growing lamina of cartilage, somewhat like the orbitar process.

344. In front of the palatine region the quadrate end of the suspensorium is short and stout, articulating with the meckelian rod, which is still short, and directed almost vertically upwards and inwards, so as to come very close to the trabecular cornu when the mouth is closed. Its proximal end is deeply and roundly notched to hinge upon the quadrate; and it is joined to its fellow by fibrous tissue.

345. These meckelian rods do not appear to be of much functional importance as yet: their direction alone indicates this. The fore part of both jaws is occupied by the upper and lower labials. The upper pair are almost vertical falcate flaps of solid hyaline cartilage, thick and heart-shaped in section, the upper thick edge being grooved to embrace the decurved end of the trabecular horn. The inferior labials are drumstick-shaped rods, with the lower end stoutest, and the *upper* end attached to the anterior face of the meckelian bar near the symphysis. They nearly meet below.

346. The oral opening is very small, nearly vertical, and suctorial: but a horny dentigerous plate is developed in the front of each jaw between the labial cartilages and the skin. The upper plate is behind the corresponding labials; the lower is below rather than behind them. The creases of the lips are also covered with small hooked teeth, in addition to the two principal plates.

Somewhat behind the mouth the oral cavity widens, and has lateral angles, as well as a deep angular fossa between the lower labials. Further backwards, the section of the cavity is simplified, becoming wider and less deep, with its lateral angles downwardly produced between the meckelian and the quadrate cartilages.

347. The hyoid arch (Fig. 35, *hy.*) remains completely below the level of the suspensorium: each moiety is a broad and massive plate, roughly four-sided, scooped within by an antero-inferior and a postero-superior fossa. At its most elevated (median) part it articulates by a shallow cup-and-ball joint with the under surface of the suspensorium beneath the hinder edge of the orbitar process. Its blunt antero-inferior end articulates with the small pisiform basihyal.

348. The four branchial arches are at their full development, and are greatly modified since the last stage. They have all coalesced together both above and below, by the formation of a continuous thin bar of cartilage. From the epibranchial bar (opposite the second branchial) there project two or three small processes. The ventral connective articulates on either side with the rounded first and the rudimentary second basibranchial. This hypobranchial tract, behind the basi-branchial, overlies the pericardium; and the arches are so much curved and thrown outwards in the side wall of the throat, that the hypobranchial and epibranchial connectives lie on nearly the same level; and the arches, which were once immediately beneath the skin and parallel with the sides of the face, now hang like hammocks obliquely across the throat. The first and fourth arches are baggy and crumpled, and extremely thin. All are covered with papillæ transversely placed, bearing the very abundant branchial tufts[1].

349. We can now review this surpassingly interesting stage. If the student will make a comparison between

---

[1] For a full description of the branchial tufts in Rana pipiens, see *Phil. Trans.* 1871, p. 156.

the form here described and the skull of the adult Frog, he cannot but be astonished at the contrast; and he may go where he will in the broad field of the morphology of the skull, and returning with added knowledge and developed thought, he will find himself capable of deeper insight attended with intensified marvel in studying this minute skull. The figures on Plates V. and VI. in the Philosophical Transactions for 1871 will be found fruitful if looked at again and again. A shallow cartilaginous boat supports the brain, and united with it behind is a pair of ball-like masses containing the organs of hearing. Diverging horns in front of the brain-case support large labial cartilages. The mouth proper is entirely *in front of* the cranial cavity: there are rudiments of the future upper and lower jaws, but placed most dissimilarly to their ultimate directions: while large transitory labials predominate among the oral tissues. The mouth is suctorial, but also has a pair of transitory horny dentary plates, with other denticles. The side of the face is traversed by a large subocular bar, confluent with cranial cartilage in front and behind, and from its forward edge a lamina arises to roof over the temporal muscle. The hyoid arch has a massive and strange form, appearing to play a very subordinate part in the structure: and the gill-arches, framed into a complex basket-work, bound together above and below, bent almost double in the sides of the throat, are at the summit of their ascending development, soon to retrograde and become obsolete.

Fourth Stage: Tadpoles with tails disappearing.

350. In this stage we consider the changes taking place in the skull during the gradual disappearance of the tail. Before the anterior limbs are seen externally, the lower jaw becomes much elongated, and the suspensorium, not increasing in size so rapidly as other structures, appears to be carried backwards and outwards, the palatine bar being greatly lengthened. By the time that the fore limbs are exposed, and the tail is reduced to half its

original size, the brain-case is complete in cartilage. When the tail has disappeared, most of the cranial bones are manifest, the sphenethmoid or "girdle-bone" being a remarkable exception: the creature has become far removed from the ichthyic type, and in many respects specially amphibian.

351. The cranial floor is composed of almost continuous cartilage, nearly uniform in level. The intertrabecular region is fully chondrified, but behind the pituitary body a triangular median fenestra remains, with the base directed forwards; it appears to be partly a remnant of the intertrabecular space, partly of the fissure between the parachordals which the notochord no longer occupies. In the basioccipital region the gelatinous remainder of the notochord lies in a deep median groove in the cartilage.

Fig. 36.

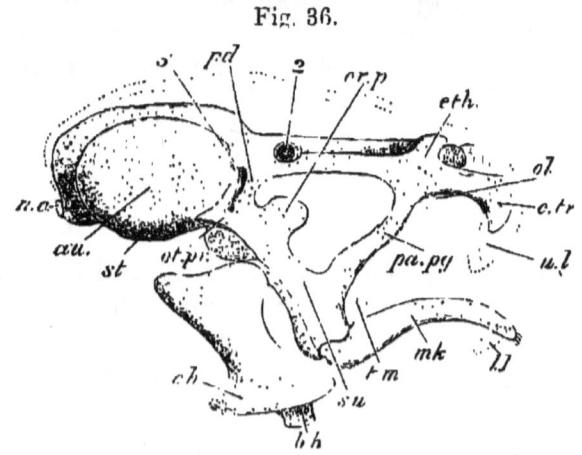

Tadpole with tail beginning to shrink; side view of skull without branchial arches.

n.c. notochord; *au.* auditory capsule; between it and *eth.* the low cranial side wall is seen; *eth.* ethmoidal region; *st.* stapes; 5, trigeminal foramen; 2, optic foramen; *ol.* olfactory capsules, both seen owing to slight tilting of the skull; *c.tr.* cornu trabeculæ: *u.l.* upper labial, in outline; *su.* suspensorium; *pd.* its pedicle; *ot.pr.* its otic process; *or.p.* its orbitar process; *t.m.* temporal muscle, indicated by dotted lines, passing beneath the orbitar process; *pa pg.* palatopterygoid bar; *mk.* meckelian cartilage; *l.l.* lower labial, in outline; *c.h.* ceratohyal; *b.h.* basihyal. The upper outline of the head is shown by dotted lines.

352. The side walls of the cranium are complete, and confluent with the floor and with the auditory capsules; above they form a slight inturned selvedge in the roof. The fore end of the skull is gently concave, pierced on either side of the middle line by the olfactory nerves. The optic nerve passes out in the side wall not far in advance of the auditory capsules. The supraoccipital cartilaginous roof extends as far as to the fore end of the ear-capsules, but is deficient on either side in its anterior half, leaving two oval parietal fontanelles. The cranium and the periotic cartilages constitute one broad continuous mass. The cranial roof is membranous in its anterior two-thirds, forming an oblong fontanelle.

353. In the earliest period of this stage the trabecular cornua remain distinct, although the hinder commissural (ethmoidal) region is more extensive than it was. The cornua become broader and more flattened, approaching the middle line; the nasal sacs gradually come to lie upon them. Anteriorly the sacs are separated by soft tissue, but behind there is a cartilaginous vertical plate, the rudimentary nasal septum, growing from the thickening ethmoidal region in front of the brain-case. The ethmoidal cartilage passes directly into the palatine connective in the antorbital region (Fig. 36).

354. Later, the septum, the cornua, and the ethmoidal region become perfectly confluent, and the original trabecular moieties can only be distinguished antero-externally where the extremities of the cornua project inwards as small rhinal processes from the front and lower wall of the nasal capsule. The latter is almost completely invested by cartilage: the septum has grown forwards to the front of the face; the floor is formed by the flattened cornua; and a continuous aliseptal roof grows outwards from the upper edge of the septum and the ethmoidal region, curving downwards into the external nasal wall.

355. The external nostrils are on the upper surface of the face, very near its fore end. They are bounded in

front by a distinct but small *prorhinal* cartilage, derived from the upper labial, the lower part of which is separated, and covered by the premaxillary; while the aliseptal lamina curves round behind, and bounds them infero-externally. In this external boundary an ectosteal *rhinal* bone arises, which extends in the outer wall down to the nasal floor. The roof of the cavity is mostly behind the external nostril, forming an alinasal pouch directed backwards and outwards, distinct from the ethmoidal cranial wall and the palatine bar. The cartilage curves downwards infero-externally as a falcate process to bound the inferior or internal nostril (which is more posterior than the external), becoming continuous with the main part of the nasal floor. The internal nostrils are separated by a comparatively wide basinasal plate.

356. In previous stages we have seen the mandibular suspensorium as a subocular bar parallel with the cranial wall, and connected with it in front by a short palatine pedicle. The latter gets gradually elongated backwards and outwards, so that the original subocular bar is at length situated entirely behind the eyeball: first passing through a stage in which it forms a V with the palatine, the angle being subocular (Fig. 36). Finally, the palatine (or palatopterygoid as it may now be called) itself becomes a subocular bar, nearly parallel with the cranial wall, but descending to a lower level behind. The suspensorium, considerably shorter than the palatopterygoid, is directed forwards, outwards, and downwards, at an angle of about 50° with the cranial axis (Fig. 37).

357. The anterior or palatine part of the subocular bar remains connected with the lateral ethmoidal region of the skull by a short transverse process beneath the alinasal cartilage; it also gives off antero-externally a short prepalatine spur. The palatine region soon becomes marked off from the pterygoid by the appearance of constriction at the anterior third of the bar. Posteriorly the pterygoid is continuous with the suspensorium at the junction of its middle and lower thirds.

358. The orbitar process, formerly so large, does not advance in size, as the suspensorium increases. The latter is lengthened in front of the orbitar process, which consequently, instead of being anterior to the eye, comes to be placed quite behind it, covering part of the hinder surface of the temporal muscle (Fig. 36, *or. p.*). It finally becomes an inconspicuous thin lamina on the fore edge of the suspensorium.

359. The upper or metapterygoid region of the suspensory cartilage is more and more differentiated into regions, though its size diminishes relatively to the quadrate or lower tract. The attachment to the cranial floor remains, but the primary posterior process is first enlarged as a triradiate lamina, and then segmented off, to be transformed into the tympanic *annulus* in a later stage (*a. t.* Fig. 38). The upper posterior angle from which this cartilage has been cut off forms a rounded *otic process* which is apposed to the outer half of the front part of the auditory capsule. The hinder edge of the main part of the suspensorium becomes thin and shell-like, convex internally and concave externally.

360. The distal extremity of the suspensorium (*su.*) is concavo-convex, fitting accurately upon the condyle of the greatly elongated sigmoid mandible (*mk.*). The latter is thicker behind, and pointed in front; its course is forwards and inwards, subparallel with the cranial floor. The lower labial becomes identified with the symphysial end of the meckelian cartilage. The facial nerve passes out of the cranium in front of the periotic cartilage, and runs downwards and forwards on the inside of the posterior margin of the suspensorium, where it divides into a chorda tympani or mandibular branch and two lesser hyoid branches.

361. The hyoid arch loses its lozenge shape, and elongates, articulating with the middle and upper part of the hinder edge of the suspensorium instead of near its distal extremity. It is expanded above and below, and constricted in the middle. Furthermore, it coalesces early with the

soft azygous basihyal; so that the arches of opposite sides are continuous below. Confluence also takes place between the basihyal and the basibranchial, thus fusing the hyoid and the branchials into one hyobranchial system. The hypobranchial regions have also lost their distinction in this broad medio-ventral plate of cartilage. The first and fourth branchial arches lose their flattened expanded form, and all become very slender (Fig. 37). But each epibranchial connective sends upwards a series of six tooth-like lobes. Later, these same spurs degenerate and pass into fibrous tissue; while a process developed from the hypobranchial region of the first and second arches elongates, diverging from the middle line.

362. Ossification is just commencing in the fibrous tissues of the cranial roof when the tail begins to diminish in size: it advances very far before the end of this stage. Bones affecting the cartilaginous cranium arise some weeks later than the earliest membrane bones, but they may be mentioned first. There are two pairs, the *prootics* and *exoccipitals* (Fig. 37, *pr o.*, *e.o.*). The prootic embraces the trigeminal nerve at its exit from the cranium, and extends beneath the fore part of the auditory capsule, and upwards in the side wall of the cranium in front of it, so as to be sickle-shaped. The exoccipital is very similar, surrounding the foramen for the vagus nerve, ossifying the base of the condyle behind and the hinder part of the periotic mass in front. These bones commence in the perichondrium, and in the superficial cells of the cartilage (*superficial endostosis*): they proceed to ossify the cartilage throughout.

363. Two pairs of parostoses arise on the cranial roof: the anterior (*frontals*) cover the part between the ear-capsules and the front end of the cranial cavity. They are much longer than the *parietals* (*pa.*) which lie close together between the fore part of the auditory capsules. Later, the parietal and the frontal of each side begin to coalesce. The frontals do not quite roof over the great fontanelle in front; the parietals seem to lie entirely upon

cartilage, but the cranial roof is imperfect beneath the centre of each bone. The *nasals* (*na.*) arise as small patches over the middle of each nasal capsule, and become crescentic bones with the concavity forwards and outwards.

Fig. 37.

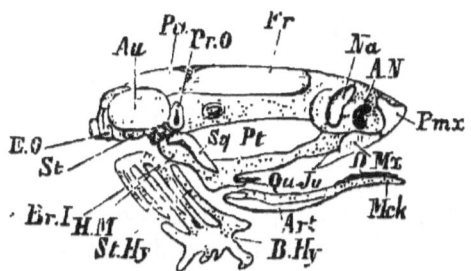

Young Frog, with tail just absorbed; side view of skull.

Au. auditory capsule; in front of it, cranial side wall; *a.n.* anterior nostril; *st.* stapes; *mck.* meckelian cartilage; *b.hy.* basihyobranchial plate; *st.hy.* stylohyal or ceratohyal; *br.* 1, first branchial arch.

Bones: *e.o.* exoccipital; *pr.o.* prootic; *pa.* parietal; *fr.* frontal; *na.* nasal; *pmx.* premaxillary; *mx.* maxillary; *pt.* pterygoid; *sq.* squamosal; *qu.ju.* quadratojugal; *art.* articular; *d*, dentary.

364. The *parasphenoid* underlies the cranial floor from the ethmoidal nearly to the end of the basioccipital region, being convex and subcarinate below, and scooped on its upper surface. It is almost as broad as the cartilaginous basis cranii, to which it is closely applied, tongue-shaped in its anterior two-thirds, and pointed behind. In the middle of the periotic region it acquires a basitemporal wing on either side, ending sharply near the foramen ovale. When the parasphenoid has assumed this complete form, the *vomers* are only small irregular patches of bone on the surface of the palate, a little internal to each inferior nostril.

365. The *premaxillaries* (*pmx.*) and *maxillaries* (*mx.*) originate before the vomers. The former appear as short thick bars laid in front of the rhinal processes, with a nasal process on the upper surface of the face. The premaxil-

laries meet in the middle line, but their nasal processes are separated by a circular membranous space. The maxillaries are styloid bones continuing the line of the premaxillaries in the upper lip, so as to form with them part of a hyperbolic curve. They have an ascending lamina on the face throughout most of their length, and end pointedly behind, outside the prepalatine spur of the palatine cartilage. They appear rather as splints to the nasal capsule than to the palatine bar. There is a ligamentous union in the upper margin of the gape between the end of the maxillary bone and the quadrate condylar region.

366. Delicate spicules of bone are found in the fibrous stroma lining the inner face of the suspensorium and of the pterygoid cartilage. The *pterygoid* lamina (*pt.*) becomes thicker close to the suspensorium, extending upwards as a needle (which expands to a spatula-form above) on its inner and anterior edge. There is a distinct ectosteal (*metapterygoid*) plate embracing the inner convex face of the hinder part of the suspensorium. Near the quadrate condyle a *quadrato-jugal* (*qu. ju.*) has appeared, extending as a small style towards the maxillary.

367. On the outer face of the suspensorium a *squamosal* parostosis early arises, and soon acquires a sigmoid shape, being broader above and pointed below. Its upper or supratemporal region becomes transversely extended, and moulded on the remnant of the cartilaginous orbitar process. The *dentary* and *articular* bones (*d., art.*) arise similarly at first, as thin fenestrate plates lying in contact with the meckelian cartilage, or only slightly separated from it by fibrous stroma. The dentary is at first on the outer surface of the fore half of the mandible, the articular on the inner and lower surface of the hinder two-thirds, not however reaching to the angle.

368. The cranial cavity is now very largely bounded by cartilage, replaced to a slight extent by bone. The nasal region has taken definite form in continuity with the

primary trabecular cornua, forming the considerable olfactory capsules, and affording a basis for the anterior jawbones. Instead of a suspensorial subocular bar, the palato-pterygoid cartilage is subocular: the suspensorium is still directed a little forwards, but also passes very largely outwards. It has become differentiated proximally, giving off the lamina for the tympanic ring, and forming a new simple otic process, as well as a rounded condyle on its inner cranial process or pedicle; the orbitar lamina is now insignificant. The transformation of the hyoid and branchial apparatus is very notable, tending to the formation of a single expanded basal plate, and a rod-like hyoid style, and to the disappearance of the branchial arches.

369. The bones existing at the close of this stage are the following: exoccipitals and prootics; parasphenoid, parietals and frontals; nasals, premaxillaries, maxillaries, vomers, and rhinals; pterygoids, metapterygoids, quadrato-jugals, and squamosals; dentaries and articulars. It is remarkable how indefinite are the boundaries between parostoses and ectostoses in the early stages of ossification in the Frog: the "membrane"-bones arise for the most part closely in contact with the face of the cartilage.

Fifth Stage; Frogs of the first summer.

370. The young Frog during the first summer increases in bulk threefold, and at the same time its skull is greatly developed, especially with regard to ossification. The cranial cavity becomes less oblong, and broader behind; the periotic masses are relatively wider and shorter. The direction of the suspensorium is altered by the gradual retraction of its lower extremity (Fig. 38).

371. The prootics and exoccipitals extend considerably into the surrounding cartilage, completely encircling the trigeminal and the vagus nerves respectively. A new bone, of great interest from its later development, makes its appearance as a narrow transverse splinter in the ethmoidal region immediately constituting the front wall

of the cranium. This bone, arising as a true ethmoid, has to be subsequently denominated *sphenethmoid (eth.)*; it possesses both ectosteal and endosteal portions, formed synchronously; at present no trace of it can be seen from the sides or from below.

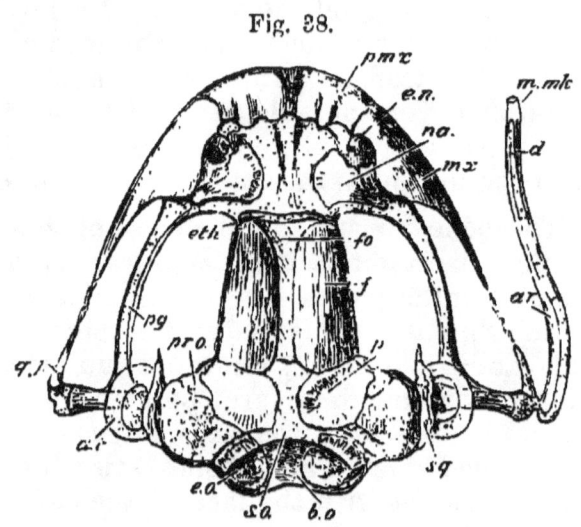

Fig. 88.

Young Frog, near end of first summer; upper view of skull, with left mandible removed, and the right extended outwards.

b.o. basioccipital tract; *s.o.* supraoccipital tract; *fo.* frontal fontanelle; *e.n.* external nostril; internal to it, internasal plate; *a.t.* tympanic annulus.

Bones: *e.o.* exoccipital; *pr.o.* prootic, partly overlapped by *p.* parietal; *f.* frontal; *eth.* rudiment of sphenethmoid; *na.* nasal; *pmx.* premaxillary; *mx.* maxillary; *pg.* pterygoid, partly ensheathing the reduced cartilage; *q.j.* quadratojugal; *sq.* squamosal; *ar.* articular; *d.* dentary; *m.mk.* mento-meckelian.

372. Other regions of the cranial cartilage begin to have their peripheral cells directly calcified: but this bony matter is not deposited in tesseræ as in the Elasmobranchs, nor throughout distinct morphological regions; the osseous granules do tend, however, to be aggregated where in another type a proper bone-territory would be established. The epiotic, supraoccipital, pterotic, basisphenoidal, and upper and lower nasal regions are those in which this superficial endostosis can be traced. The

bony deposit is in very small grains, in semilunes extending half round a cell, and in separate and connected rings.

373. The frontal and parietal of the same side still retain a partial suture between them: the parietal overlaps the ascending part of the prootic; the frontal extends to the fore end of the great cranial fontanelle (*fo.*), just behind the commencing sphenethmoid; the parietofrontals of opposite sides do not yet meet by a considerable space. The nasals (*na.*) are much altered in shape; the upper part is now a broad subquadrate plate, with a small spur running downwards and outwards on the ethmoidal cartilage: the two nasals are widely separated from one another. The rhinal bone is seen as a notched and grooved plate lying on the floor of each external nostril.

374. The premaxillaries and maxillaries (*pmx., mx.*) are much advanced, and dentigerous. Their nasal processes are considerably enlarged: those of the premaxillaries are completely apposed, and extend pretty evenly to the level of each external nostril; those of the maxillaries rise to the outer boundary of the nostrils, and are higher behind, towards the ethmoidal spur of the nasals. Behind this region the nasal plate rapidly sinks, and the maxillary is a mere style passing backwards to within a short distance of the quadratojugal. The palatal laminæ of the premaxillaries and maxillaries are of less extent than their nasal processes; they are thin shelves which die away behind. The parasphenoid is broader in front and much more dense than formerly; it does not reach relatively so far back as it did, being rather emarginate than pointed. Its basitemporal wings are wider and broader, corresponding with the increased breadth of the auditory region; their hinder edge is conterminous with the fore margin of the exoccipitals. The vomers have developed into very irregular triradiate bones, mainly related to the inner margin of the inferior nostrils, but sending a spur backwards and inwards towards the parasphenoid.

375. The great subocular space between the cranial wall and the palatopterygoid bar is now nearly oblong. The prepalatine spur at the antero-external angle is strongly marked. Beneath the transverse part of the cartilage between this region and the ethmopalatine junction a delicate slightly-curved narrow *palatine* bone is applied. It rather overlaps, but does not touch the styliform ectosteal pterygoid (*pg.*) which lies on the inner face of the whole longitudinal part of the cartilage, and tends to ensheath it above and below. Posteriorly the pterygoid sends off at an obtuse angle a spur which underlies two-thirds of the suspensorium: above, it becomes applied to the small metapterygoid and partly fused with it.

376. During the first summer the suspensorium, which was still turned forwards when the young frog left the water, becomes progressively retracted so as to form more than a right angle with the fore part of the skull. The pedicle where it had coalesced with the cranial cartilage is converted into fibrous tissue, and the boss below is a distinct broad condyle articulating with the antero-lateral angle of the auditory cartilage below. The secondary otic process has completely fused with the supero-lateral edge of the auditory capsule, and overhangs the tympanic cavity.

377. The meckelian and labial cartilages are replaced at the symphysis by a small solid *mento-meckelian* bone (*m. mk.*) on each side; and the very delicate dentary, not half the length of the ramus, has coalesced with it. The main part of the bar lies in a trough-like ectosteal bone, the articular (*ar.*).

378. The stages by which the parts of the middle ear, so late in their development, become manifest, may here be recounted[1]. In tadpoles we have described merely a simple stapes, having between it and the continuous skin a small space, a relic of the first visceral cleft. In

---

[1] See Huxley; "On Representatives of Malleus and Incus," *P. Z. S.* 1869, pp. 391—407.

young frogs that have taken to the land, a definite tract of granular tissue can be traced from the stapes forwards and outwards to the upper part of the suspensorium beneath the otic process; the facial nerve passes over this tract. In the course of rather more than a month a delicate clubbed rod of cartilage is formed in this tissue, its clubbed end fitting into the narrow anterior part of the fenestra ovalis, which has not become completely filled by the stapes. The cartilage cells in the pointed anterior end of this rudimentary *columella* pass insensibly into fusiform connective-tissue cells.

Fig. 39.

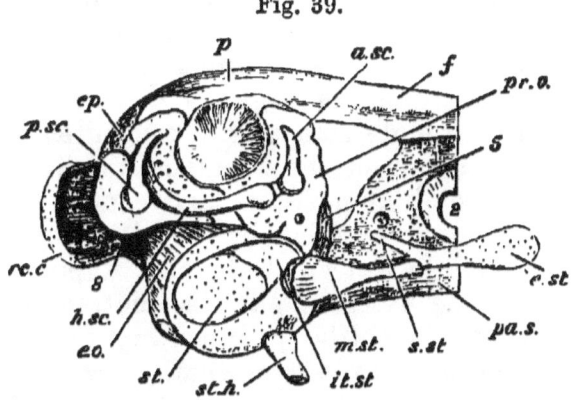

Adult Frog, side view of auditory region, with semicircular canals and parts of middle ear displayed.

oc.c. occipital condyle; *ep.* epiotic region; *a.sc., h.sc., p.sc.* anterior, horizontal, and posterior semicircular canals; 5. trigeminal foramen; 2. optic foramen, surrounded by a ring of membrane; *st.* stapes; *it.st.* interstapedial; *m.st.* mediostapedial (bone); *e.st.* extrastapedial; *s.st.* suprastapedial; *st.h.* stylohyal cartilage cut off short.

Bones: *e.o.* exoccipital; *pr.o.* prootic; *p.* parietal; *f.* frontal; *pa.s.* parasphenoid; *m.st.* mediostapedial.

379. About two months later, towards autumn, the columella is complete (see Fig. 39). The posterior extremity is segmented off as an oval mass of cartilage about half the size of the stapes; this is the *interstapedial* (*it. st.*). The middle third, or *mediostapedial* (*m. st.*), is more slender, and is invested by a bony sheath. The anterior third curves outwards, applying itself to the

back of the upper part of the suspensorium; this *extrastapedial* (*e. st.*) is broader and flatter than the middle part, and becomes involved in the fibres of the tympanic membrane. At its proximal end the extrastapedial gives off a small rounded cartilaginous process upwards and backwards at an acute angle with the middle piece; this *suprastapedial* (*s. st.*) is attached by ligament to the side of the auditory capsule.

380. By this time the little cartilage which was detached from the postero-superior angle of the suspensorium forms a tympanic annulus, constituting more than three-fourths of a circle. It is a broad band, thick on the exterior, with a thin inner edge. Its extremities lie upwards, and are strongly attached to the outer edge of the supratemporal portion of the squamosal, hiding the upper half of the descending part of that bone. The circular space thus enclosed is filled by a thick web of fibrous tissue, to the inner face of which the extrastapedial cartilage is attached. The external skin is continuous over this layer of fibrous tissue (the rudimentary tympanic membrane), dipping down to it somewhat.

381. The lower part of the hyoid arch is a styloid cartilage loosely connected to the skull and suspensorium by ligaments. Ventrally it sends forward a little hypohyal lobe, and is fused with the broad shield-shaped basibranchial plate. Posteriorly this plate bears remains of the hypobranchial portions of two branchial arches: the anterior are slender and unossified; the posterior pair are ossified, and embrace the larynx. The other regions of the branchial arches have disappeared.

382. The principal points to be borne in mind with respect to this period, in addition to the growth of bones previously existing and the moulding of the skull into its adult form, are the development of the ethmoid and mento-meckelian bones, the retraction of the suspensorium and of the lower jaw, the completion of the columella and the tympanic annulus, and the disappearance of the branchial

arches. The facial nerve passes over the columella in the same position as it passes over the upper member of the hyoid arch where that is early and fully developed.

The Skull of the Adult Frog.

383. The skull of the frog is a flat semielliptical structure, broadest in the hinder region, where it is formed of continuous cartilage and bone. In front of this expansion, which occupies about one-fifth of the length of the skull, the narrow axial parts enclosing

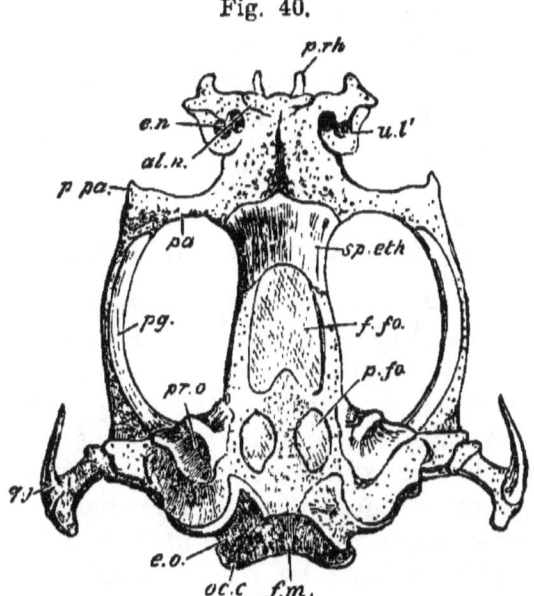

Fig. 40.

Adult Frog: upper view of skull with investing bones and lower jaw removed.

f.m. foramen magnum; *oc.c.* occipital condyle; *p.fo.* parietal fontanelle; *f.fo.* frontal fontanelle; *p.rh.* prorhinal cartilage; *e.n.* external nostril; *al.n.* alinasal cartilage; *u.l'.* modified upper labial; *p.pa.* prepalatine spur.

Bones: *e.o.* exoccipital; *pr.o.* prootic; *sp.eth.* sphenethmoid; *pa.* palatine; *pg.* pterygoid; *q.j.* quadrato-jugal.

the brain are separated from the lateral boundaries for about half the length of the head, by a large ovoid tract, occupied by the eyeballs and by muscle. In the anterior part of the skull the axis and the lateral bars again unite in a precranial expansion. The external bars we have mentioned are carried at a level below that of the cranial floor.

384. When the investing bones of the skull are stripped off, the underlying cartilage and its proper bones are seen to present a somewhat different appearance. The continuous elliptical outline is lost, and the nasal cartilages appear as a very distinct mass attached to the axial or proper cranial region.

385. The cranial cavity is enclosed by continuous cartilage and cartilage-bone, except where nervous structures pass out of it, and where the roof presents membranous fontanelles. Behind, there is an occipital region mostly posterior to the periotic masses. The auditory region is broad, and there is no distinction between periotic and cranial cartilage. The preauditory part of the cranial cavity is three-fifths of its length.

386. The brain-case is nearly twice as deep behind as it is in front (Fig. 41), and considerably wider above than below. Consequently its side walls in front of the ear-masses slope outwards as they ascend. With these peculiarities, the cranial cavity of the frog is of a very smooth and symmetrical type, without marked prominences or irregularities. The base is very flat beneath, and emarginate in the middle line behind. The condyles, largely ossified by the exoccipital bones, stand out on either side above the base (Fig. 41, *occ. c.*), having about half the height of the skull. The exoccipitals extend nearly to the middle line of the cranial floor, and approach it almost as closely above; the supraoccipital region has a thin film of superficial endostosis.

387. The exoccipitals have a large extension into the auditory masses, which are flattened and broad, with a curved outline externally. They are very considerably unossified, possessing indefinite films of superficial

THE ADULT FROG.

endostosis, and a pair of large anterior bones, the prootics (*pr.o.*). These occupy the antero-internal regions of the periotic masses, are deeply notched below where the trigeminal and facial nerves pass out in front of the auditory tract, and extend forwards above these nerves to ossify a considerable part of the alisphenoidal region. The anterior canal and the fore end of the horizontal are especially girt by the prootic; while the exoccipital is related to the posterior canal, and the hinder end of the horizontal. There is a pterotic ridge or tegmen tympani, external to the horizontal canal, and overhanging the stapes and tympanum. The exoccipitals contain the foramina for the glossopharyngeal and vagus nerves, at the junction of their auditory with their occipital regions.

388. The continuous cranial floor in front of the exoccipitals has some slight endostosis in its substance, but not sufficient to be denominated a bone. The cartilaginous side walls of the skull present nothing remark-

Fig. 41.

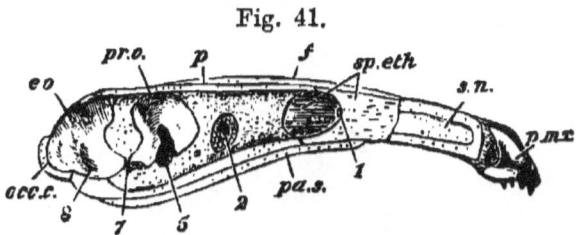

Adult Frog; median longitudinal section of skull, lower jaw removed.

occ.c. occipital condyle; *s.n.* nasal septum; 1, olfactory, 2, optic, 5, trigeminal, 7, facial, 8, glossopharyngeal and vagus foramina.
Bones: *e.o.* exoccipital; *pr.o.* prootic; *p.* parietal; *f.* frontal; *sp.eth.* sphenethmoid; *pmx.* premaxillary; *pa.s.* parasphenoid.

able, except the very posterior position of the foramina for the optic nerves: two-fifths of the length of the cranial cavity is in front of them. The fontanelles on its upper surface consist of a pair of small oval parietal openings (Fig. 40, *p. fo.*) in the region between the prootic bones; and of one frontal (*f. fo.*) much larger, oblong in outline, with a triangular spur of cartilage projecting forwards from the posterior boundary. The cartilaginous roof between the frontal and the parietal fontanelles is feebly ossified by endostosis, a rudiment of a supra-sphenoidal bone.

389. The floor, side walls, and roof of the cranial cavity are ossified anteriorly by one bone, the sphen-ethmoid (Figs. 40, 41, *sp. eth.*). This bone extends further backwards in the side walls than in either the floor or the roof of the cranium. Below, it has a nearly transverse posterior margin; in the roof it is concave behind, embracing the fore end of the frontal fontanelle. The olfactory crura pass through its cranial concavity to gain the nasal cavities; a slight vertical ridge projects between the crura into the cranium, like a rudimentary *crista galli*. The sphenethmoid also extends forwards above and below in the ethmoidal region, aborting the cartilage, and ending by a blunt median projection in each case; which point corresponds to the fore end of a median vertical plate connecting the upper and lower laminæ: this is a true perpendicular ethmoidal (mes-ethmoid) plate (Fig. 41).

Fig. 42.

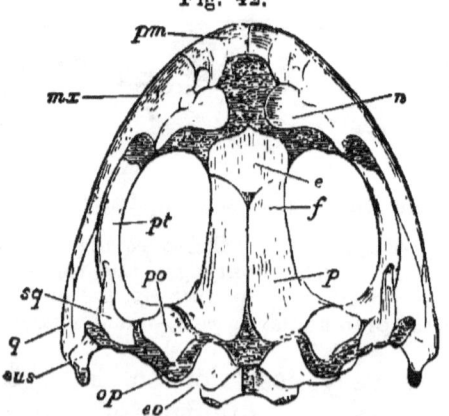

Adult Frog; upper view of skull, with lower jaw removed.

e.o. exoccipital; *p.* parietal; *f.* frontal; *e.* sphenethmoid; *n.* nasal; *pm.* premaxillary; *mx.* maxillary; *q.* quadrato-jugal; *pt.* pterygoid; *sus.* suspensorium; *p.o.* prootic; *op.* opisthotic.

390. The investing bones of the cranium proper are very simple. There is a pair of parieto-frontals, joined

by suture, above, and a parasphenoid below. The parieto-frontals extend from the sphenethmoid, whose lateral regions they partially cover, to the exoccipitals. They diverge from each other behind, exposing the cartilaginous cranial roof, and also in front. They are thick and smooth, and dip towards their sagittal suture: their parietal portions are conterminous with the prootics, and in front of them the bones dip over the edges of the cranial roof, investing its upper side walls for a small depth.

391. The parasphenoid underlies the whole cranial floor (separated from it by a perichondrial membrane) except in the occipital region, even sending a small spur backward between the exoccipitals. Its oblong basitemporal wings are very wide, undergirding most of the width of the auditory masses, and lying between the prootics and the exoccipitals. In front of these wings the bone is very regularly linear, and subcarinate.

392. Beyond the sphenethmoid the whole of the precranial cartilages are unossified. A rather low septum (*s. n.* Fig. 41), thicker above, runs forwards to the end of the snout. The nasal floor formed by the coalescence of the hinder half of the trabecular cornua is a broad thin flap, separated by a notch from the antorbital or ethmopalatine bar, extending outwards. Anteriorly the cartilage is expanded outwards and curved round the inner nostril on its outer side, so as nearly to enclose it. The broad anterior end of the cornu has sent forward in the middle a slender *prorhinal* process, and at its outer angle is bilobate (Fig. 40. *p. rh.*).

393. The roof of the capsule is broadest at its posterior region, where it has completely coalesced with the ethmopalatine bar. Both above and below, the cartilage is partially hardened by endosteal bone: one more notable patch than the rest is in the ectethmoid or antorbital region. The external nostrils are almost directly above the internal, and are protected by a lunule of cartilage which lies over the opening. This cartilage

is the upper two-thirds of the original upper labial (Fig. 40, *u. l'.*). The lower third has become a rounded nucleus which lies between the premaxillary and the nasal capsule in front. Within the nasal opening the

Fig. 43.

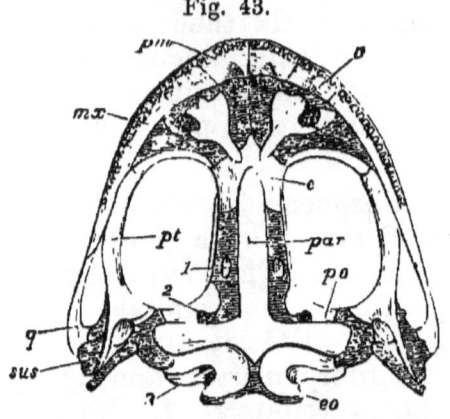

Adult Frog: under view of skull, with lower jaw removed.

e.o. exoccipital; *p.o.* prootic; *par.* parasphenoid; *e.* sphenethmoid; *v.* vomer; *pm.* premaxillary; *m.x.* maxillary; *q.* quadrato-jugal: *sus.* suspensorium; *pt.* pterygoid; 1, optic foramen; 2, trigeminal foramen; 3, glossopharyngeal and vagus foramen.

septum sends out from its upper part a broad thin lamina of cartilage, which runs outwards to the outer and anterior wall of the nose, and unites with the subnasal floor; half way outwards it becomes subdivided into two laminæ, so that at this region the nasal cavity has three horizontal floors. The nasal passage, from external to internal nostril, is completely lined by the little subcylindrical septo-maxillary.

394. On either side of the cranium is a large oval suborbital membranous space. It is bounded in front by the flat ethmopalatine bar, passing directly outwards and sending forwards a prepalatine spur (Fig. 40, *p. pa*), while posteriorly it is continued into the slender pterygoid bar. Behind, this passes into the mandibular suspensorium, marked by three principal regions, (1) an antero-inferior, the *pedicle,* directed inwards, and articulating by a

thickened condyle with the auditory capsule in front of the fenestra ovalis; (2) a supero-posterior, the otic process, attached to the fore part of the tegmen tympani; (3) the quadrate or main suspensorium, directed backwards, downwards, and outwards, so as to reach nearly as far back as do the occipital condyles.

Fig. 44.

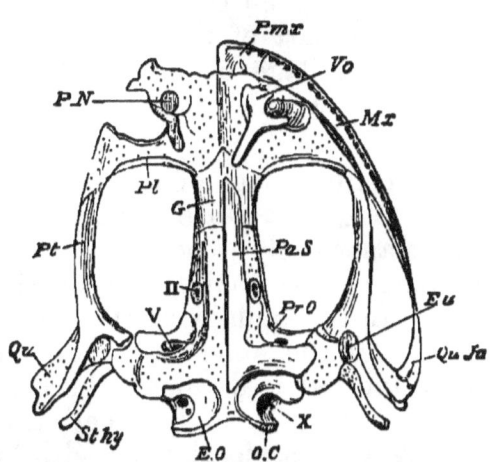

Adult (edible) Frog: under view of skull; investing bones removed on right side. (Huxley.)

o.c. occipital condyle; *p.n.* posterior (inferior) nostril; *qu.* quadrate tract; *st. hy.* stylohyal, partially shown; *eu.* eustachian opening; II. optic foramen; V. trigeminal foramen; X. glossopharyngeal and vagus foramen.

Bones: *e.o.* exoccipital; *pr.o.* prootic; *pa.s.* parasphenoid (left half); *g.* sphenethmoid; *vo.* vomer; *pmx.* premaxillary; *mx.* maxillary; *pl.* palatine; *pt.* pterygoid; *qu.ju.* quadrato-jugal.

395. The palatine or antorbital bone invests the hinder surface of the transverse ethmopalatine bar; it is slender and almost crescentic, and slightly underlies the sphenethmoid. The pterygoid is elongated and more massive; it has transformed much of the cartilage of the subocular bar, and just touches the palatine bone in front. Behind, it is forked, passing inwards by a square (metapterygoid) process to support the pedicle, and backwards as a narrow parosteal spur to clamp the under surface of the suspensorium almost as far as the condyle. The

quadrate cartilaginous region is only very slightly ossified by the hinder end of the quadrato-jugal.

396. The T-shaped squamosal (*sq.* Fig. 45) lies on the suspensorium and the external auditory region. Its horizontal tract is outside the prootic behind, and sends forwards over the pterygoid a considerable postorbital spur. At more than a right angle with this is set the descending portion, lying on the suspensorium and overlapping the quadrate part of the quadrato-jugal. It is separated from the cartilage by perichondrium, and from the quadrate by periosteum. It lies beneath the tympanic annulus.

397. The investing bones of the nasal and maxillary tracts have now to be described. The nasals are considerable bones, transversely placed on the nasal roofs, broad within, narrow externally, not nearly meeting in the middle line. In the middle of their anterior margin they are notched for the external nostril, and then extend backwards and outwards along the nasal lamina of the maxillary to end in a sharp preorbital spur. The premaxillaries are of considerable breadth, are apposed but unanchylosed in the middle line, and possess both dentary and nasal plates. The latter project upwards and backwards by a curved spur in their middle region, touching the nasals. Mesially a tract of cartilage is uncovered by bone between the sphenethmoid, the nasals, and the nasal plates of the premaxillaries (see Fig. 42). The palatal plates of the latter are thin and quadrate, with a somewhat irregular surface.

398. The maxillaries are elongated, with extensive nasal laminæ in their anterior third; the septo-maxillary (*s.mx.* Fig. 45) appears between the inner and anterior angle of this plate and the nasal process of the premaxillary; the external nostril notches it behind this point. The nasal plate of the maxillary dies away, at first by a sudden decrease in the antorbital region, and then gradually, the bone becoming a fine pointed style, largely overlapping the quadrato-jugal. It is adjacent to the anterior

third of the pterygoid, but diverges outwards from it, while the pterygoid curves inwards (Fig. 44). The palatal plate of the maxillary is but slight and ranges with that of the premaxillary, soon becoming insignificant.

399. The vomers are elegant trifoliate plates, wide apart like the nasals. The middle leaf, which is emarginate, and the narrow falcate posterior leaf, together largely surround the internal nostril. The pointed anterior leaf nearly reaches the suture between the maxillary and the premaxillary. The rounded stalk converges towards its fellow, lies partially under the sphenethmoid, and almost touches the parasphenoid; this tract of the vomer is dentigerous. The septo-maxillary can just be seen between the middle and posterior leaflets of the vomer.

Fig. 45.

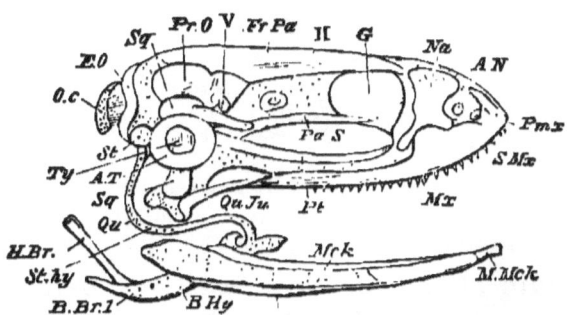

Adult Frog: side view of skull, dissected.

o.c. occipital condyle; *st.* stapes; *ty.* tympanic membrane; *a.t.* tympanic annulus; *a.n.* anterior nostril; *qu.* quadrate condyle; *mck.* meckelian cartilage; *st.hy.* stylohyal cartilage; *b.hy.*, *b.br.*1. basi-hyobranchial plate; II. optic foramen; V. trigeminal foramen.

Bones: *e.o.* exoccipital; *sq.* squamosal, partly covered by tympanic annulus; *pr.o.* prootic; *fr.pa.* parietofrontal; *g.* sphenethmoid; *na.* nasal; *pmx.* premaxillary; *s.mx.* septomaxillary; *mx.* maxillary; *pa.s.* parasphenoid; *pt.* pterygoid; *qu.ju.* quadrato-jugal; *m.mck.* mento-meckelian; *h.br.* hypobranchial or thyrohyal.

400. The lower jaw (Fig. 45) is comparatively simple, and very elongated, with its rami widely separated behind, and approaching each other rapidly in front. The meckelian cartilage persists except in the symphysial region,

where it is replaced by a small cylinder of bone, the mento-meckelian, which however is partially derived from the remains of the lower labial cartilage. The articular region has an elevated coronoid part; the condyle is a smooth egg-like mass, with its long axis longitudinal; it plays very freely beneath the smoothly-scooped base of the quadrate. The dentary, bearing no teeth, and continuous with the mento-meckelian, runs back two-fifths of the length of the jaw, surrounding the cartilage; the articular[1] ensheaths the inner side completely, but very much of the cartilage is bare of bone externally.

401. It appears unnecessary again to describe the columella and tympanic annulus: the essential arrangement is given in §§ 379, 380. The anterior boundary of the tympanic cavity is traversed by the main facial nerve, which lies above the pedicle. The stapes itself is attached

Fig. 46.

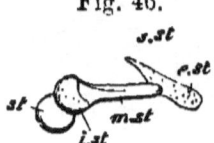

Adult Frog: columella auris.

st. stapes; *i.st.* interstapedial; *m.st.* mediostapedial; *e.st.* extrastapedial; *s.st.* suprastapedial.

to the margin of the fenestral fossa by a delicate band of fibrous tissue; but much of the inner face is in immediate contact with the cavity of the vestibule. This fenestral fossa is worthy of some note; it is an egg-shaped fossa of considerable depth, into which the prootic extends slightly; the exoccipital does not reach to its posterior rim. The fenestra ovalis merely occupies the postero-inferior third of the bottom of this fossa; it is uniform, with the concave edge looking upwards and forwards; the main otoconial mass can be seen through it. At an earlier period the fenestra was as large as the stapes, but the

---

[1] Prof. Huxley says this bone represents the angular, coronary, and splenial elements; but it is here named articular from a consideration of transitional cases. See Huxley, art. 'Amphibia,' *Encyc. Brit.* Vol. I. p. 755.

latter has grown, and the surrounding cartilage has enlarged to embrace it.

402. The slender curved hyoid arch (*st. hy.*) is attached to the periotic capsule a little way below the rim of the stapedial fossa at its anterior third. It has changed little since the Frog was two or three months old; it has not become ossified. Ventrally it passes into the anteroexternal angle of the broad hyobranchial plate, which is a great cartilaginous escutcheon with irregular patches of superficial endostosis on its surface (Fig. 47). It has lost

Fig. 47.

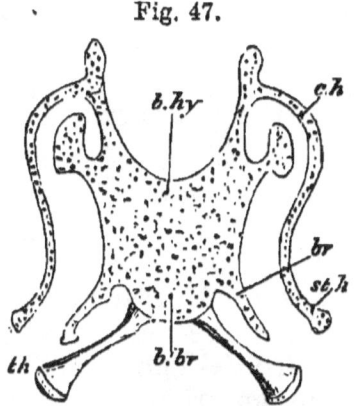

Adult Frog : hyobranchial plate.

st.h. hyoid arch or stylohyal; *c.h.* ceratohyal tract; *b.hy.* basihyal; *b.br.* basihyobranchial; *br.* remains of third branchial arch; *th.* fourth branchial or thyrohyal.

the symmetrical foramina it formerly had; and its various outgrowths are unossified. Its upper surface is gently concave; and it is rendered exquisitely mobile by reason of its suspension by the sigmoid hyoid bars. The anteroexternal angle is produced into a leafy hypohyal lobe in front of the attachment of the hyoid cornu; there is another broad lamina, the remnant of the first two branchial arches, just behind this attachment, and a third more elongated process (*br.*), representing the third branchial arch, extends almost from the postero-external extremity of the plate. The fourth branchials (*th.*) are ossified as shaft-bones, a core of cartilage remaining, and attached near one another at the posterior extremity of the hyobranchial plate; they are slender in the middle and expanded behind.

403. SUMMARY. The primordial skull of the Frog, arising very early in development, possesses the typical elements quite distinct from one another; but there is nothing to represent the palatopterygoid arch, and the suctorial mouth is situated far forwards. The mesocephalic flexure is early overcome, and at a subsequent period considerable labial cartilages form the functional jaws. When the external gills are at their fullest development, the face is enlarged proportionally to the brain-case, and the head is comparatively flat, the parts of the brain being arranged almost in a straight line.

404. By the time the tadpoles are half an inch long, extensive coalescence of the primitively distinct elements has occurred. A cranial floor is formed by the union of parachordals and trabeculæ; the latter give rise both to an internasal plate and to nasal floors. The mandibular arch unites with the trabecular cartilage both in front of and behind the orbit. The meckelian bar is cut off, and the definite suspensorium has articulated to it the unusually massive hyoid bar, which possesses no dorsal region. Subsequent metamorphosis puts the tadpole of an inch long in possession of a boat-shaped cartilaginous cranium united behind with the ball-like otic masses; while the mouth remains entirely in front of the brain-case, the suspensorium of the mandible (still functionally in abeyance) is a large subocular bar, bearing a superior lamina roofing the temporal muscle. The branchial arches have coalesced into a most remarkable basket-work, which is folded almost double in the sides of the throat.

405. As the tail is disappearing the cranium is found much more extensively chondrified, and nasal capsules are well formed. The character of the mouth is changed; the labial cartilages become subordinate, and the two jaws functional. The mouth is elongated by the gradual

retraction of the suspensorium, concurrently with a growth of (palatopterygoid) cartilage from the anterior confluence of the suspensorium with the trabecular cartilage, and the meckelian bar is correspondingly lengthened. The hyoid and branchial arches gradually dwindle, and ultimately form a rod-like arch and a broad basal plate.

406. During this great transformation most of the membrane-bones of the skull appear, and also the ex-occipitals and prootics. Some most characteristic bones are particularly late in development, for instance the sphenethmoid and the mento-meckelian. The series of parts which is identified as the upper part of the hyoid arch, forming the columella, does not arise till after the young Frogs have taken to the land; while another structure, which is related to the organs of hearing, the tympanic annulus, becomes connected with them, having been originally a postero-superior leaf cut off from the mandibular suspensorium.

407. The contrasts between the Elasmobranch skull, destitute of bones, the Salmon's, with its multitude of osseous plates little compacted together, and the Frog's, so highly specialised, and with bones so accurately moulded on cartilage, are very obvious, and need not be detailed here. But it may be serviceable to emphasize some of the features in which the Frog is unlike the Axolotl, both having skulls presenting at first sight many resemblances. The Axolotl's suspensorium is not a subocular arch at first; its palatopterygoid cartilage is comparatively scanty, and very slight in the adult. The Axolotl has no labials, no orbitar process, no columella. The Frog developes a more complex branchial framework, which is more completely absorbed than in the Axolotl. The retraction of the suspensorium is a special feature in the Frog, as well as its giving rise to a tympanic annulus. The early history of the trabeculæ differs much in the two types, especially as regards the parachordal extension, so marked in the Axolotl, and the trabecular cornua, which are early formed and of notable size in the Frog.

408. The skull of the Axolotl is ossified by more bones, appearing earlier, and has less of permanent cartilage than the Frog. Among bones found in the former and not in the latter are the pterotic, opisthotic, and ectethmoid. The mento-meckelian and columellar ossifications of the Frog however are not represented in the Axolotl.

APPENDIX ON THE SKULLS OF AMPHIBIA[1].

409. The skull of *Rana pipiens* (the Bull-frog) presents marked distinctions from that of the common Frog. It is both highly generalized, having more features looking back to the Fishes, and very specialised, for no creature turns its suspensorium so far back. The larvæ are much more Petromyzine in aspect than those of the Frog. The cranial box is very narrow in proportion to the width of the skull. There are many additional bones, frequently minute, and having the character of ganoid scales. Multiplication of bones takes place in several of the best defined tracts; the metapterygoid and mesopterygoid are distinct from the pterygoid; the palatine tract is divided into two bones; there is an anterior parasphenoid distinct from the vomers; a series of small bones represents the septo-maxillary, and another series covers the suspensorium. The tympanic annulus is of great size, and forms a complete circle.

410. In the small Frog, *Pseudis paradoxa*, the skull is proportionate to the body, and its details are of marvellous interest. In one stage of development the whole of the occipital and auditory regions is ossified by one continuous bone on either side. Later, each occipito-otic is cut into two bones; and subsequently the hinder pair of bones (mainly occipital) unite into one, while the anterior pair remain separate as prootics. The parasphenoid is split down the middle in front, and has no posterior handle (or projecting spur behind the basitemporal wings). The parieto-frontals arise as one bone on either side, and are subsequently segmented into parietal and frontal. The internasal cartilage is of great extent, but

[1] For much valuable information see Prof. Huxley, art. 'Amphibia,' *Encyc. Brit.*, ninth edition, Vol. ɪ.

the nasal cavities are small. The prepalatine region is cons'derable; the antorbital cartilage is almost severed from its ethmoidal attachment. There are large suctorial cartilages, or labials.

411. The skull of the Toad (*Bufo vulgaris*) differs from that of the common Frog in the following respects. The orbitar process in the Tadpole forms a perfect arch over the temporal muscle; the second upper labial coalesces in the adult with the internasal plate in front, and with the antorbital bar behind; the palatine cartilage is segmented off both from the antorbital and from the pterygoid; the upper part of the suspensorium in the adult rises above the "condyle of the pedicle;" there is no separate metapterygoid bone; the columellar cartilage is not segmented, subdivision taking place by the formation of two separate bony shafts, whose meeting-point is distal, not near the stapes; the extrastapedial is a large semioval leaf, and the suprastapedial coalesces with the tegmen tympani; the lower hyoid bar coalesces with the auditory capsule; the cartilaginous tympanic annulus forms only two-thirds of a circle; the continuous nasal roof is only one-fourth the size of the same tract in the Frog; the prorhinals are much smaller.

The Skull of Dactylethra[1].

412. The skull in this remarkable form, the aglossal Cape Toad, is as notable in its history and adult structure as any among the Anura. The larvæ are very divergent in character from those of ordinary Frogs and Toads; they present much general resemblance in outward form to the extinct *Coccosteus* and *Pterichthys*. The mouth is suctorial, and small, but very wide, like that of Siluroid F.shes and *Lophius*. The lower jaw protrudes beyond the upper; a very long tentacle proceeds on either side from the upper lip, and there is no trace of the horny plates possessed in early life by most Frogs and Toads. In conformity with these characters the whole head is extremely flat or depressed, instead of being high and thick. There are no claspers beneath the chin. The branchial orifice is not confined to the left side, but exists also on the right. The tail, like the skull, is very Chimæroid, the whole caudal region being narrow and elongated, and terminating

[1] See *Phil. Trans.* 1876 for figures and detailed description.

in a long, thin, pointed lash. The fore limbs are not hidden beneath the opercular fold.

413. The youngest larvæ examined were an inch and a quarter in length, three-fifths of this being caudal. The fore and hind limbs at this stage are nearly equal in size, although subsequent growth makes them very unequal. The grade of development of the skull corresponds to that described in the third stage of the Frog. The cranial structures are much more flattened even than an external view would suggest; for the outer skin is very loose, and the subcutaneous stroma copious and gelatinous. Unlike the larvæ of the ordinary Frogs and Toads, the tissues are very transparent, although richly supplied with brown pigment.

414. The cartilaginous cranial floor has a very astonishing aspect; as in the Skate's skull, there is a long, common, cranio-nasal valley, formed behind of the parachordal cartilages and notochord, and in front, for twice the extent, of the coalesced trabeculæ, to whose transversely extended anterior edge a transverse labial cartilage is united. The periotic masses are confluent with the parachordal and trabecular cartilage, but the cranial floor extends considerably behind them, and gives rise to exoccipital walls. The trabeculæ form blunt side walls to the cranium in front of the auditory masses, and these elevations converge, narrowing the cranial cavity. The brain proper ends at a point in front of the periotic cartilages only one-fourth of their length, and the diverging olfactory crura are as long as the brain. The precranial valley is embanked by the continued trabecular elevations, which again recede from one another, forming no transverse ethmoidal wall. The trabecular laminæ in the sides of the precranial valley are but thin, and arched outwards over a considerable infero-lateral concavity on either side. The precranial valley is filled with a watery tissue through which the olfactory crura run.

415. In the auditory capsule the fenestra ovalis is opening, but is merely filled with indifferent tissue. The tegmen tympani projects as a considerable lobe from the hinder part of the capsule, is narrower laterally, and then extends broadly from the whole anterior and antero-lateral margin. It runs forwards and outwards as a broad ribbon, equal in length and breadth to the capsule, and then is confluent with a footshaped flap having a pointed end forwards,

nearly reaching the orbitar plate of the suspensorium. This lobe is to be transformed into the tympanic annulus. The facial nerve runs beneath the tegmen.

416. The mandibular suspensorium is broadly confluent behind with the trabecular cartilage in the whole of that tract where it forms a wall for the brain, and then diverges gently from it so as to leave a narrow falcate membranous space, the subocular fenestra, rounded in front, pointed behind. The suspensorium broadens as it passes forwards, and has a good-sized orbitar process, in a region not far behind the nasal openings. The quadrate condyle is just at the same level as these apertures; and at that region the suspensorium is confluent with the axial cartilage by a broad postnarial or palatine lamina and by a narrower prenarial lamina, which takes the place of the ligament uniting the cornu trabeculæ of the Toad to the quadrate. The eyeball rests on the suspensorium somewhat outside the subocular fenestra.

417. The meckelian cartilage is small as compared with the existing portion of the hyoid arch. The symphysial end is pointed, and between the opposite points of the two bars is found a pair of short inferior labial cartilages continuing their line (very unlike the lower labials of the Frog's tadpole). The upper labials, as has been mentioned, are very different; the moieties have coalesced with each other, and to some extent with the anterior trabecular edge; but on either side there is a sharp fissure between the labial and the cornu trabeculæ, and the labial runs, as a gradually attenuating thread of true cartilage, to the extremity of the long tentacle, so that at this stage it reaches to the end of the abdominal cavity.

418. The hyoid arch shows a ceratohyal piece, with no segmentation; it is broad and massive, yet sinuous, and articulates definitely with the suspensorium beneath the orbitar process. A basi-hyobranchial plate separates the two hyoid moieties. There is no upper or columellar region at present. There are four branchial arches, all confluent above and below, separated laterally by three narrow oblique slits. The first arch is a large wide bag of cartilage, with sinuous walls, through which can be seen radiating rows of internal tufted gills. The second and third arches are comparatively narrow; the fourth is much wider and somewhat like the first.

12—2

Through the clefts can be seen the interdigitating papillæ from which the dendritic gills grow.

419. In Tadpoles of Dactylethra at their largest size the proportionate width of the hinder part of the skull has greatly increased; the precranial valley has become filled with cartilage in which the olfactory crura run; and from the auditory to the nasal sacs the floor of the skull is one wide sheet of cartilage (as in Sharks and Rays), gently *convex* in the middle, concave near the margins. Far forwards there is a small nasal septum. The labial tentacles are one-third shorter: but there are two new pairs of labials around the inferior and two around the superior nares. The quadrate condyle has not relatively receded. There are two apposed frontals covering the supracranial fontanelle and bearing on their hinder surface a pair of small parietals. The parasphenoid (the first bone to appear) is dagger-shaped; the cranial cartilage is affected by one continuous bone on either side, occupying the exoccipital and prootic tracts.

420. In Dactylethras with large legs and diminishing tail one of the most striking changes is the retraction of the quadrate condyle for a distance equal to one-third the length of the skull, it being now opposite the fore margin of the frontal. The precranial ethmoidal cartilage is still of great width, as also is the suspensorium; a much narrower palatopterygoid pedicle has arisen, and the arcade is disengaged from the trabecular cornu. The tympanic annulus is definitely segmented off, and has become semicircular. The stapes is well chondrified, but there is no columella. The parasphenoid has grown forwards to the subnasal region, and bears a small vomer. The frontals have coalesced. Nasals, maxillaries, and premaxillaries have appeared. The meckelian cartilage has grown very much, with a thick proximal end, an ectosteal articular within, and a small dentary in front. The mouth is no longer Siluroid, but Batrachian. The hyoid arch is still very large and strongly marked, but somewhat disengaged from its suspensorial articulation.

421. In the adult Dactylethra an enormous transformation has taken place, for we find a narrow elongated cranial cavity, quadrate condyles at the level of the periotic masses, and a very large subocular fenestra separating the cranium from the palatopterygoid bar. The part of the cranial walls and floor between the nasal and

the auditory regions is nearly all solid bone, but this sphenoidal bone scarcely extends into the ethmoidal region. The long superior fontanelle is covered by the large parieto-frontal. There is a triradiate supraethmoid upon the cartilage in front of the cranium, its median ray overlying the nasal septum. The nasal region, with its cornua, abbreviated first labials, and numerous pairs of valvular secondary labials, presents many features of resemblance to the Elasmobranchs.

422. The parasphenoid reaches nearly to the extremity of the snout, the single vomer lying beneath it behind its termination. There is no palatine and no jugal; and the hinder end of the maxillary is far distant from the quadrate. The large pterygoid behind underlies the suspensorium, extending inwards to touch the short basitemporal wings of the parasphenoid. The suspensorium is not quite directly attached to the cranial wall, but is confluent with the auditory cartilage without dividing into very marked processes. The elongated mandible presents much cartilage, and has but two bones, the articular and dentary. There is a very large auditory columella, in which the cartilage is not segmented, but there are two shaft-bones of nearly equal length, the inter- and mediostapedial. The extrastapedial is a large cartilaginous plate applied over the outer face of the squamosal, quadrato-jugal, and quadrate, and embedded in the tympanic membrane; around it is the attenuated tympanic annulus, whose upper deficiency is supplied by two small tympanic bones. The lower hyoid tract, from being massive, has become very slender; it is not closely related to the auditory capsule, but is suspended from it behind and below the attachment of the suspensorium by a ligament of some length. Medioventrally the two bars are fused with one another and with the basibranchial tract. The fore part of the original branchial pouch has become a long and broad plate on either side. From the basibranchial tract project backwards a pair of branchials ("thyrohyals") ossified as shaft-bones, and supporting the larynx.

The Skull of Pipa.

423. Embryos of the remarkable Surinam Toad, *Pipa monstrosa*, taken from the dorsal pouches when 9 lines long, have a skull as highly developed as the earliest described larvæ of Dactylethra,

while as yet they are coiled upon a little differentiated yelk-mass, the tail almost meeting the chin. The limbs are well developed and free externally, and yet the cerebral vesicles are very bulbous, and the opercular flap is free and small. There are no horny jaws, no labial cartilages, no traces of branchiæ; although the branchial and aortic arches are formed. The ear-sacs with the hindbrain are very large, taking up half the head. In many respects, tadpoles of the common Frog and Toad only half the length of these embryos are more advanced; they are free and active at one third their length. But in Pipa the embryo is very unlike a tadpole: it appears never to have gills; and the metamorphosis is perfect before extrusion from the dorsal pouch.

424. The skull is essentially very comparable to that of Dactylethra, with considerable differences in relative size of parts. The hinder region is predominant, with a long cephalic notochord; the brain reaches nearly to the anterior extremity of the head. The cartilage has not united around the notochord, but the intertrabecular space is already filled in with cartilage, forming a broad trabecular plate, continued outwards in the mandibular suspensoria; the subocular fenestra is a mere chink. There is a well marked otic process of the suspensorium. The tegmen tympani is not so conspicuous a feature as in Dactylethra. The fenestra ovalis is forming. The trabecular cornua are free. The hyoid and branchial cartilages are on the whole very similar to those of Dactylethra, although feebler. This is their highest state of development. The aortic arches show no secondary branches.

425. Metamorphosis is complete while the larva remains in the maternal pouch. In creatures still rolled up, and little more than half an inch long, as well as in others showing their faces at the mouths of the dorsal pouches, seven and a half lines long, the ossification is already intense. The head and body are only one-fourth of an inch longer than when the yelk-mass was but little lessened; the original ground-plan, so to speak, is filled in with but little modification in the size of the territories.

426. Two of the most remarkable features of the chondrocranium are, the absorption of most of the cartilage in the floor of the skull, including some of the postpituitary tract, and the very great increase in the subocular fenestra. The palatopterygoid is

a slender rod of cartilage. The columella is complete at this early period, contrasting with its late development in the Frog. The ceratohyal has become entirely absorbed; the branchial arches are reduced to a form substantially agreeing with their condition in Dactylethra.

427. The notochord is but little diminished relatively: it lies mainly *upon* the basilar plate behind, and over the parasphenoid in front, *free* from relation to cartilage. On either side behind, a fissure marks the division between the almost absorbed auditory cartilage and the basilar plate; in front of that chink the auditory, basilar, trabecular, and suspensorial cartilages are united, the right and left tracts being widely separated. About the region of the optic nerve the trabecula is reduced to a normal size. From thence each bar rapidly widens, running inwards to unite with its fellow and form the ethmoidal floor, upon which lie the olfactory lobes, and the anterior fourth of the cerebral hemispheres. The cartilaginous cranial wall, continuous with the trabecular base, is a very thin vertical lamina extending from the auditory capsule to the ethmoidal region, united with the ethmoidal floor just described, and forming a narrow anterior ethmoidal wall to the cranium: all this is entirely unossified. It is continued forwards into a comparatively narrow nasal septum with recurrent cornua trabeculæ, and is laterally produced into the almost transverse antorbital and palatine tract. There is a small alinasal roof growing on either side from the top of the septum; and the nasal investment is completed by three pairs of upper labial cartilages, two of which extend into the upper lip.

428. There is a strong prepalatine spur from the ethmopalatine bar, running forwards parallel with each trabecular cornu. Behind this the main bar is very elongated, joining the fore face of the suspensorium just within the condyle. The suspensorium is very different from its primary condition: the pedicle in front of the auditory capsule is narrow, but joins broadly with the base and side wall of the cranium; its direction is *forwards* and inwards from the quadrate condyle; formerly it was directed backwards. The otic process almost continues the line of the pedicle backwards: it is not fused with the periotic mass; it is directly behind the palatopterygoid bar. The quadrate tract is little extended: it

furnishes a bilobed condyle. Instead of being situated at the level of the anterior end of the subocular fenestra, it is now below its posterior angle. The columella extends forwards over the otic process to the palatopterygoid bar. The process of the suspensorium bearing the condyle for the ceratohyal has disappeared in correlation to the absorption of that cartilaginous plate.

429. The cranial bones present remarkable features. There is an occipito-otic mass, possessing small exoccipital tracts, nearly meeting below in the middle line a strong supraoccipital bridge, beneath which a cartilaginous plate remains, and very complete otic investments, which have eaten away most of the cartilage below, leaving it more extensively above. The extreme external and anterior regions of the capsule remain unaffected by bone. The very large parasphenoid greatly resembles that of an adult perennibranchiate or a young caducibranchiate Urodele. It ends behind about the middle of the auditory region by a sharp spur; and in front beneath the nasal septum by another sharp spine; it is broadest in the ethmoidal region; there is no vomer. The cranial fontanelle is more than covered by a great single parieto-frontal bone, extending backwards to the supraoccipital: an anterior median spur is due to a supraethmoid ectosteal plate.

430. The nasals are semicircular bones notched in front for the external nares: in front of each is a small preorbital and a smaller septo-maxillary. The premaxillaries and maxillaries are flat, simple bones, lying entirely beneath the anterior cartilaginous structures. The maxillaries are falcate, ending behind in sharp points, a little in front of the pterygoids; the palatopterygoid cartilage is bowed farther out than the maxillary bone. A delicate styloid palatine bone extends between the parasphenoid and the maxillary. The pterygoid lies underneath the corresponding cartilage and the suspensorium, almost touching the parasphenoid: its transverse tract is much the larger. There is no quadrate ossification; the meckelian cartilage is largely invested by the ectosteal articular, reaching nearly to the symphysis; the dentary is largest at the symphysis. The squamosal consists of a lamina on the suspensorium, hidden by the columella; and a horizontal extension at the side of the auditory capsule. The columella is fastened to the simple unossified stapes; the elongated cartilage is not

segmented, but it is invested by an interstapedial and a mediostapedial shaft-bone, and is continued forwards into the enlarged extrastapedial cartilage, surrounded by the U-shaped tympanic annulus: there is no suprastapedial.

431. The skull of the adult Pipa, although presenting no structures which cannot be interpreted by the last described form, is one of the most remarkable pieces of osseous architecture that can be imagined. Cartilage is almost entirely absorbed; the bony tracts have become greatly extended and massive, and by the alteration of their relative size the skull wears an extremely different aspect. Its breadth has become half as great again as its length, the auditory capsules being thrust out on long stalks. The quadrate condyles are retracted to a position parallel with the occipital condyles, though at a lower level. The cranium is excessively flattened, and the bones of the roof and floor meet laterally. The maxillaries combine with the pterygoids to bound the upper jaw.

432. The posterior view of the skull presents a bony arc, convex upwards, bearing numerous bosses for muscular attachment, and consisting almost exclusively of the occipito-otic mass. In the basi-occipital region a small median tract is unossified: the rest of the foramen magnum is encircled by bone, and no cartilage is left. The occipital mass extends outwards to the squamosal and quadrate bones, and nearly to the quadrate condyle. The occipital condyles are very outwardly directed. The prootic tract of the occipito-otic mass reaches out to the tympanic apparatus, which is close to the quadrate condyle. The parasphenoid is a broad oblong bone, posteriorly filling up an angle between the right and left occipito-otic masses where they nearly touch below, anteriorly presenting a median spur beneath the nasal septum, and a pair of palatine wings reaching to the maxillaries. There is no vomer. The massive fronto-parietal is twice as broad in the ethmoidal region as it is behind, where it overlaps the supraoccipital tract. It has a median supraethmoidal spur, and large lateral wings extending on either side down to the maxillary. There is no cartilage in the cranial walls: the deficiency is supplied by the bevelling of the parasphenoid upwards on either side to meet a long orbital plate of the parietofrontal. There is no sphenethmoid.

433. Cartilage is persistent in the ethmoidal, nasal, and ethmopalatine tracts. The prepalatine spur is retained, and also a slender part of the palatopterygoid bar, not nearly reaching the suspensorium or the antorbital region. The nasal septum remains, but is very low; the alinasal cartilage is small; the recurrent trabecular cornua are long, slender, and straight, diverging from the septum at about 45^0; a small prenasal spur projects forwards from the septum; all but one pair of labial cartilages have vanished. The large triangular nasals extend forwards to the extremity of the snout, where they meet; they are notched for the external nares, and overlap the parietofrontal. Below the narial opening are the preorbital and septo-maxillary.

434. The premaxillaries and maxillaries are edentulous palatal plates. The latter are much the larger, have a spur within the internal nostrils, and extend backwards under the pterygoid bones for half their length. The anterior part of the pterygoid is of considerable breadth, bounding the subocular space, which is wider than the cranium; while the eyeball is quite small, and rests against the anterior external part of the pterygoid. Posteriorly, the pterygoid is transversely extended beneath the otic mass so as to reach the parasphenoid and come very near the middle line; there is also a smaller process extending outwards and backwards beneath the quadrate. The suspensorium is undistinguishable from the otic mass (although some cartilage remains) except by the fact that the quadrate is a distinct considerable bone. The squamosal clasps the outer edge of the otic mass and the upper and hinder face of the quadrate. The columella has its extrastapedial relatively dilated, and surrounded by an almost perfect ring of cartilage, the tympanic annulus. The stapes and the fenestra ovalis are surrounded by an annular elevation of cartilage. The facial nerve emerges from the *upper* surface of the otic mass about half way towards the quadrate condyle; it passes *over* the columella. The moieties of the edentulous lower jaw have a fibrous symphysis, and are very rib-like; the articular bone runs nearly to the chin, and is covered by the dentary, which is but two-thirds its length.

CHAPTER VI.

THE SKULL OF THE COMMON SNAKE.

(Tropidonotus natrix.)

435. THE early phases of the development of the skull in this form were long ago observed by Rathke, and his account of them is not by any means superseded. The first two stages given below are abridged from Rathke's *Entwickelungsgeschichte der Natter*, 1839[1].

First Stage.

436. The trabeculæ at their first appearance form two narrow bands, consisting of the same gelatinous substance as that constituting the whole investment of the notochord, not sharply differentiated, but only thickened and more solid parts of that half of the cranial floor which lies under the anterior cerebral vesicle. Posteriorly they are separated by a small interval, and thence sweep in an arch to about the middle of their length, separating as they pass forwards; afterwards they converge, so as to approach one another very closely, or even to come into contact. Altogether they form, as it were, two horns, into which the investing mass of the notochord (parachordal tract) is continued forwards. The elongated

---

[1] See Huxley, *Proc. Roy. Soc.*, Nov. 1858, p. 56, and *Elements of Comparative Anatomy*, p. 287.

(intertrabecular) space between them, moderately wide in the middle, is occupied by softer formative substance, upon which rests the infundibulum and the forebrain. Anteriorly the trabeculæ reach to the front of the head, and bend slightly upwards into the frontal wall: but there is also a small lateral projection or cornu passing outwards on either side.

437. The middle trabecula[1] grows, with the brain, further and further into the cranial cavity, and as the dura mater begins to be now distinguishable, it becomes more readily obvious than before, that the middle trabecula raises up a transverse fold of it, which traverses the cranial cavity transversely. The fold itself passes laterally into the cranial wall; it is highest in the middle, where it encloses the median trabecula, and becomes lower externally, where it forms, as it were, a short ala proceeding from the trabecula. With increasing elongation, the (middle) trabecula becomes broader and broader towards its free end, and, for a short time, its thickness increases. After this, however, it gradually becomes thinner, without any change in its tissue, till, at the end of its second period, it is only a thin lamella, and after a short time (in the third period) entirely disappears.

438. In Mammals, Birds, and Lizards, that is, in those animals in general, in which the middle cerebral vesicle is very strongly bent up and forms a protuberance, while the base of the brain exhibits a deep fold between the infundibulum and the posterior cerebral vesicle, a similar part to this median trabecula of the skull is found. In these animals, also, at a certain very early period of embryonic life, it elevates a fold of the dura mater, which passes from one future petrous (prootic) bone to the other, and after a certain time projects strongly into the cranial cavity. Somewhat later, however, it diminishes in height and thickness, until at last it disappears entirely, the two layers of the fold which it had raised up coming into contact. When this has happened, the fold diminishes in height and eventually vanishes almost completely.

[1] This account of Rathke's "middle trabecula" is given at length in order that its significance may be prominently brought forward.

Second Stage.

439. The trabeculæ attain more solidity, acquire greater distinctness from the surrounding parts, and assume a more determinate form, becoming in fact filiform, so that the further forward, the thinner they appear. They increase very little in thickness, but far more in length, during the growth of the head. Quite anteriorly, they coalesce with one another, forming an internasal tract. As soon as the olfactory organs increase markedly in size, the septal region is moderately elongated and thickened, without becoming so dense as the hinder part of the trabeculæ. The lateral prolongations which now proceed from the internasal tract become little denser, although considerably enlarged.

440. The lateral parts and upper wall of the cranium, with the exception of the auditory capsules, remain merely membranous, consisting in fact only of the cutaneous covering, the dura mater, and a little interposed blastema, which is hardly perceptible in the upper part, but increases in the lateral walls, towards the base of the skull.

Third Stage: Embryo Snakes about an inch and three-quarters in length.

441. These embryos have recovered from the mesocephalic flexure; but the visceral clefts remain open, three fairly distinct, the fourth obscure. The head (Fig. 48) has a very monstrous appearance, being covered with a series of bulbous protuberances. The nasal region is rounded and beaked: the olfactory cavities are formed, and the nasal meatus is pervious. The eyes (*e.*) are very large, being twice the diameter of the auditory and nasal sacs. Although the mesocephalic flexure is lost, there is a remarkable bending of the head upon the neck.

442. The angle of the mouth is directly below the anterior margin of the auditory sac. The inner nares

are formed, opening together in one common depression on the mid line of the palate. The maxillopalatine cheek-mass, from the nasal sacs to the angle of the mouth, is distinguishable into two tracts, which are better seen when the lower jaw is removed. There is an outer or maxillary, and an inner palatopterygoid tract. The lower jaw is of equivalent length and thickness to the upper; and between its rami can be seen a pair of rod-like prominences containing the rudiment of the tongue and its sheath.

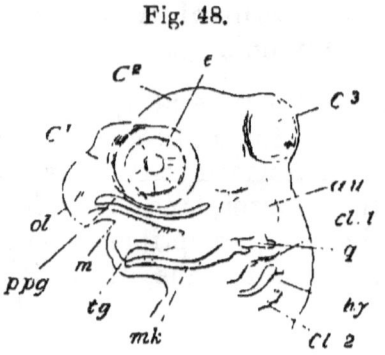

Fig. 48.

Embryo Snake 1¾ inch long; side view of head with facial arches seen through.

C 1, 2, 3, cerebral vesicles; $ol.$ nasal sac; $e.$ eyeball; $au.$ auditory mass; $m.$ mouth; $ppg.$ palatopterygoid tract; $mk.$ meckelian cartilage; $q.$ quadrate; $tg.$ rudiment of tongue; $hy.$ hyoid arch; $cl.$ 1, 2, visceral clefts.

443. At this stage the chondrocranium is already well formed. The cranial notochord ($ch.$ Fig. 49) is short, and twisted upon itself, and does not reach more than two-fifths of the distance between the occipital end of the skull and the pituitary body. The parachordal cartilages are united at two points behind the pituitary body and in front of the notochord. There is one broad bridge immediately in front of the notochord; the other is narrower, and lies immediately behind the pituitary body. Between these bridges there is a large oval space vacant of cartilage ($p.\ b.\ c.\ f.$). The anterior part of the parachordal cartilage is turned outwards opposite the postpituitary bridge, to form a hooked process, with a hinder concave margin. Just outside the same bridge the carotid artery pierces the cartilage on either side: this point marks the junction of the parachordal with the trabecular cartilages.

Fig. 49.

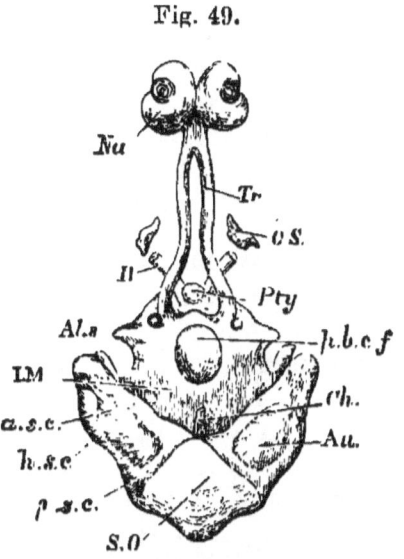

Embryo Snake, about 1¾ inch long; chondrocranium seen from above, the brain and jaws having been removed.

s.o. supraoccipital tract; *au.* auditory capsule; *a.s.c., h.s.c., p.s.c.* anterior, horizontal, and posterior semicircular canals; *ch.* notochord; *i.m.* basilar plate; *p.b.c.f.* posterior basicranial fontanelle; *al.s.* alisphenoid cartilage; *pty.* pituitary body; *tr.* trabecula; *o.s.* orbitosphenoid cartilage; II. optic nerve; *na.* nasal capsular cartilage.

444. The trabeculæ (*tr.*) pass forwards as rounded rods, wide apart opposite the pituitary body, rapidly converging in front, but remaining separated by a distance almost equal to their own width as far as the nasal region. The tract in which the trabeculæ remain distinct is about one-third the length of the skull. In front the trabeculæ coalesce, forming a considerable internasal plate; and develope a low nasal septum which is continuous at its upper edge with the nasal capsules. There are lateral

processes at the sides of the fore end of the septum—
the trabecular horns; and a median anterior downgrowth,
the prenasal cartilage.

445. The nasal glands lie right and left of the septum,
but on a lower plane, in the anterior part of the palate.
The whole nasal wall has chondrified; seen from above
each nasal roof (*na.*) appears as a reniform cartilage, the
two being in contact back to back, and the external nasal
opening occupying the hilus. The nasal cartilages are
relatively large as compared with the trabeculæ, and have
coalesced with nearly the whole extent of the nasal septum.

446. The posterior region of the basicranial cartilage
has two mammillary condyloid projections, one on either
side of the notochord; and a little in front of these are
the anterior and posterior (or rather external and internal)
condyloid foramina. The chondrocranium is completed
in the occipital region laterally and superiorly. The
supraoccipital region (*s. o.*) forms a prominent boss in the
middle line behind, and anteriorly is shaped like a
rectangular wedge fitting between the auditory capsules;
but the union of the original supraoccipital moieties is
not yet complete, and their lateral union with the auditory
capsules is also imperfect.

447. The periotic masses have a remarkable triangular shape; the long base of the triangle is external,
while its apex is situated internally, at the fore part
of the supraoccipital wedge: the two capsules nearly
meet in the cranial roof. The front part of the capsule
diverges much farther from the middle line than the
hinder portion: it is also longer and reaches to the level
of the middle of the posterior basicranial fontanelle. The
semicircular canals are relatively very large and distinct;
the anterior and posterior meet by less than a right
angle at the inner (supraoccipital) part of the capsule;
and the ampullæ are all well seen from above. The
great supracranial fontanelle is margined behind almost
entirely by the somewhat concave inner and upper edges

of the auditory capsules, along which margin the large anterior canals run.

448. Opposite the apex of the notochord in the lower edge of each capsule the membranous labyrinth is budding into a rudimentary cochlea lying just mesiad of the external canal. Behind and above this swelling, on the infero-lateral face of the capsule, is the fenestra ovalis, containing a membranous plate which is chondrifying continuously with the hyoid arch. A diverticulum can be seen growing backwards from the fenestra ovalis; this is the commencement of the fenestra rotunda, which afterwards becomes separated from the former by a bar of cartilage. Behind these and internal to them the glossopharyngeal and vagus nerves are seen passing out in the boundary between the otic mass and the occipital cartilage.

449. Between the out-turned anterior termination of the basilar cartilage and the side-wall of the otic mass is a large rounded notch, just opposite to the postpituitary fontanelle and of similar size. Part of this notch is occupied by a small independent ear-shaped cartilage, having a narrow end forwards, and its edge convex inwards; this is the alisphenoid cartilage (*al. s.*). The orbitonasal branch of the trigeminal nerve passes out between this cartilage and the auditory capsule, and then forwards into the orbit. The second and third divisions of the trigeminal also pass out behind this alisphenoid cartilage.

450. The only chondrified tracts in the visceral arches are found in the mandibular and upper part of the hyoid (Fig. 48). The quadrate (*q.*) is segmented from the meckelian rod, and lies loosely external to the upper and outer side of the auditory capsule, its posterior otic process, which is sub-bifid, nearly reaching to the posterior canal. The part answering to the pedicle in the frog is a short rounded lobe. The shaft of the quadrate is directed forwards and downwards, and the condylar articulation is on as high a level as the external semicircular canal. The mandible (*mk.*) is a slender sigmoid rod, continuing the line of the quadrate shaft, and having a long angular

process ascending behind the condyle. The two mandibular bars nearly meet in the chin.

451. The whole of the hyoid arch ($hy.$) is only about half the length of the meckelian cartilage. It is a small rib-like bar, with a bilobate proximal extremity similar to that of the quadrate; and a curved distal arc turned backwards. The proximal part is closely applied to the infero-lateral surface of the auditory capsule, and the hinder and inner of the two lobes coalesces with the stapedial plate filling the fenestra ovalis. The rounded outer tubercle is free from the otic mass and is directed forwards. The posterior region of the hinder part of the hyoid is ready to be cut off as a free piece homologous with the mandible; it is separate a little later, and diverges further backwards.

452. At this period the chondrocranium of the Snake is fairly complete, without bony deposit. The notochord occupies only the hinder part of its original territory; the trabeculæ and parachordals have united, and while union has taken place at two places behind the pituitary body, there is a considerable posterior basicranial fontanelle. An occipital arch and alisphenoid cartilages complete the list of proper cranial elements. Anteriorly the trabeculæ are parallel for a considerable distance without uniting, but in the nasal region they coalesce with each other and with the cartilages of the nasal capsules: cornual and prenasal tracts are likewise present. The auditory capsules have a notable triangular form, and already possess a rudimentary cochlea. The mandibular arch is divided into upper and lower pieces; the tissue in the maxillo-palatine process has not chrondrified. The palatal surface possesses median, submedian, and lateral longitudinal thickenings, which are nearly prepared for osseous deposit. The simple hyoid bar is continuous with the stapedial plate in the fenestra ovalis.

Fourth Stage: Embryos about two and a half inches long.

453. The head is more elongated, and much less monstrous. The cerebral vesicles are still well marked from above, but not nearly so protuberant as before (Fig. 50). The jaws have lengthened correspondingly with the rest of the head. The visceral clefts are entirely obliterated. Ossification is considerably advanced in several regions.

454. The general form of the basis cranii has altered very little from that of the more advanced individuals of the last stage. The fontanelle behind the pituitary body is larger, and the space between it and the end of the notochord is less than it was. The two moieties of the occipital roof have completely coalesced. The cranial part of the notochord lies on the united parachordals or basilar plate, and is surrounded by a bony sheath which is spreading into the substance of the cartilage right and left, forming the *basioccipital* (*b. o.*). In the cartilage on either side of this is a bone (*e. o.*) pierced by the foramen for the hypoglossal nerve and extending as far as that for the vagus, which it is beginning to enclose. These *exoccipitals* can be seen postero-laterally on the occipital roof. The auditory capsules have altered their external shape very little; they are still unossified. The diverticulum to form the cochlea is a rounded bud extending downwards and forwards below and in front of the fenestra ovalis. It has no definite fenestra (rotunda) in the auditory wall as yet.

455. The small alisphenoid cartilage (*al. s.*) is still earshaped, and is placed somewhat longitudinally. Its hinder dilated extremity is applied to the front of the auditory capsule below, and its anterior end lies along the edge of the basilar plate, behind the region of the pituitary body. A notch in the hinder part of the alisphenoid embraces the posterior division of the trigeminal nerve (5), which passes backwards; and the remainder of the trigeminal emerges from the cranial cavity in the space between the concave notch of the front aspect of the

13—2

basilar plate and the concave edge of the alisphenoid looking inwards.

456. The postpituitary bridge of cartilage separating the pituitary space from the basicranial fontanelle remains as it was in the last stage; the external carotid artery pierces the cartilage on either side, and on the outer edge of this region is a bony rudiment ossifying the cartilage,

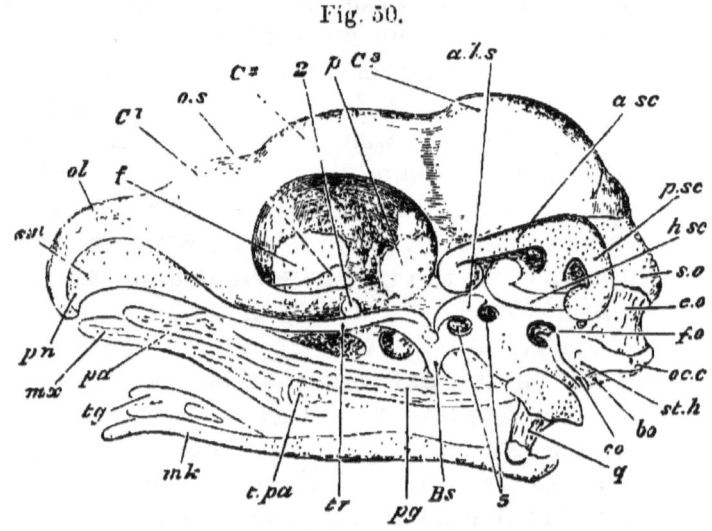

Fig. 50.

Embryo Snake, about 2½ inches long; side view of head, dissected; several membrane bones having been removed.

C 1, 2, 3, fore-, mid-, and hind-brain; *oc.c.* occipital condyle; *s.o.* supraoccipital cartilage; *a.sc.*, *h.sc.*, *p.sc.* anterior, horizontal and posterior semicircular canals; *al.s.* alisphenoid cartilage; *o.s.* orbitosphenoid cartilage; *tr.* trabecula; *s.n.* nasal septum; *pn.* prenasal cartilage; *ol.* upper part of nasal capsule; 2, optic foramen; 5, trigeminal foramina; *f.o.* fenestra ovalis; *st.h.* stylohyal cartilage; *co.* columella; *mk.* meckelian cartilage; *tg.* tongue.

Bones: *b.o.* basioccipital; *e.o.* exoccipital; *b.s.* basisphenoid; *p.* parietal; *f.* frontal; *mx.* maxillary; *pa.* palatine; *pg.* pterygoid; *t.pa.* transpalatine; *q.* quadrate.

the two ossifications coalescing later to form the *basisphenoid* ($b.s.$). The long uncoalesced portions of the trabeculæ (*tr.*) in front of the pituitary body remain closely apposed; and between them is appearing a delicate styloid bone, enlarged into a spatulate plate just in front of the pituitary body, beneath the broader part of the intertrabecular space. This is the origin of the *parasphenoid*. The small infero-lateral granular tracts of the cranial (orbitosphenoidal) wall above and in front of the optic foramina have not yet chondrified. The nasal cartilages have not changed in form, but the septum (*s. n.*) is higher. The trabecular cornua are now indistinguishable from the nasal capsules; and the prenasal spur (*p. n.*) is curved downwards and backwards in front of them.

457. On either side of the midbrain in its *lower* region a thin film of bony matter is giving rise to the *parietal* (*p.*); and the *frontals* (*f.*) are similar bony plates on the sides of the forebrain, nearly meeting *beneath* the cranial cavity, above the trabeculæ. Also on the inner face of each olfactory roof there is a thin shell of bone, the *nasal;* and these two shells lie back to back above the low septum. In front of the nasal cavities a bilunate plate of bone has appeared, the azygous *premaxillary*.

458. On the side and below the level of the nasal septum is a large reniform nasal gland with a duct which passes downwards and outwards. This gland is covered above by a delicate film of bone, which enlarges inwards into a vertical plate near the septum; this is the *septomaxillary*. Mesiad of the duct, the lower surface of the gland on either side is invested by a bone, the *vomer*, having a convex surface downwards. Two small (labial) cartilages are attached to the duct of each nasal gland.

459. Along the anterior region of the palate is a pair of bony styles lying external to the trabeculæ, and partly underneath the nasal cavities. In its middle region each of these *palatine* bones (*pa.*) sends inwards a process (the ethmopalatine) towards the trabeculæ, just opposite the pointed end of the parasphenoid: it curves over the posterior nasal passage, which opens in the middle of the palate. A pair of slender gently-curved bones begin immediately beneath the hinder end of the palatines, and pass outwards and backwards to the inner face of the quadrate cartilage. These are the *pterygoids* (*pg.*). Neither

palatines nor pterygoids arise in cartilage. In the margin of the cheek and upper lip there is a splinter of bone reaching from the outer angle of the premaxillary to opposite the posterior end of the palatine—the *maxillary* (*mx.*). Between the posterior end of the maxillary and the middle of the pterygoid, lying very obliquely, is another slender bone with a broad end towards the maxillary: this is the *transpalatine* (*t. pa.*).

460. The quadrate, instead of being a rib-shaped cartilage in the same line as the meckelian, passing downwards and forwards, is broad and fan-shaped, and its narrow lower part is rapidly ossifying as the *quadrate* bone (*q.*). The curved upper edge is applied loosely to the side of the auditory capsule, and is separated from it already by a delicate sickle-shaped plate, the *squamosal*, whose pointed anterior end is attached to the fore part of the capsule.

461. The meckelian cartilage, now very elongated, is gently sigmoid, and prolonged into an angular knob behind the quadrate articulation. The rods of opposite sides are widely separated by connective tissue. Films of parosteal bone have appeared in relation to it; the *dentary* in front and outside, forking behind, and exposing a considerable tract of cartilage; behind it, applied to the outer face of the jaw, is the *surangular*, reaching to the condyle. On the inner surface there are three splints, enlarging from before backwards—the *splenial*, the *coronoid*, the *angular*, the latter reaching nearly to the angle of the jaw. Ectostosis has arisen in relation to the cartilage near the articulation, and extending into the angular process; this is the *articular*.

462. The backward direction of the hyoid arch is now much increased, and the rod has become slender and less curved (*co.*). The part resembling the tubercle of a rib, instead of turning forwards, is directed backwards, and is applying itself to the posterior edge of the stapedial plate. A lamina has been separated from the hinder and upper edge of the bar, forming a distinct stylohyal cartilage (*st. h.*), which is somewhat heart-shaped with the apex directed backwards.

463. We have here no record of great transformation in cartilaginous parts; and the principal changes of relation are those by which the upper mandibular segment and the hyoid arch are carried backwards, the former becoming at the same time more loosely connected with the brain-case. The bones which have appeared in cartilage are the basioccipital, exoccipital, basisphenoid, quadrate, and articular. The principal membrane-bones are also present, the low lateral position of the frontals and parietals, the azygous condition of the premaxillary, and the occurrence of the septo-maxillary and transpalatine being points of interest. The auditory capsule is as yet unossified.

Fifth Stage: Embryo Snakes near and up to the time of Hatching.

464. The skull has made great progress towards its adult conformation; the brain-case is very straight, and the original protuberances are scarcely distinguishable on the upper surface. Ossification has extended rapidly, but is still far from complete: many new centres are arising.

465. The basioccipital has become a large heart-shaped bone, the notochord with its bony sheath (cephalostyle) still lying above it, though united with it; the bone is separated by considerable tracts of cartilage from the surrounding elements. The pointed end of the basioccipital lies backwards, and is flanked on either side by a tubercle of cartilage, forming the two somewhat projecting lobes of the median condyle. In front the bone is extending beyond the original cartilage into the posterior basicranial fontanelle. The paired basisphenoidal ossifications described in the last stage have coalesced into one bone, ossifying the prechordal (postpituitary) bridge of cartilage, and beginning to extend backwards into the fontanelle. It passes forwards also on either side of the pituitary body, as far as the posterior end of the parasphenoid, with which it ultimately coalesces. The internal carotid arteries perforate

the bone. There is no ossification of the trabecular cartilage in front of the basisphenoid.

466. The exoccipitals have extended, so as to meet by suture over the foramen magnum. The remainder of the supraoccipital region is ossified by a distinct *supra-occipital*, which is five-sided, broadest transversely between the highest part of the auditory capsules. Its concave anterior margin is the hinder border of the great supracranial fontanelle: posteriorly it is conterminous with the exoccipitals, and laterally with the epiotics. The glossopharyngeal nerve pierces the hinder end of the ear-cartilage; the vagus passes out between the capsule and the exoccipital; the hypoglossal pierces the exoccipital.

467. The auditory capsule has acquired its three most constant ossifications. The *epiotic* lies at the side of the supraoccipital, and occupies the most elevated part of the capsule, where the anterior and posterior canals unite: it is the smallest of the periotic bones. The *opisthotic* is related to the posterior part of the horizontal canal and the lower part, with its ampulla, of the posterior canal. It is of considerable vertical extent, and passes downwards as a wedge between the fenestra ovalis and the fenestra rotunda (of the cochlea), also passing behind the latter so as almost to encircle it. Its anterior process nearly meets the prootic beneath the fenestra ovalis, over the neck of the cochlea, whose bulb is still unossified.

468. The rest of the capsule is enclosed by the *prootic*, which includes most of the anterior and horizontal canals with their ampullæ, and surrounds the meatus internus. This opening lies in front of the lower ray of a triradiate synchondrosis between the three periotic bones. Beneath the two ampullæ the prootic sends down a wedge-like process into the floor of the cranium between the basioccipital and the basisphenoid. In this wedge there are three foramina, two large ones, anterior and inferior, giving passage to the anterior and the posterior divisions of the trigeminal nerve; and a smaller above

between them and the meatus internus, for the facial nerve.

469. External to this lower part of the prootic, standing on the oblique, hinder, and outer edge of the basisphenoid, and underpropping the anterior part of the prootic, is a small four-sided bone which has ossified the alisphenoid cartilage, as well as an additional band of membrane bridging over the notch which previously existed in the cartilage, (§ 455, p. 195). Through the foramen thus formed the anterior portion of the trigeminal nerve passes after its exit from the prootic. The concave hinder part of the *alisphenoid* is in front of the posterior division of the trigeminal, which passes almost directly backwards: the facial nerve is immediately above this. There is one other cartilage bone in the cranial wall, the *orbitosphenoid*; it is triradiate, wedged in between the frontal and parietal, above and in front of the optic foramen.

470. The parietals, very solid and thick, now occupy a large portion of the side-walls of the brain-case, and reach inferiorly to the lateral edges of the basisphenoid (anterior part) and the front of the alisphenoid and prootic; they send backwards a horn along the upper edge of each prootic and epiotic, reaching the anterolateral angle of the supraoccipital. In front of the anterior ampulla the parietals also send a vertical ridge inwards as a partial boundary between the midbrain and the hindbrain. At present they have not roofed over the brain-case, but simply lie in its sides. The frontals are more complete above, and below nearly meet, above the parasphenoid. Their side-wall is concave externally, the parietals being convex.

471. The nasals extend much further over the nasal capsules than they did; and the vomers and septomaxillaries are becoming solid, and taking the shape which they possess in the adult. The small azygous (edentulous) premaxillary has developed two distinct palatal processes, one on either side of the median pre-

nasal spur. The parasphenoid has grown large after the pattern described in the last stage.

472. The palatines are hatchet-shaped: the blade of the hatchet is very concave below where the bone arches over the posterior nasal meatus. There is a small external and a larger anterior process; all these parts are related to the ethmoidal region of the skull, (between the eye and the nose). The handle of the palatine passes directly backwards, and lies above the fore end of the pterygoid. This latter is very much enlarged; it is a spindle-shaped bone extending somewhat outwards and far backwards to the inner side of the quadrate; its middle region is broad, and supports the curious transpalatine, which is forked anteriorly, receiving the posterior end of the maxillary. This is now much elongated and dentigerous, bounding the anterior half of the gape, and towards its hinder end having a horizontal expansion which supports the inner fork of the transpalatine.

473. The squamosal is an oblong splint-like bone attached loosely by its fore end to the auditory capsule over the anterior ampulla; behind, it diverges backwards and outwards. On its inner face is a much smaller splint, the *supratemporal*. The handle of the fan-shaped quadrate is much longer than it was, and its extended upper edge is still cartilaginous: it applies itself over and outside the posterior end of the squamosal. Thus it has lost all direct connexion with the cranium, and is also directed definitely backwards. The quadrate condyle is now behind the level of the basioccipital region; it forms a hinge-joint, with an angular process (like an olecranon) behind, and a raised rim in front.

474. The posterior part of the meckelian cartilage is well ossified by the articular; anteriorly the cartilage persists to the front of the jaw, covered on its outer side by the dentary. In front of the middle of the jaw the broad ends of two bones meet in a vertical line: one of these, directed forwards, is the splenial, the other, lying behind, is the coronoid; both end by a point. A

narrow angular invests the lower edge of much of the articular, extending back to the region of the joint. The surangular is on the outer side of the same region, but is twice as large.

475. The columella is now composed of a bony stapedial plate and a shaft continuous with it, all but the distal third of the original cartilage being ossified. The stapedial plate itself is nearly straight behind, and convex in front; convex externally and scooped within. The ascending (tubercular) process of the shaft is above and behind, the capitulum below and in front. To the curved posterior half of the sigmoid shaft a thick subcrescentic (*stylohyal*) bone is attached. It has ossified all but the extremities and the free under edge of its cartilage: and it is very tubercular and rough. Its outer face is attached to the inner surface of the quadrate, and both it and the columella extend behind the posterior edge of that bone.

476. The condition of the skull above described requires careful attention to details for its full comprehension; the growth of the parietals and frontals is especially noteworthy. The anterior halves of the trabeculæ remain unossified; but supraoccipital, pro-, epi-, and opisthotic, alisphenoid, and orbitosphenoid centres have appeared. The quadrate condyle is now carried behind the basioccipital region, and it is connected with the side of the cranium by the intervention of the squamosal,—a very novel feature. The columella and the stylohyal have become ossified.

The Skull of the Adult Snake.

477. The cranial investment constitutes an exceedingly strong box, the bones being united by even sutures, sometimes slightly squamous. The bony substance is very dense, almost like ivory, with very little diploë; and a large number of the sutures that were present at the time of hatching have disappeared. The bones of the face are also dense, but instead of being welded together, they are

for the most part loosely connected with one another, as well as with the brain-case, by elastic ligaments constituting the most mobile facial apparatus to be found among Vertebrates. The axial parts are condensed into a comparatively small space, while the jaws even in their most contracted state are widely extended on either side of and behind the rest of the skull. The brain rests upon the almost flat cranial floor, without being elevated upon an interorbital septum. The hyoid apparatus is as insignificant as in any vertebrate type, while branchial arches are non-existent.

478. The cranial box is of tolerably even breadth, for the ear-masses do not project laterally as in other types, but are rather elongated. The occipital ring is much flattened, and is continuous to a great extent with the auditory capsule. The basioccipital is a broadly heart-shaped plate (*b.o.* Fig. 52), its rounded posterior apex constituting the median transversely-oval occipital condyle. A considerable part of each exoccipital (*e.o.*) appears in the cranial floor, and an equal portion in the cranial roof; and the postero-superior edge of each overlaps the corresponding part of the atlas. The exoccipitals meet above (Fig. 53), and are separated by the condyle below: each has coalesced with the contiguous opisthotic. The supraoccipital is now apparently larger than before, for it has annexed to itself the two epiotics.

479. The largest of the periotic bones, the prootic (Fig. 51, *pr.o.*), is apparently quite distinct; but it has coalesced by its outer face in front with the little four-sided perforate alisphenoid. Internally the prootic rests equally upon the antero-external edge of the basioccipital and the outer edge of the basisphenoid. Anteriorly it is in contact with the thick hinder edge of the lateral plate of the parietal, with its inturned thick vertical crest. Superiorly at its fore part it adjoins the postero-lateral edge of the parietal, and the epiotic behind. Here, as in Reptiles generally, a Y-shaped suture is persistent between the three periotic bones, which never unite with

each other, but always with some adjacent bone. (When there is no alisphenoid, the prootic remains distinct from any other element.) The nerve foramina do not need further description beyond that given in the last stage.

480. The large irregularly four-sided basisphenoid has coalesced with the parasphenoid, to form a very curiously-shaped bone (Fig. 52). There is a posterior clinoid wall arching over the hinder part of the pituitary body. The latter rests upon a bony floor formed by the backward growth of the parasphenoid and by a concomitant growth of bone (basisphenoidal) from the whole margin of the intertrabecular space. Thus a sort of subcranial hollow is formed, which besides being partially roofed by the posterior clinoid wall, is covered at either side by a shelf passing inwards from the base of the parietal, on which the brain very largely rests. The subcranial hollow contains the pituitary body, a quantity of fibrous tissue, and the internal carotid arteries, which pass into it laterally beneath the parietal shelf, having previously perforated the basisphenoid.

481. The basisphenoidal ossification proceeds forwards on either side of the intertrabecular space to a point just under the very large optic foramen. Anteriorly to this point the trabeculæ are entirely unossified, and lie in a pair of almost closed channels, on either side of the parasphenoid. The latter has a rather broad base and a crested summit which is wedged in between the frontal bones. Anteriorly to the brain-case the parasphenoid becomes compressed and knife-like, wedging in between the hinder ends of the vomers.

482. The two parietals have completely anchylosed into one, with no crest along the line of union. The bone is somewhat pentagonal above, partly overlapping the supraoccipital behind, and the epiotics laterally. Just in front of its projecting postorbital angle the parietal is overlaid by a small oblong but slightly curved postorbital membrane bone. The front edge of the parietal recedes at the mid-line, and is slightly overlapped by the frontals.

483. The side-wall of the cranial cavity is provided by three bones, prootic, parietal, and frontal, the larger

Fig. 51.

Adult Snake; side view of skull, with jaws removed.
tr. trabecula; *a.n.* anterior nostril; II. optic foramen.

Bones: *b o.* basioccipital; *e.o.* exoccipital; *s.o.* supraoccipital; *pr.o.* prootic; *b.s.* basisphenoid; *al.s.* alisphenoid; *pa.* parietal; *pa'.* lateral plate of parietal; *fr.* frontal; *fr'.* lateral plate of frontal; *o.s.* orbitosphenoid; *pa.s.* parasphenoid; *vo.* vomer; *na.* nasal; *pmx.* premaxillary; *l.* septo-maxillary; *a.or.* antorbital or prefrontal; *p.or.* postorbital or postfrontal; *col.* columella.

third being afforded by the parietal (Fig. 51). The alisphenoid does not appear in the cranial wall at all, being thrust out by the parietal and prootic. The inner face of the parietal wall is deeply scooped for lodging the optic lobes; and the external face is correspondingly convex. The hinder edge of the hollow is very thick and almost vertical, projecting inwards as far as the prootic does; from the lower edge of the bone on either side the shelf over the subcranial cavity proceeds inwards.

484. The optic foramen is much larger than the optic nerve which passes through its centre; its boundaries are, behind, the antero-inferior part of the parietal wall; in front and below, the parasphenoid; above, a fan-shaped fenestra between the anterior edge of the parietal and the posterior edge of the frontal wall. The anterior part or handle of this space is occupied by a very small trilobate bone, the orbitosphenoid.

485. The frontals are very solid bones, not anchylosed to one another, but coapplied so as to form a single small tubular cavity. The inferior meeting-place is in the cranial floor above the parasphenoid; supero-externally they expand to form a supraorbital ridge over each orbit. In the extreme anterior part of the cranium a median wall grows downwards from each frontal and separates the fore part of the cerebral hemispheres, which are somewhat divergent in front; but even these partitions, though in contact, are unanchylosed to one another. The frontal floor is here more expanded than behind; while at the same time the outer wall is deficient infero-laterally, the space being filled in by fibrous tissue, supported by an ingrowth of the prefrontal. This membrane-bone (*a. or.*) lies at the side of the fore end of the frontal, and bounds the orbit anteriorly; its upper angle wedging in between the frontal and the nasal.

486. The trabeculæ have united underneath the fore part of the frontals, and become compressed into a vertical ethmoidal plate passing on into the nasal septum. The hinder part of the nasal capsules are also wedged in beneath the fore part of the frontals. The nasal septum is highest behind, but nowhere very high, for the nasal roofs continuous with it diverge very gently at first and are highly arched. The nasal bones (*na.*) lie like shells upon the cartilages, dipping down between the two almost to the septum. Externally they narrow as they descend towards the maxillaries: in front of these they are concave for the passage of the external nostrils. Anteriorly to the nasals the walls of the capsules dip gradually and are confluent, the trabecular cornu forming a bulging front-wall, resting on the premaxillary, and scooped by the external nostril. A very small spur of cartilage (prenasal) grows backwards between and underneath the extreme fore end of the capsules.

487. In the fore part of the nasal floor, on either side of the vomers, is a small tongue of cartilage, which may be connected on one side with a trabecular cornu, and which passes more directly outwards behind towards the outer process of the vomer. At the angle where the outward divergence takes place is another small piece of cartilage,

which converges a little towards its fellow of the opposite side, and lies in a rounded notch of the vomer. These are two upper labial cartilages.

Fig. 52.

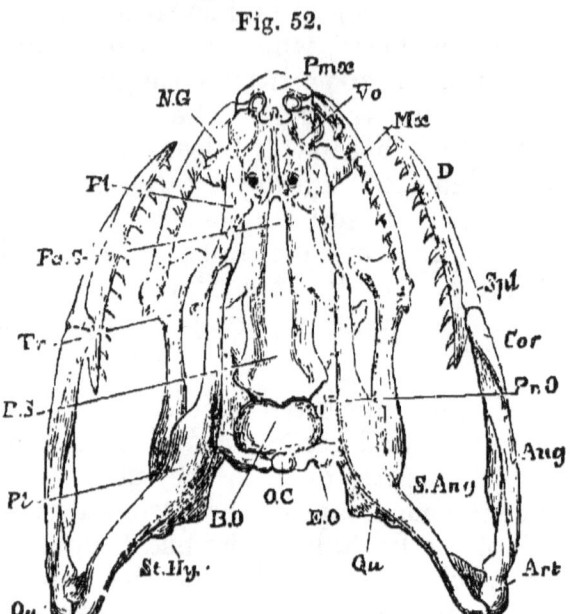

Adult Snake; under view of skull.

o.c. occipital condyle; *n.g.* nasal gland.

Bones: *b.o.* basioccipital; *e.o.* exoccipital; *pr.o.* prootic; *b.s.* basisphenoid; *pa.s.* parasphenoid; *vo.* vomer; *pmx.* premaxillary; *mx.* maxillary; *pl.* palatine; *pt.* pterygoid; *tr.* transpalatine; *qu.* quadrate; *st.hy.* small stylohyal on quadrate; *art.* articular; *ang.* angular; *s.ang.* surangular; *cor.* coronoid; *spl.* splenial; *d.* dentary.

488. The greater part of each nasal floor is occupied by the bone called septo-maxillary, which is vertically extended along the low nasal septum, and after underlying the cavity extends upwards for some distance in the outer nasal wall towards the nasal. The transverse plate of the septo-maxillary which is a floor to the nasal cavity is a roof to the large reniform nasal gland, which itself is bedded in a hollow of the vomer. The latter bone (*vo.*) has a longitudinal plate applied to its fellow beneath the nasal septum, and a transverse wing which is cupped to receive the nasal gland, and notched antero-laterally to give exit

to its duct. The inner (posterior) labial applies itself closely to this notch and extends inwards in the hilus of the gland, expanding upon its duct.

489. The azygous premaxillary (*pmx.*) is a small triradiate edentulous bone, with a short median nasal process above, and a longer double palatine process below. On either side it reaches the maxillary, which is a rather long slightly-curved dentigerous rod, thicker in front and flatter behind. It is attached by fibrous tissue at its anterior end to the premaxillary, at its anterior third (internally) to the descending edge of the prefrontal, and behind passes under and is supported by the broad transpalatine. It occupies less than half of the gape, the rest of which in the upper jaw is margined merely by fibrous tissue, there being neither jugal nor quadratojugal.

490. The palatine (*pl.*) is a small dentigerous rod of bone not half as long or wide as the pterygoid: in front of the middle it sends inwards a small plate beneath the posterior nasal passage. Its anterior third lies beneath the prefrontal: behind this it diverges gently outwards, not lying at all under the brain-case, and is attached to the outer side of the fore end of the pterygoid. The latter (*pt.*) is a large somewhat falcate bar, thick in the middle, and nearly as long as the distance from occiput to premaxillary; it reaches from the palatine behind the occiput to an extent nearly equivalent to half the length of the skull; it obliquely crosses the inner face of the suspensorium, and extends even behind the quadrate condyle to the most posterior part of the angle of the mandible. The transpalatine (*tr.*) is a curious hatchet-shaped bone, passing obliquely from the posterior end of the maxillary to rest upon the broadest part of the pterygoid. Its anteroexternal process touches the palatine and is attached to it.

491. The squamosal (*sq.*, Fig. 53), the uppermost bone of the suspensorium, is almost horizontally placed, passing also a little downwards; it is oblong, rounded in front, and wedge-shaped behind, where its downward face is covered

with articular cartilage. It lies on and is attached by ligament to the anterior part of the prootic, and is slightly in contact with the temporal region of the parietal. The quadrate (*qu.*) is obliquely extended above, with a cartilaginous face lying over the bevelled hinder and external

Fig. 53.

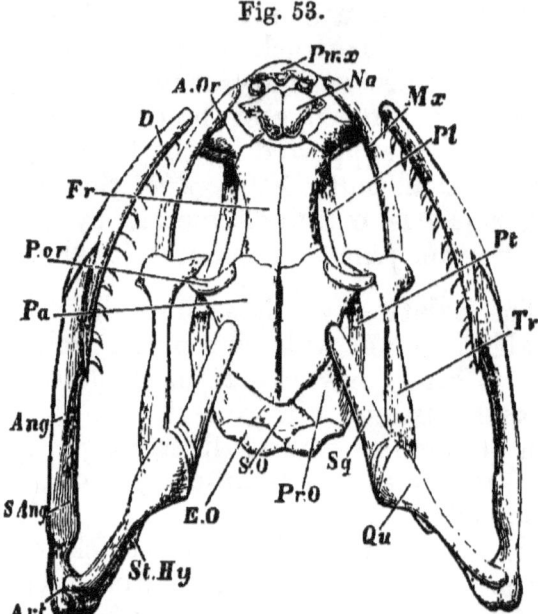

Adult Snake: skull seen from above.

Bones: *s.o.* supraoccipital; *e.o.* exoccipital; *pr.o.* prootic; *pa.* parietal; *fr.* frontal; *na.* nasal; *pmx.* premaxillary; *a.or.* antorbital or prefrontal; *p.or.* postorbital or postfrontal; *pl.* palatine; *pt.* pterygoid; *tr.* transpalatine; *mx.* maxillary; *sq.* squamosal; *qu.* quadrate; *st.hy.* minute stylohyal on quadrate; *art.* articular; *s.ang.* surangular; *ang.* angular; *d.* dentary.

edge of the squamosal: there is a synovial joint between the two bones. The quadrate becomes more slender and thicker below, and bears a rounded condyle for a perfect hinge joint. Its direction is outwards and backwards at an angle of about 45° with the middle line.

492. The lower jaw reaches as far as to the front of the maxillary bone, and is gently arcuate: its dentigerous

dentary (*d.*) is just half the length of the ramus. The bones are on the whole unchanged from the last stage. The broad junction of the coronoid and splenial is at two-fifths the length of the jaw from its anterior extremity. The angular (*ang.*) and surangular (*s. ang.*) are of great length, and are in some degree consolidated with the articular (*art.*); the meckelian cartilage is almost entirely absorbed.

493. The columella with its stapedial plate lies under the horizontal canal, in a recess which leads to the fenestræ ovalis and rotunda; it is relatively much diminished, is entirely bony, and the sigmoid mediostapedial part is now no longer than the long axis of the stylohyal. This latter is a little scale-like bone, partly anchylosed with the posterior edge and inner face of the quadrate a little above its middle. By the growth of the suspensorium it is carried away from the columella to a distance of four times its own length. The small supratemporal has completely coalesced with the squamosal. A small pair of cartilages is sometimes found at the sides of the larynx, which may represent the distal part of an arch (Huxley).

494. SUMMARY. In this type the non-coalescence of the trabeculæ in their postnasal tract, combined with their late persistence, is a feature of striking interest. Furthermore, we note a marked diminution in the relative proportion of the cartilaginous structures to the rest of the skull. The cranium is only posteriorly roofed by cartilage; the anterior part of its lateral wall also acquires very little cartilage, and the orbito- and alisphenoids are very feeble, and contribute but slightly to the cranial investment. The occipital and auditory bones are very complete, showing much more likeness to Osseous Fishes than to Amphibians; the basisphenoid is ossified by two centres, and then coalesces with the large parasphenoid which by its median upward growth separates the trabeculæ. But the fore part of the cranium has yet more notable features, in the first appearance of the frontals and parietals being low

down laterally, in the perfect coalescence of the parietals superiorly, and the formation of inferior plates, above the trabeculæ and supporting the brain.

495. The palate is of a distinctive "cleft" type, and the palatal bones present an extraordinary elongation and high specialisation of parts. A new bone, the transpalatine, is added, extending between the posterior end of the maxillary and the middle of the pterygoid. The latter, as in forms previously described, is connected with the quadrate, but it has a perfection of structure not before manifested. The premaxillary is azygous; the septomaxillaries are very perfect and interesting; "labial" cartilages are again found in relation to the posterior nares. The suspensorium of the mandible is notable (1) as giving rise to but one bone, the quadrate, which ossifies it entirely; (2) for altering its direction during development from forwards to backwards; (3) in being quite dissociated from contact with the cranial wall, and loosely united with the long squamosal, both bones being backwardly directed. The lower jaw has a very full number of distinct ossific centres. The side wall of the skull possesses membranous antorbital (prefrontal) and postorbital (postfrontal) bones. The small representative of the hyoid arch is divided into two parts, of which one persists as the simple columella, separated by some distance from the minute stylohyal piece.

496. There are but few points of special community between this skull and the types examined in the Amphibia, while there are many features which remind us of the Osseous Fish and the Bird. While the specialisation of membrane-bones and diminution of cartilage and cartilage-bones is far beyond what we have seen in the Salmon, yet in the occipital and otic bones, the parasphenoid, and the palato-mandibular structures there are many resemblances between these two forms. But there is lost almost entirely the conformation of parts necessitated by the function of breathing in water; and the remnants of the postmandibular arches very early after

their appearance take up the relations to which they are adapted in the adult. In many respects we are on the high plane of vertebrate development; the membrane bones are co-applied and united with the cartilage-bones to form a strong defence for the brain; the bones of the palate have a form and relations which need but little modification to present the Avian type; the suspensorium is one definite bone, the quadrate. From the consideration of this skull alone, the association of Amphibians and Reptiles in a group distinguished on one hand from Fishes, and from Birds on the other, becomes a contradiction to anatomical truth.

APPENDIX ON THE SKULLS OF REPTILES.

497. The skulls of venomous snakes vary in their palatal structures in a manner which finds its extreme expression in *Crotalus* (the Rattlesnake). Here the shortening and specialisation of the maxillary, bearing the poison-fang, combined with great mobility of the other palatal bones, are the distinctive characters. The premaxillaries are small; and the maxillaries are capable of being moved from their normal position so as to erect the poison-fangs, by the same muscular action which opens the mouth[1]. In *Typhlops* the palatines are slender, and transversely placed, behind the posterior nares. The pterygoids lie beneath the cranial floor, and do not extend to the quadrates. In *Tortrix* the quadrate is short, and directly connected with the cranium, the squamosal being insignificant.

The Skull of the Turtle.

498. In the Turtle (*Chelone*) at the time of hatching[2], the chondrocranium is much more developed than in the Snake. The occipital roof is extensive, reaching beneath the hinder fourth of

---

[1] See Huxley, Anatomy of Vertebrates, p. 239.

[2] This description will explain most of the structure of the adult skull.

the parietal bone; yet the supracranial fontanelle is large. The floor of the cavity presents cartilage almost perfectly continuous, the posterior clinoid ridge being prominent; but there is a posterior basicranial fontanelle. The ossifications of the floor of the cranium are a large basioccipital, and a basisphenoid arising by three centres, two paired and lateral, behind the pituitary body, as in the Snake, and one anterior and median, representing the rostrum of the same animal. In front of this the presphenoidal cartilage is low at first, and then becomes suddenly compressed vertically, forming an interorbital septum beneath the optic nerve. This cartilage attains its greatest height as mesethmoid, and then is gently lowered where it constitutes the nasal septum, finally terminating antero-inferiorly in the median prenasal cartilage.

499. From the whole of the upper edge of the interorbital septum arises on either side a large semilunar orbitosphenoid cartilage, which ends posteriorly just above the optic nerve. The space behind this in the cranial wall is largely ossified by a descending plate of the parietal, there being no alisphenoid cartilage. The nasal cartilages form very perfect roofs and side-walls to their capsules, and, curving inwards laterally, give rise to rudimentary turbinal plates. Posteriorly the nasal wall constitutes a steep antorbital plate. In its hinder part the capsule is almost entirely invested by the large ethmo-nasal bone. Anteriorly the lateral wall is continuous in cartilage with the nasal floor, so that the capsules form two perfect cylinders.

500. The exoccipitals are vertical and oblong, perforated by the vagus and hypoglossal nerves. The supraoccipital region is large, with a great crest or spine growing backwards; it possesses one considerable ectostosis. There are three auditory centres, but much of the capsule remains cartilaginous. Anteriorly is the prootic; supero-posteriorly the epiotic, passing forwards and inwards, and uniting with the supraoccipital; and postero-inferiorly the opisthotic, remaining permanently distinct. Between these bones is a triradiate tract of cartilage, the remains of which persist through life as a suture. There are large parietals (with lateral cranial plates), and frontals.

501. The quadrate has a large swollen otic region closely applied to the auditory capsule, and an anterior (orbital) process growing

forwards and inwards to the outer side of the descending plate of the parietal. This is not ossified in the adult. The pterygoids are long and broad, and unite with one another in the middle line beneath the cranial floor. Each pterygoid has an angular ascending process applied to the inner face of the quadrate, and also abutting against the outer face of the parietal beneath the orbital process of the quadrate; it is ossified from membrane. In a groove on this osseous process is found a tract of cartilage, whose apex nearly reaches the apex of the orbital process, forming a right angle with it. It becomes invested with a bony shaft, and constitutes a small *epipterygoid;* it is subsequently developed into an oblong plate of bone wedged in between the parietal and the ascending process of the pterygoid, and has been mistaken for an alisphenoid.

502. The palatines are united with the pterygoids behind, and with the median vomer in front and above. The latter bone is expanded beneath, on a level with the small inferior palatal plates of the palatines, which are here seen for the first time, tending to carry backwards the orifice of the posterior nares in the palate. There is no transpalatine, and no septo-maxillary. The premaxillaries are small; the maxillaries large, and jugals and quadrato-jugals continue the series to the quadrate. There is a postfrontal (membrane bone) behind and above the orbit ; and further back, a squamosal at the sides of the auditory capsule and above the quadrate. The temporal fossa is roofed over by the junction of the postfrontals and squamosals internally with an external plate of the parietals, and inferiorly with the jugals and quadrato-jugals. The dentaries are anchylosed in the adult into one bone. The postoral arches present much more development than in the Snake. There is a broad basihyal with anterior (larger) ceratohyal cornua and posterior smaller (branchial or "thyro-hyal") processes. These cornua are all ossified. They are not directly connected with the axial parts; but there is a long columella auris, ending above in the stapes.

The Skull of Lizards.

503. In a typical Lacertilian (as Varanus, Iguana) the hinder region of the skull is very completely ossified. Basi- and exoccipitals enter into the composition of the single condyle. The basioccipital is broad, the exoccipitals ascend to two-thirds of the height of the

foramen magnum, which is completed by the large supraoccipital. The auditory capsules, together with the exoccipitals, are extended widely outwards; the opisthotic unites with the exoccipital; the epiotic with the supraoccipital; the prootic remains distinct.

504. The basisphenoid is ossified from a pair of ectosteal centres behind the pituitary body; and sends out on either side a large basipterygoid process to articulate with the inner aspect of the pterygoid. The basisphenoid extends forwards for some little distance into the trabecular cartilage on either side of the pituitary space, but in front of this the cartilage is unossified and passes into the extensive interorbital septum, which again is continued into the nasal septum. The membranous floor of the pituitary space is supported by the posterior broad end of the delicate styliform parasphenoid, which underlies the posterior two-thirds of the interorbital septum. The latter is largely fenestrate, and but slightly ossified. There is no alisphenoid, but a membrane occupies the cranial wall from a little behind the optic foramen to the front margin of the prootic (which is notched for the trigeminal nerve). The parietal descends slightly into this tract. The orbitosphenoid cartilage is extensive, but very largely fenestrate; the cartilage extends for some distance behind the optic foramen. There is an orbitosphenoid ossification below the fenestra. The nasal septum is unossified: the nasal passage is floored on either side by a large septo-maxillary, which ascends as a small vertical plate against the septum.

505. The premaxillary as a rule is azygous, having a long nasal process. The nasals may be large as in Iguana, or very small as in Monitors. There are two membrane-bones in the antorbital region, both of which may be perforated; the upper or prefrontal is the larger, and has an orbital plate; it narrows into an arcuate spur between the orbits, uniting by suture with the frontal. There is one supraorbital bone in Monitor, Iguana has two or three, the Scincoids a considerable series. The frontals are simple superior bones, united with the parietals by a straight coronal suture, whose ends abut on the postorbitals (postfrontals). These membrane-bones (peculiar to reptiles) are always large, and rest on the squamosal behind, and sometimes on the jugal in front. As in Snakes, the parietals are always densely anchylosed, except for the small parietal foramen, so characteristic of these forms, which may be behind the

coronal suture, in its course, or slightly in front of it. Posteriorly a large horn is sent out on each side from the parietal, passing backwards to the ear-capsule and the parotic process of the exoccipital. The jugal is sometimes large, reaching to the postfrontal and parietal. The quadrato-jugal is absent except in Hatteria. The squamosal is connected with the posterior spur of the parietal, and may be also united with the postfrontal. There is a supratemporal behind the squamosal. The fossæ which lie between the external bones of a Lizard's skull may be distinguished as supratemporal, between the parietal, postfrontal, and squamosal; post-temporal, between the parietal, supraoccipital, and auditory prolongation; and lateral temporal, between the squamosal and postfrontal above, the jugal and quadrate in front and behind, and the quadrato-jugal ligament below.

506. In the *Geckos* the external lateral bones of the skull are small or absent, so that the upper and lower lateral bars are not constituted. In *Sphenodon* both arcades are complete, and the quadrate is anchylosed to the squamosal, quadrato-jugal and pterygoid, and the front of the ear-capsule. There are two premaxillaries. In *Mosasaurus* the pterygoids articulate with one another in the middle line behind the posterior nares. In *Chamæleon* the skull has an interorbital septum, but no epipterygoid. The quadrate bones in *Amphisbæna* are directed downwards and forwards, and there is no interorbital septum. In Chamæleon there is a backwardly produced median crest from the occipital and parietal bones, joined by a pair of lateral prolongations from the squamosals. The orbit is closed by a process of the jugal, but the lower arcade is not completed by a quadrato-jugal. The quadrate is articulated with the squamosal and the auditory capsule, but the pterygoid does not reach it.

The Skull of Crocodiles.

507. The hinder region of the skull is very complete, and there are large alisphenoids. There is an interorbital septum; and the orbitosphenoids are little developed. There is no parietal foramen. The bones anterior to the cranial cavity are greatly elongated, and the palatal plates of the maxillaries and palatines are very large, meeting in the middle line, and forming a secondary floor to the nasal passages; this palate is carried far backwards by the development of a palatal plate on the pterygoids. The latter also extend

upwards to the base of the cranium, being suturally united, and are united externally with the upper and inner surface of the quadrates. The vomers are usually concealed by the palatal plates. The large quadrate is firmly fixed in the side of the skull, though not anchylosed. The tympanic cavity is bounded by almost all the bones of the hinder part of the skull, the epiotic and supraoccipital excepted. The opisthotic is confluent with the exoccipital. The lateral bony arches of the skull are very complete and strong, forming considerable fossæ resembling those of the Lacertilia. The Eustachian tubes are very complex.

CHAPTER VII.

THE SKULL OF THE COMMON FOWL.

First Stage: Embryo at the end of the fifth day of incubation.

508. THE head of the embryo chick has at this period become well-defined, with notable bulbous protuberances of the cranial vesicles, and very large eyeballs. The mesocephalic flexure is at its greatest intensity; the head is bent over so as nearly to touch the neck; the mouth faces the chest. The cerebral hemispheres are already well-developed on the inferior aspect of the head (Fig. 54, *c.h.*); the midbrain is prominent in front of the cranial bend; while the hindbrain is elongated and comparatively small. The olfactory pits (*n.*) are little depressions below and behind the forebrain, and just internal to the fore part of the eyeballs. The auditory vesicles are some way above and a little behind the eyeballs, at the sides of the posterior region of the hindbrain.

509. The mouth (*m.*), beneath and behind the forebrain, has an oblong aperture, bounded in front by the nasofrontal process (*n.f.*), which separates the two olfactory depressions; behind by the first visceral arch, and laterally by the maxillopalatine processes (*s.m.*) growing forwards from the sides of that arch, and lying underneath the eyeballs. The nasal pits, as they deepen, acquire a break in their rim, giving rise to a groove directed obliquely downwards towards the cavity of the mouth. The naso-

frontal and maxillopalatine processes abutting on this groove on either side, tend to deepen it: at a later period they form a bridge over it, and thus enclose a channel from the nasal pits to the mouth, opening on its roof.

Fig. 54.

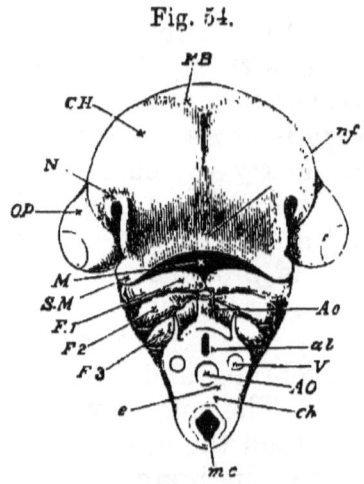

Embryo Chick of the fourth day of incubation; head viewed from below as an opaque object (Foster and Balfour). The neck has been cut across between the third and fourth visceral folds.

c.h. cerebral hemispheres; *f.b.* first cerebral vesicle; *op.* eyeball; *n.f.* nasofrontal process; *m.* mouth cavity; *s.m.* superior maxillary or maxillopalatine process from *f.* 1, the first visceral (mandibular) fold; *f.* 2, *f.* 3, second and third visceral folds; *n.* nasal pit.

In the section of the neck, *al.* alimentary canal; *m.c.* neural canal; *ch.* notochord; *ao.* below, dorsal aorta; *v.* vertebral vein; *ao.* above, bulbus arteriosus.

510. Three visceral folds (Fig. 54, *f.* 1, 2, 3) exist in the side-walls of the throat, the hinder being the smaller; and they are most prominent ventrally. Behind each fold there is a cleft on either side, opening into the throat.

511. The primary parts of the cranio-facial skeleton, from being constituted of stellate cells, are mostly passing into hyaline cartilage. The cranial part of the notochord (Fig. 55, *n.c.*), slightly constricted in two places, extends forwards nearly to the pituitary region (*pt.s.*). It is flanked on either side by the *parachordals* (*pa.ch.*): there

VII.] THE FOWL: FIRST STAGE. 221

is a prechordal bridge, and from this region the *trabeculæ* (*tr.*) bend downwards and diverge to surround the pituitary space. A discontinuity of these primary elements has not been discovered in any earlier stage. No separate cartilaginous auditory capsule is formed in the chick; but

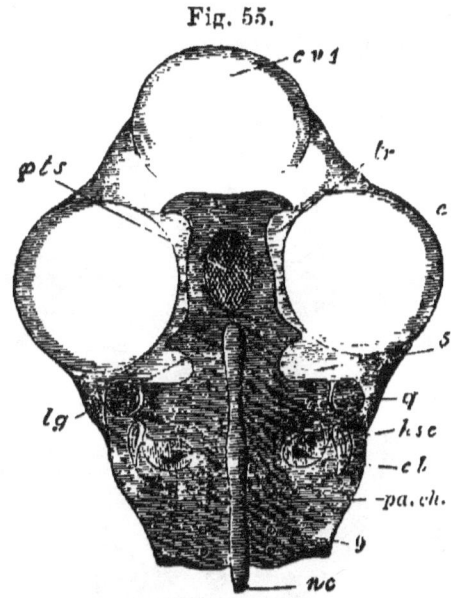

Fig. 55.

Embryo Chick, fifth day of incubation; view of cranial structures from above, the upper part of the head having been sliced away horizontally; part of the first vesicle and the lower part of each eyeball are left. The cartilaginous parts are those indicated by dark horizontal shading.

c.v. I, forebrain; *e.* eyeball; *tr.* trabecula (the line stops short of the part indicated); *pt.s.* pituitary or intertrabecular space; 5, notch for trigeminal nerve; *l.g.* process which becomes the lingula sphenoidalis; *q.* quadrate cartilage; *h.s.c.* horizontal semicircular canal; *cl.* cochlea; *pa.ch.* parachordal cartilage; *n.c.* notochord; 9, foramen for hypoglossal nerve.

each parachordal is from the first continuous in the middle of its outer aspect with the chondrified wall of the ear-sac, in which the cochlea (*cl.*) and semicircular canals already begin to be evident. A little way in front of the ear-mass the parachordals are suddenly narrowed, forming a notch for the passage of the trigeminal nerve (5). In front of this there is a slight angular expansion, on each side of the front end of the notochord; and at this point a small transverse crest exists, marking out exactly the line about which the mesocephalic flexure has taken place.

512. The space enclosed by the descending trabeculæ is longitudinally oval: below this region the paired rods coalesce to form a broad *internasal plate*, lying behind the

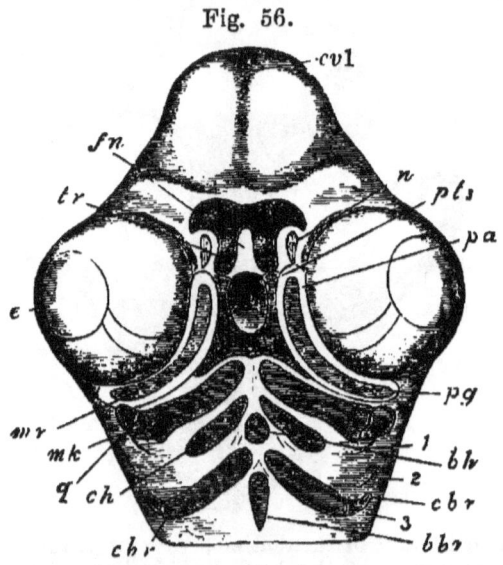

Fig. 56.

Embryo Chick, fifth day of incubation; head viewed from below, with skeletal parts seen through.

*c.v.*1, forebrain; *e.* eyeball; *n.* nasal pit; *f.n.* nasofrontal process; *tr.* points to one of the trabecular horns, recurrent in the nasofrontal process beneath the level of the main part of the trabeculæ, which are seen on either side of *pt.s.* the pituitary space; *mr.* subocular bar, in which *pa.* palatine and *pg.* pterygoid tracts are indicated; *q.* quadrate; *mk.* meckelian cartilage; *b.h.* basihyal; *c.h.* ceratohyal; *b.br.* basibranchial; *c.br.* ceratobranchial; *e.br.* epibranchial; 1, 2, 3, visceral clefts.

forebrain. It is emarginate in the middle line, and extended into a horn-shaped process on each side. These *cornua* partially arch round the external nasal openings. The inferior part of the internasal plate is curved somewhat backwards upon the remainder, so that the cornua look directly backwards (see Fig. 56).

513. In the inner (palatine) region of the maxillo-palatine process, the curved *palatopterygoid* bar (subocular in position) is beginning to be established, but it is considerably later in solidifying than the mandibular arch (Fig. 56, *pa.*, *pg.*). The latter is already segmented into two pieces; (1) a smaller tuberous *quadrate* cartilage (*q.*) lying behind the eyeball, and closely apposed to the antero-external angle of the parachordal and otic cartilage behind the exit of the trigeminal nerve; and (2) a meckelian rod (*mk.*) lying in the first visceral fold, having a sinuous articular surface for the quadrate and a slight angular outgrowth behind. The meckelian bars nearly meet in the middle line. The first visceral cleft (1), behind the quadrate, is commencing its development into the complex tympanic cavity.

514. The hyoid arch contains ventrally a pair of small ill-differentiated rods (*c. h.*) and an azygous piece or *basihyal* (*b. h.*). The second visceral cleft is fast closing, and behind it the first branchial arch has a pair of well-defined *ceratobranchial* rods (*c. br.*) nearly meeting in the middle line, surmounted by very small distinct *epibranchial* pieces. A *basibranchial* (*b. br.*) is forming in the middle line behind.

At this early stage the hyoid arch has begun to lie within the mandibular, just as the first branchial comes to lie within the massive hyoid arch in the Osseous Fish. Already the third visceral arch of the chick has outgrown the second,— a state of things universal in the class of Birds.

515. We here see that a grade of structure is reached in this skull as soon as its rudiments can be made out, which is only attained by a series of changes in the skulls of inferior types. The basicranial elements are continuous notwithstanding the mesocephalic flexure; the internasal plate and cornua are equally complete and continuous. The otic masses are also coalesced from the first with the special cranial structures. The separate development of the upper piece of the mandibular arch is a feature of high specialisation; the future palatopterygoid, so important

in the adult, is already distinguishable. The slightness of the skeletal parts in the hinder arches is related to the absence of any necessity for breathing in water.

Second Stage: Sixth and Seventh Days of Incubation.

516. The head, still disproportionately large, is now five lines long, and the rudimentary cranium is well chondrified. The mesocephalic flexure has greatly diminished, and the head is more squared and less bulbous (Fig. 57). The forebrain occupies nearly the front half of the cranial cavity; the midbrain lies behind; while the hindbrain, considerably smaller, is entirely below the hinder part of the midbrain. The mouth, instead of being totally inferior, is now a space, the axis of which points more forwards than downwards.

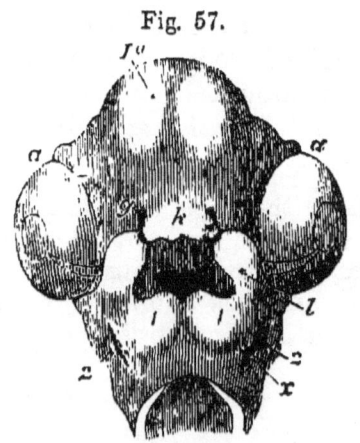

Fig. 57.

Embryo Chick, sixth day of incubation; head seen from below. (After Huxley.)

1*a*. cerebral hemispheres; *a*. eyeball; *g*. nasal pit; *k*. nasofrontal process; *l*. maxillopalatine process; 1, mandibular arch; 2, hyoid arch; *x*. first visceral cleft. The cavity of the mouth is seen between *k* and 1; the darkest part indicates the opening into the throat.

517. The trabecular region of the cranium has become longer than the parachordal. At the angle formed by the bending of the cranial floor, marked by a slight crest in the first stage, there has arisen the high ridge,

directed upwards and somewhat backwards, known as the posterior clinoid wall, protecting the pituitary body behind. The internal carotid arteries enter the cranial cavity in front of this wall. The parachordal cartilages are well consolidated, although the notochord is not obliterated. Posteriorly the cartilage grows outwards and upwards around the neural tube, forming exoccipital regions, and tending to complete the occipital ring. An external growth of the exoccipital behind the ear-mass (*tympanic wing*) comes to overarch the tympanic cavity (*e.o.* Fig. 58).

518. The trabecular rods at the sides of the pituitary body appear slightly bulged out and more distinct from the parachordals than in the first stage, though perfectly continuous with them. In older embryos there is an *alisphenoid* (lateral) tract of cartilage (*a.s.*) partially completing the base of the cranium, separated from the otic mass by the trigeminal nerve below, but united with it outside the nerve. Externally this alisphenoid cartilage has an oblique projecting ridge, partially bounding the orbit behind, and overarching the quadrate. The trabecular cartilage remains low for a little distance, as far as the exit of the optic nerves. Underneath the fore part of the basilar region the fibrous stroma is thickening to receive the basitemporal ossification.

519. In front of the optic foramen and rising considerably above its level, the brain has become elevated upon a vertical cartilaginous internasal and interorbital plate, oblong in side view, thinner above, but thick below where it is directly formed by the coalesced trabeculæ. It extends continuously to the fore extremity of the head, and its front margin is almost vertical; it is continued below into a rounded knob, directed a little backwards; and its upper edge is united on each side with the small cartilaginous nasal roof.

520. Within a very short time in incubation the anterior part of this vertical plate grows forwards below, so as to form the axis of a beak projecting downwards and

forwards: the beaked part (prenasal cartilage) curving below the level of the basicranial axis. Thus we have an oblong *interorbital plate* (Fig. 58, *p.s.*), with a little

Fig. 58.

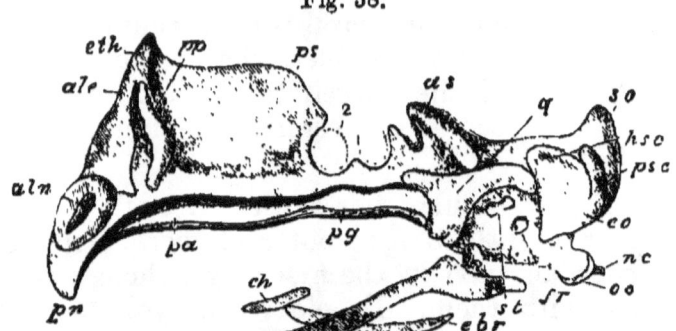

Embryo Chick, seventh day of incubation; side view of skull.

o.c. occipital condyle; *nc.* notochord; *e.o.* exoccipital tract, with its tympanic wing; *s.o.* supraoccipital tract; *h.s.c.* horizontal semicircular canal, *p.s.c.* posterior semicircular canal, both seen through the cartilage; *a.s.* alisphenoid cartilage, with its external ridge; 2, optic foramen; *p.s.* presphenoidal region; below it, interorbital septum; *p.p.* antorbital plate; *eth.* ethmoidal region; *al.e.* aliethmoidal cartilage; *al.n.* alinasal cartilage; *p.n.* prenasal cartilage; *pa.* palatine, *pg.* pterygoid tract; *q.* quadrate; *st.* stapes; *f.r.* fenestra rotunda; *mk.* meckelian bar; *c.h.* ceratohyal; *b.h.* basihyal; *b.br.* basibranchial; *c.br.* ceratobranchial; *e.br.* epibranchial.

spine projecting upwards in the front wall of the cranial cavity; and a triangular *internasal plate*, with its apex directed downwards and forwards. This internasal plate rapidly becomes differentiated into three regions, distinguished by the nature of the nasal outgrowths continuous with its upper and anterior edge. In front there is the *alinasal* valvular structure (*al.n.*), protecting the external nostril; behind this the *aliseptal* roof in the proper septal region, with soft coiling growths already projecting downwards and inwards from it; and finally the *aliethmoidal* roof (*al.e.*) growing from the highest part of the vertical plate, coiling to form an upper turbinal, and also developing a transverse vertical partition (*p.p.*) between the nasal cavities and the orbits (*antorbital* plate).

521. The arcuate rod in the maxillopalatine process (*pa.*, *pg.*) is elongated to reach from the front of the quadrate cartilage to the middle of the nasal septum. In section these rods and the thickenings in which they lie appear as the walls of the cleft palate. The quadrate (Fig. 58, *q.*) is now a very definite triradiate cartilage; its antero-internal process passes under the trigeminal nerve, but does not reach the cranial wall. The postero-external or otic process forms a large hook extending along and applied to the side of the ear-mass, and reaching to and partially overarched by the exoccipital (tympanic) wing. This otic process overarches the tympanic cavity, and the fenestræ ovalis and rotunda (*f. r.*) are on the wall of the capsule beneath and within it. The quadrate condyle for the mandible, looking forwards and downwards, is grooved so as to be double. The meckelian rods (*mk.*) are long sinuous cartilages approximated anteriorly, extending as far forwards as the beak; thickest behind, where they fork into two processes, the inner and the posterior angular processes.

522. A small piece of cartilage (stapedial or *columellar*), at first continuous with the ear-capsule, has become detached from it below the otic process of the quadrate, and consequently its thick base fills up an oval fenestra (*ovalis*), behind and below which is another membranous fenestra (*rotunda*), in the wall of the rudimentary cochlea (Fig. 58, *st.*, *f. r.*).

523. The history of this part has been observed more completely in the House-Martin (*Chelidon urbica*) where the upper end has been observed at first continuous with the auditory cartilage, but dilated at its free end into a fan-shaped *extrastapedial* plate which is pointed above and below, the upper process being the rudimentary *suprastapedial*, and the lower being continuous with a delicate ligamentous tract, which chondrifying afterwards becomes the *infrastapedial* (see Fig. 74). From the lower end of this ligament there passes backwards at a right angle a small flat oblong cartilage: this exactly corresponds to

the stylohyal which Prof. Huxley described in the young Crocodile (see his paper on the Representatives of the Incus and Malleus, P.Z.S.). The dilated extrastapedial part has the same shape as described by the same author in Hatteria.

524. The basal hyoid and branchial cartilages are the same in number as in the last stage, but are more solid, and the lateral moieties are directed backwards.

525. Comparing the skull at this stage with its primary condition, we see that a marvellous advance has taken place, elevating it from an elementary grade to one characterized by the notable interorbital septum. Already the skull is definitely Sauropsidan, and in several points Struthious. The hinder part of the cranial investment is proceeding to form its occipital ring; the anterior has developed the main outline, not only of the interorbital framework, but also of the beak which distinguishes the Bird. The prenasal cartilage is a very fit model on which the premaxillaries of any type of bird might be formed; in no bird whatever, let the shape of its face be what it may, is the type of the first model wholly lost, although in every case the primordial structure very early undergoes wasting. The palate is cleft as in Struthious Birds; the present articulation of the quadrate is like that which is persistent in them. Likewise the temporary condition of the branchial arches in this stage agrees with their form in adult Struthionidæ. The angular processes of the jaw already indicate the type of the animal, for they have a peculiar development in the whole Gallo-Anserine series, attaining their most extraordinary form in the Tetraonine group.

Third Stage: Embryos in the middle of the second week of incubation, with heads from eight to nine lines in length.

526. The skull has made great progress towards a bird-like form. In this stage it is in some respects parallel

with the adult skull in certain Teleostean Fishes, where there is a free development of the cartilaginous framework, which is comparatively little ossified; whilst the parostoses are delicate, although numerous. The Lumpfish (Cyclopterus) and the Salmon possess such skulls.

527. When the axial parts of the skull are dissected out, they are seen to be very easily divisible into two main regions: a large posterior rounded expansion, more than hemispherical, and an elongated vertically-compressed plate, lying between the orbits and between the olfactory organs, and flanked anteriorly by moderate-sized appendicular cartilages belonging to the nasal sacs. The posterior expansion is remarkable for the curious way in which the periotic masses are compressed and curved so as to be moulded into the rounded form of the cranium (see Fig. 60). Thus a considerable portion (the *cochlear*) of the ear-mass is in the cranial floor, very median in position and far forwards: but the region containing the canals is much compressed and tilted backwards, rising high in the side wall, where it becomes postero-external, to accommodate the large brain; so that all the semicircular canals are visible in a posterior view.

528. The notochord has relatively retired, and is separated from the pituitary body by a distinct bridge of cartilage; this forms the somewhat elevated posterior clinoid wall, and the cartilage shelves downwards on either side of it. The cranial notochord is slightly constricted in two places, thus presenting three subequal regions, the anterior cylindrical, the other two fusiform; the hinder of these is enclosed in a bony sheath, and at the posterior extremity of the skull by a cartilaginous bridge above and below. This is the only point where the parachordal cartilages have as yet united; but here they completely invest the notochord, and form the single rounded occipital condyle, the two mammillæ which first appear becoming one. Anteriorly indeed the parachordals have somewhat withdrawn from the middle line, leaving a

lanceolate chink in which the notochord lies: this is the posterior basicranial fontanelle (*p.b.f.* Fig. 62).

529. The occipital ring of cartilage is complete; the foramen magnum is obliquely directed backwards (*f. m.* Fig. 59), and the supraoccipital region is almost vertical; the latter is emarginate in the middle line, both in front and behind. A pair of triangular *exoccipitals* (*e.o.*) are ossifying; they extend from the lateral margin and base of the foramen magnum forwards to the foramen for the vagus nerve. The exoccipital region sends a lamina of cartilage (tympanic wing of the exoccipital) around the side and base of the hinder part of the ear-capsule. This wing is the posterior boundary of a scooped antero-inferior hollow in the periotic mass, which is to become the tympanic cavity (Fig. 59, *ty.*); at present it is filled by a soft flocculent stroma, soon to be absorbed. There is but a slight roof to this cavity (tegmen tympani or pterotic ridge).

530. Mesiad of this tympanic hollow, and extending in front of it, is a thick rounded boss of cartilage, protruding both within the cranial cavity and on its inferior aspect, and containing the finger-like cochlear process of the ear-cavity (*cl.* Fig. 62). These cochlear protuberances are remarkably near to the median line of the cranial floor, at the sides of the middle region of the notochord. The whole of the periotic capsule is unossified at present.

531. The circumpituitary or basisphenoidal region is perforated by a nearly circular pituitary hole, into which the pituitary body passes, and through which the internal carotid arteries enter the cranial cavity. The axial cartilage is narrow in the posterior clinoid region, then a little expanded outwards and much extended vertically; and projecting from the circumpituitary cartilage antero-laterally is a curved process or *lingula* of cartilage, extending downwards and backwards for a short distance. From the hinder half of the basisphenoidal region a considerable elongated *alisphenoidal* lamina arises (*a. s.* Fig. 59), passing outwards and backwards on either side in the cranial floor, and ascending a little into the side

wall; coalescing at its extremity with the periotic mass, but leaving an elongated space unfilled by cartilage between the two tracts. The cranial surface of the alisphenoid conforms closely with the concave curvature of the hinder part of the cranium (see Fig. 62).

532. In the cartilage of the antero-lateral boundary of the pituitary fossa is a pair of small bony centres; they are continuous, however, with the parosteal ossification of the rostrum to be hereafter mentioned. Just in front of these basisphenoidal centres, where the compression of the basicranial cartilage to form the interorbital septum begins, is a pair of small distinct cartilaginous plates, affording articulating surfaces for similar cartilages developed on the pterygoid bones (indicated in Fig. 59). The interorbital septum is now of much greater depth, carrying the fore part of the brain up to a high level. The prenasal continuation of the nasal septum is thickened above, and almost straight below, though slightly downbent at its extremity (*p.n.* Fig. 59). The trabecular cornua have become lost in the lower anterior part of the alinasal cartilage.

533. The frontal, parietal, and nasal bones have appeared; they need not be described in the present stage. The bones of the floor of the cranium and of the palate must be noticed more fully, because of the great interest of their history. The subcranial region which in the Frog is ossified by the basitemporal wings of the parasphenoid, is here supplied with a pair of distinct and large *basitemporal* bones (*b.t.* Fig. 59), which extend from near the median line, beneath the cochlea, and so far outwards as to constitute a floor for the tympanic cavity: their anterior limit is near the fore margin of the alisphenoid cartilage. These ossifications arise in a thick web of fibrous tissue in the hinder part of the palate; and the matrix is abundant in the middle line, extending forwards to the bone next to be described. The Eustachian tubes run forwards and inwards above the anterior edge of these bones, and meet in the middle line, beneath the pituitary fossa.

Fig. 59.

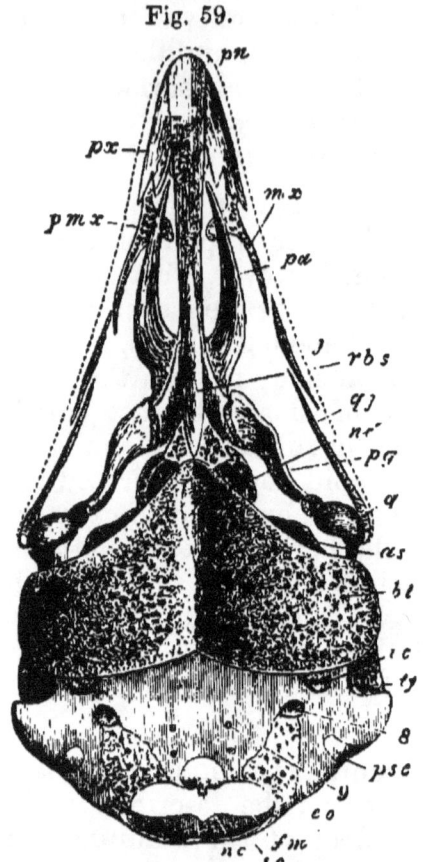

Embryo Chick, middle of second week of incubation; under view of skull, with arches removed.

nc. notochord; *f.m.* foramen magnum; *s.o.* supraoccipital tract; 9, foramen for hypoglossal nerve; *p.s.c.* posterior semicircular canal; 8, foramen for glossopharyngeal and vagus nerves; *ty.* tympanum; *i.c.* internal carotid artery passing above the basitemporal bone; *a.s.* alisphenoid cartilage; *nc'.* notochord diagrammatically seen in the posterior basicranial fontanelle; *p.n.* prenasal cartilage.

Bones: *e.o.* exoccipital; *b.t.* basitemporal; *r.b.s.* rostrum; at its hinder end on either side the basisphenoid centres; *pa.* palatine; *pg.* pterygoid; *px.* premaxillary; *mx.* maxillary; *pmx.* maxillopalatine plate of maxillary; *j.* jugal; *q.j.* quadrato-jugal; *q.* quadrate.

534. The interorbital septum rests upon a grooved fusiform rod of bone (*r.b.s.*), representing the anterior part of the parasphenoid of the Salmon, the Axolotl, and

the Frog, and the whole of that of the Snake and the Lizard. This ossification in Birds receives the special name of *rostrum*; its posterior extremity, at the anterior part of the pituitary space, has crept into the perichondrium and the superjacent cartilage on either side.

535. The *premaxillaries* (*px.*) are formed as double laminæ (nasal and palatal), united at the outer edge: above, they nearly meet by their nasal processes, although the prenasal cartilage protrudes between them in front: below, they lie outside it. The posterior extremity of each is forked, the outer being the dentary process, and the inner the palatal. From each premaxillary along the margin of the upper jaw to the quadrate condyle there are found three subequal styliform parostoses, *maxillary* (*mx.*), *jugal* (*j.*), and *quadrato-jugal* (*q.j.*). The maxillary at its anterior third sends inwards an earshaped process over the palatine bone; this is the maxillopalatine process (*pmx.*).

536. The palatine bones (*pa.* Fig. 59) are flat, pointed in front, where they pass between the palatal processes of the premaxillaries and the prenasal cartilage; flattened behind and somewhat arcuate outwards, leaving a pair of long palatonasal spaces between them and the nasal septum and the rostrum. Posteriorly they lie broadly against the rostrum, and then turn a little outwards, becoming pointed where they articulate with the pterygoids. The latter (*pg.*) are sigmoid clubs, flattened in front where they articulate with the cranial axis and the points of the palatines; and subcylindrical behind, articulating by a cupped face with a small knob on the quadrate cartilage, just above its principal condyle. The articulation of the pterygoid with the basis cranii is by the intervention of a distinct plate of cartilage playing against the basipterygoid cranial plate formerly mentioned (§ 532). The tissue in which the palatines and pterygoids are formed does not become true hyaline cartilage before being ossified, although fast tending towards that condition; and more or less cartilage may appear in different parts of these tracts in various birds.

537. The ascending portion of the quadrate cartilage is enveloped in an ectosteal sheath (*q.*): the mandible is becoming covered with its splints, the *dentary, splenial, surangular,* and *angular* pairs (Fig. 63, p. 242): there is no coronoid. The stapes or columella is fully formed, but not ossified. The two small ceratohyals have united in the greater part of their extent, forming a small arrow-head-shaped lingual cartilage. Shaftbones are appearing in the ceratobranchials.

538. This stage is marked by the more perfect moulding of the occipital region, and the great development of the nasal parts. The constriction of the notochord and the ossification surrounding its posterior portion are of great interest. The tympanic cavity is plainly indicated, and the Eustachian tubes acquire a notable position above a pair of bones, the basitemporals. Attention should be especially directed to the origin of these bones and of the rostrum. Few ectosteal centres have appeared; exoccipitals, basisphenoids, quadrates, and ceratobranchials. The principal membrane-bones of the roof of the skull, the palate, and the jaws are now found; but the vomer is not yet in existence.

Fourth Stage: Embryo at end of second and beginning of third week of incubation: head about an inch long.

539. The chondrocranium is rapidly becoming ossified, its general form remaining substantially the same. The notochordal ossification has extended on each side into the parachordal cartilage, but the condyle is still unossified. The basioccipital bone thus originating is broadest in the middle, and is most prominent above along the notochordal line (*b.o.* Fig. 62). The posterior basicranial fontanelle (*p.b.f.*) has become a wide space between the clinoid wall and the basioccipital. The exoccipitals (*e.o.*) have increased in size, and possess a process directed outwards in the tympanic wing; and paired

supraoccipitals (Fig. 60, *s.o.*) have appeared, ossifying most of the cartilage above the foramen magnum.

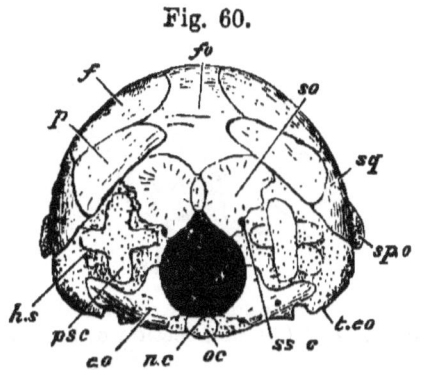

Fig. 60.

Embryo Chick, end of second week of incubation; posterior view of cranium.

o.c. occipital condyle; *nc.* indicates the superior position of the notochordal remnant; *p.s.c.* posterior semicircular canal; *h.s.* horizontal semicircular canal; *t.eo.* tympanic wing of exoccipital cartilage; *sp.o.* sphenotic process; *fo.* supracranial fontanelle.

Bones: *e.o.* exoccipital; *s.o.* supraoccipital; *sq.* squamosal; *f.* frontal; *p.* parietal.

540. The anterior part of the otic cartilage, in the cochlear region and below the anterior canal, is ossified as *prootic* (*pr.o.* Fig. 61), and perforated by the auditory nerve. The outer and hinder view of the much tilted auditory mass shows nothing but cartilage; a small *opisthotic* osseous wedge, however, has appeared in the edge of the capsule adjacent to the exoccipital (*op.*). Already the supraoccipital is ossifying the cartilage around the posterior part of the anterior canal: and the exoccipital is enclosing the posterior canal below.

541. The rostrum, the basisphenoidal ossifications connected with it, and the basitemporals are gradually becoming one bone, together with another considerable pair of osseous tracts (*pretemporal wings*) developed in stroma laid down upon the lingulæ formerly mentioned (§ 531), and extending outwards and backwards to the upper extremity of the quadrate along the anterior margin of the basitemporal (*p.t.p.* Fig. 65, p. 246). The

basitemporals, now uniting, include between them and the floor of the skull, a space in which the Eustachian tube and the carotid artery of either side run forwards

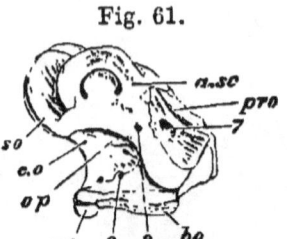

Fig. 61.

Embryo Chick, end of second week of incubation; inner view of hinder part of cranium, with median section of basilar and supraoccipital regions.

o.c. occipital condyle; *b.o.* basioccipital; *pr.o.* prootic; *a.sc.* anterior semicircular canal; *s.o.* supraoccipital; *op.* opisthotic; *e.o.* exoccipital; 7, foramen for facial nerve; 8, foramen for glossopharyngeal and vagus; 9, foramen for hypoglossal.

and inwards. The pretemporal wings and the basitemporal plate meet at an angle, and enclose a considerable cavity anteriorly, continuous with the tympanic; these are the anterior tympanic recesses.

542. The proper basisphenoidal ossification has extended in an annular form round the pituitary space (Fig. 62), and into the clinoid wall and the margin of the posterior fontanelle. The alisphenoid cartilage (*a.s.* Fig. 62) has become complicated in its relationships, being united with the fore part of the ear-mass above and below (the main part of the trigeminal nerve issuing between them); and also continuous by its base with the posterior clinoid wall and the side of the pituitary region. It has an external and posterior sphenotic process (*sp.o.*), and is also considerably curved forwards. A membranous fontanelle (*a.s.f.*) arises in its centre; and both in front and behind this, distinct ossific centres appear in the cartilage: these afterwards unite to form one *alisphenoid* bone.

543. The remainder of the chondrocranium in front of the pituitary body is unossified. The anterior part of

the brain is borne laterally upon membrane and covered by membrane and membrane-bones (see Fig. 63, p. 242); it is supported mesially by the elevated cartilaginous interorbital septum, which has a convex superior edge and a concave posterior margin, on either side of which the optic nerves diverge. The septum has a somewhat thickened lower edge where it rests upon the rostrum. The continuity of the interorbital with the nasal septum is interrupted to a great extent by the development of a vertical hourglass-shaped *craniofacial fenestra* (see Fig. 66, p. 250, in front of *m.e.*) in the cartilage just in front of the termination of the parasphenoidal rostrum. At the same spot the cartilage is expanded into an oval plate at the base, corresponding to the primary breadth of the coalesced trabeculæ.

544. There is another fenestra of great interest in the posterior and upper region of the interorbital septum; it is as large as the last-named, and may be termed interorbital fenestra; its backward termination comes very near to the limit of the cartilage (*i.o.f.* Fig. 63). Yet one other feature of interest must be mentioned; the median cartilage at the extreme anterior end of the brain-cavity sends backwards and upwards a sharp spike in its roof, the sole representative in the Fowl of the ichthyic tegmen: close beneath it the olfactory nerves emerge.

545. There is nothing especially to be noted about the proper nasal septum, except that it is grooved in front of the craniofacial fenestra for a branch of the orbitonasal nerve, which passes from the upper wall of the orbit over the lateral ethmoidal plate into the nasal roof, and then bends downwards and inwards in front of the fenestra, grooving the septum. When it has reached nearly the bottom of the septum, the nerve passes under a small cartilaginous bridge, and turns forwards along the base of the septum to its peripheral distribution. The almost cylindrical prenasal cartilage, to which the septum slopes, is at its highest relative development (*p.n.* Fig. 62).

546. The top of the ethmoidal and nasal septum is

continuous with various outgrowths on either side, arching over and extending into the different regions of the labyrinth. The nasal roof is complete on each side along its

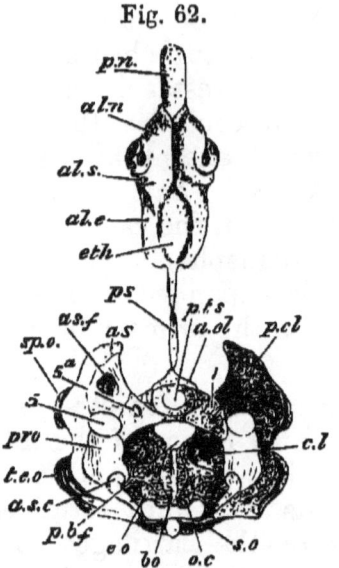

Fig. 62.

Embryo Chick,* end of second week of incubation; upper view of skull, the brain and parostoses being removed.

o.c. occipital condyle; *a.s.c.* anterior semicircular canal; *t.e.o.* tympanic wing of exoccipital; *cl.* cochlea; *p.b.f.* posterior basicranial fontanelle; *pt.s.* pituitary space; *a.cl.* anterior clinoid ridge; *p.cl.* posterior clinoid ridge, outer part; 5, trigeminal foramen; 5a. foramen for its orbitonasal branch; *as.f.* alisphenoid fenestra; *sp.o.* sphenotic process; *p.s.* presphenoidal region, or top of the interorbital septum, the great depth of which is not perceptible; *eth.* ethmoidal region; *al.e.* aliethmoidal cartilage; *al.s.* aliseptal cartilage; *al.n.* alinasal cartilage; *p.n.* prenasal cartilage.

Bones: *b.o.* basioccipital; *e.o.* exoccipital; *s.o.* supraoccipital; *pr.o.* prootic; *a.s.* alisphenoid; around *pt.s.* basisphenoid.

whole length (Fig. 62), but the side-wall growing from it is deficient in certain regions, exposing externally the superior or proper upper turbinal coil and the inferior or lower turbinal of Anthropotomy. There is no "middle turbinal." The lateral ethmoidal region furnishes a large vertical subquadrate antorbital plate, separating the orbit from the nasal cavity. The distribution of the olfactory nerve is over a bag-like involution of the aliethmoidal roof, continuous behind with the antorbital wall. The inferior turbinal coil is united with the postero-inferior angle of the same wall, but is elevated anteriorly. Further forward, in the alinasal region (*al.n.*) there is a curiously sinuous wall, with an alinasal turbinal depending from the roof and curved in correspondence with the sinuosity of the wall. The nasal opening in the cartilage is a long notch passing backwards and upwards, and rounded behind, the alinasal cartilage being enfolded around it everywhere and passing into the alinasal turbinal (see Figs. 69—71, p. 252).

547. The cranial cavity is now partially invested by membrane-bones above, but they are principally lateral, leaving a large median fontanelle. The parietals (*p.* Fig. 60) are directly in front of the ear-capsules and the supraoccipital moieties; they are expanded below, where they are overlapped by the squamosals, and narrower above. The squamosals (*sq.*) are broad squarish plates overlying the lower part of the parietals and frontals, and reaching down to the postfrontal or sphenotic process of the alisphenoid and the top of the quadrate. The frontals (*f.*) are thin hollow shells of bone, covering a large part of the crown of the head; they are bent suddenly inwards along the upper edge of the orbit, forming a considerable orbital plate in each case, but leaving a large membranous space between it and the interorbital septum (Fig. 64). They become sharp-pointed anteriorly over the ethmoidal region.

548. The nasals are flat narrow plates lying over the frontals behind, and on the aliseptal region of the nasal cartilage; they fork anteriorly behind the alinasal region, sending an upper process along the nasal process of the premaxillary, and a lower process down to the maxillary (*na.* Fig. 64). Outside the hinder part of the nasals is a rather broad *supraorbital* ossification (*lch.* Fig. 64), with a downward imperforate spur occupying the lachrymal region and applied to the outer edge of the antorbital plate.

549. The premaxillaries and the rest of the bones of the upper jaw and palate, and also the quadrate and the lower jaw, are further advanced in development, becoming more solidified, and perfecting the various relations already described. A new bone has arisen in the palate, viz. the *vomer* (*v.* Fig. 65, p. 246), which is however but a very small style, lying under the nasal septum, behind the points of maxillopalatine processes of the maxillaries.

550. In several respects the skull at this period reminds the skilled observer of conditions found in the Tinamous. Cartilage-bones have increased in number; there are paired supraoccipitals, prootics, opisthotics, and alisphenoids. The remarkable manner in which the basis cranii is compounded will attract attention; the fenestration of the orbitonasal septum is of scarcely inferior interest. The cartilages of the nasal labyrinth are now modelled as to their main outlines. The supraorbital and the vomerine ossifications are the principal new formations in membrane.

Fifth Stage: The Chick about the second day after hatching.

551. The skull of the ripe chick is interesting in many respects; it is profitable for comparison with the skulls of the Fish, the Reptile, and the immature Mammal. Looked at generally, it comes much closer to the nearest congeners of the typical Fowl than does that of the adult bird. The drawings of this form might, with very little modification, serve as diagrams to illustrate the structure of the skull in any Bird above the Struthionidæ; compared with the Aves Præcoces, the skull is like theirs at the same period; but it corresponds to that of nestlings of the Aves Altrices at about the end of the first week after hatching.

552. The period of a week has sufficed for very great changes in chondrous and fibrous tracts; and now is the best time for catching the true form of many of the osseous

territories, although some have already lost their distinctness, whilst others have not yet appeared. The ectosteal tracts have set up endostosis in the cartilage lying between their inner and outer plates, so that they are being enlarged intrinsically as well as from the immediately overlying (perichondrial) fibrous tissue. Save in the instance of the early-grafted parasphenoidal elements, the parosteal tracts or splint-bones are still altogether free from union with the endoskeletal parts, whether bony or cartilaginous. The hyaline cartilage furnishing the interspaces and headlands of the ectosteal plates has become very dense through the abundance of the cheese-like intercellular substance; it is semitransparent when thick, and for the most part of a lilac colour. The splint-bones are still fibrous, but are beginning to become smooth through the continual ossification of aponeurotic layers. As a correlate of this exogenous growth, the first deposit of bony matter is being absorbed in many places, so as to form diploë; this process is most advanced in the basitemporals and squamosals.

553. The extensive occipital plane, swelling backwards above, is largely ossified, although there are considerable chondrous tracts remaining. The basioccipital extends into the occipital condyle, and it is considerably underfloored by the basitemporal plate in front (Fig. 63). The exoccipitals ossify the lower half of the sides of the foramen magnum. They are very irregular in shape, extending considerably into the ear-cartilage: they are perforated by the vagus and hypoglossal nerves. The supraoccipital centres have coalesced almost completely to form a large bone bounding the upper half of the foramen magnum, which is pointed above; superiorly, the margin of the bone is curved like a fan, and abuts on the parietals.

554. Looking at the internal surface of the cranium, it will be seen that nearly all of the arch of the anterior semicircular canal (*a. s. c.*) lies in the supraoccipital (*s. o.*), being supero-posterior in position, owing to the backward-

tilting of the whole auditory capsule. Nearly the whole of the posterior canal and part of the horizontal lie in the exoccipital.

555. Below and in front of the supraoccipital, the prootic (*pr. o.*) ossifies most of the remainder of the auditory mass: but there is a small *epiotic* lying attached to the prootic in the anterior part of the recess for the

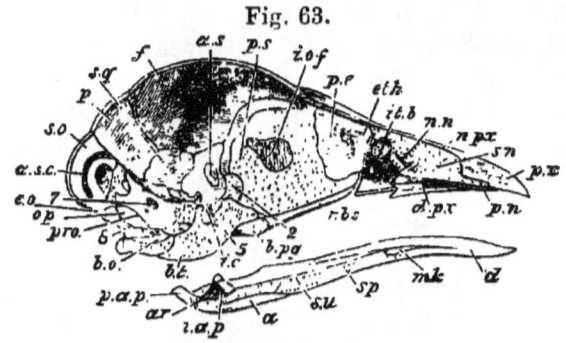

Fig. 63.

Chick two days old; median longitudinal section of skull, the brain being removed.

a.s.c. anterior semicircular canal in supraoccipital bone; *i.c.* foramen for internal carotid artery; *b.pg.* basipterygoid process; *p.s.* presphenoidal region; *i.o.f.* interorbital fenestra; *i.t.b.* inferior turbinal; *n.n.* nasal nerve passing downwards and forwards; *s.n.* nasal septum; *pn.* prenasal cartilage; 2, optic foramen; 5, trigeminal foramen; 7, foramen for facial nerve; 8, foramen for glossopharyngeal and vagus nerves; *mk.* meckelian cartilage, which is also indicated in outline beneath the splenial bone; *p.a.p.* posterior angular process.

Bones: *b.o.* basioccipital; *e.o.* exoccipital; *s.o.* supraoccipital; *op.* opisthotic; *pr.o.* prootic; *b.t.* basitemporal conjoined with basisphenoid; *r.b.s.* rostrum; *p.* parietal; *sq.* squamosal; *f.* frontal; *a.s.* alisphenoid; *eth.* ethmoid; *p.e.* mesetbmoid; *p.x.* premaxillary; *n.px.* its nasal process; *d.px.* its palatal plate; *ar.* articular; *i.a.p.* its internal angular process; *a.* angular; *su.* surangular; *sp.* splenial; *d.* dentary.

flocculus. Between the prootic and the exoccipital is a wedge of bone, the opisthotic, visible internally (*op.*): the vagus nerve (8) enters its foramen between this and the exoccipital. On the outer aspect of the skull the opisthotic can be seen separating the fenestra ovalis from the fenestra rotunda, also growing round the latter. It does not reach the posterior occipital wall. The prootic has a scooped anterior crest, continued on to the ali- and basisphenoid, and helping to form the resting place for the midbrain:

the crest is notched in front for the trigeminal nerve. The large backwardly-projecting plate of the prootic is pierced about its middle by the facial and auditory nerves (7).

556. Immediately in front of the basioccipital and as far as the rostrum the cranial floor is a thick mass of bone, for the coalesced basitemporals have united with the outspread pretemporal wings, and a section shows that their diploë is continuous (Fig. 63, *b.t.*). The basitemporal plate presents a pair of large convexities behind, and its anterior very contracted extremity is notched beneath the junction of the Eustachian tubes, which run forwards in its substance, as before described (§ 541, p. 285). Immediately in front of this region are the two oval cartilaginous basipterygoid articular plates, looking outwards and downwards on the side of the basisphenoid, which is slightly pedunculate on either aspect to carry them. The alisphenoid (*a.s.*) which arose from two centres is now one bone on each side, distinct from the basisphenoid, and retaining its fenestra; it forms a slightly bulging wall to the lower part of the back of the orbit, and is almost transverse in position: it is applied to the prootic behind, and its upper surface articulates with the large orbital plate of the frontal.

557. The rounded rostrum (*r.b.s.*), grooved above for the interorbital cartilage, is still perfectly distinct, reaching to the cranio-facial fenestra, where it curves upwards in front of the base of the ethmoidal cartilage. The latter is now completely separated from the nasal septum below—the fenestra has become a large notch. The interorbital septum and fenestra remain the same as before, but the anterior (ethmoidal) portion of the perpendicular plate is ossified by a double ectosteal lamina (*mesethmoid*), not reaching to the bottom of the septum. This bone (*p.e.*) is thick and solid anteriorly where it bounds the cranio-facial notch, and at its upper angle; behind, it does not nearly reach the interorbital fenestra.

558. From the upper edge of the mesethmoid the cartilage curves down laterally and inwards to form the

swollen upper turbinal; the antorbital wall is as before. The nasal septum has a posterior angulated margin separated by a considerable distance from the fore end of the mesethmoid; but above, there is continuous cartilage, forming a permanent isthmus. The septum sinks antero-inferiorly into the prenasal cartilage (*p.n.*), which within the last three days has become reduced to about a fifth of its former bulk and half its length: this condition is retained till about the middle of the first winter. The nasal septum and turbinal coils are unossified.

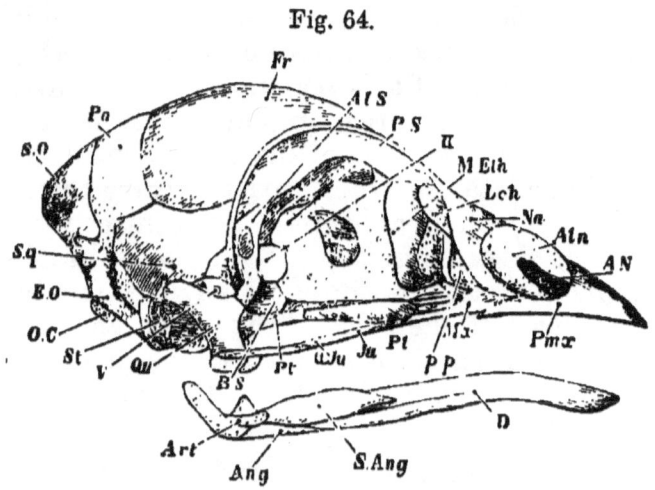

Fig. 64.

Chick two days old; external lateral view of skull.

o.c. occipital condyle; *p.s.* presphenoid region; in front of it, interorbital fenestra; *p.p.* antorbital plate, *al.n.* alinasal cartilage; *a.n.* external nostril; *II.* optic foramen; *V.* trigeminal foramen: *art.* articular surface of lower jaw.

Bones: *e.o.* exoccipital; *s.o.* supraoccipital; *sq.* squamosal; *pa.* parietal; *fr.* frontal; *al.s.* alisphenoid; *b.s.* basisphenoid; *m.eth.* mesethmoid; *lch.* lachrymal; *na.* nasal; *pmx.* premaxillary; *mx.* maxillary; *pl.* palatine; *pt.* pterygoid; *ju.* jugal; *q.ju.* quadrato-jugal; *st.* stapes; *qu.* quadrate; *ang.* angular; *s.ang.* surangular; *d.* dentary.

559. The roof- and wall-bones of the cranium, the frontals and parietals, have grown rapidly, obliterating much of the great upper fontanelle. The parietals (Fig. 64, *pa.*) are narrow transversely-placed oblong plates, fitting accurately to the posterior margin of the frontals

(which they partially overlap), and the anterior margin of the supraoccipital. The parietals are flanked below by a pair of bones larger than themselves, but more irregular, the squamosals (*sq.*) More than half the squamosal can be seen on the inner surface of the cranium, between the oblique lower edge of the parietal and the outer (hinder) edge of the alisphenoid (*sq.* Fig. 63); externally it clamps the sphenotic cartilaginous process of the alisphenoid, and overarches the quadrate and the tympanic cavity.

560. The frontals (*fr.*) are the largest cranial bones; they are very convex above, and their extensive incurved orbital plate is bounded by a sharp semicircular supraorbital ridge. The line formed by this ridge is continued farther down in front by the suture of the nasal with the lachrymal, and behind by the junction of the squamosal with the sphenotic process, thus bounding the very regular socket for the large eyeball. A membranous region remains in the cranial floor between the edge of the orbital plate of the frontal, the alisphenoid, and the interorbital septum, from the ethmoidal region to the common optic foramen. The frontals are approximated anteriorly to the coronal (parieto-frontal) suture, and then again diverge and become pointed in front, leaving the top of the mesethmoid uncovered,—a persistent condition.

561. The broad posterior edges of the nasals (*na.*) cover the pointed anterior extremities of the frontals: they are twisted upon themselves, and extend downwards and outwards and then forwards below the alinasal cartilages (*al. n.*); they also send a sharp fork above the same, by the side of the nasal processes of the premaxillary. The supraorbital plate of the lachrymal (*lch.*) is large and oval; its anterior portion remains slender.

562. The premaxillaries (*pmx.*) have coalesced into a single strong triangular bone, which has a solid rostral portion in front of the nasal septum and equal to it in length. Its nasal processes are united where they pass between the nares, but permanently distinct behind and above, where they reach the mesethmoid. These

processes are longer than the inferior sharp dentary margin (Fig. 65, *d.px.*), which is one-fourth longer than the palatal processes (*p.px.*). Each palatal process passes under the apex of the palatine bone. The axial part of the premaxillary is grooved below, and the hinder half of the groove is filled by the small prenasal cartilage (*p.n.* Fig. 63).

563. The slender maxillary (Figs. 64, 65) now rises in front into the angle between the descending crus of

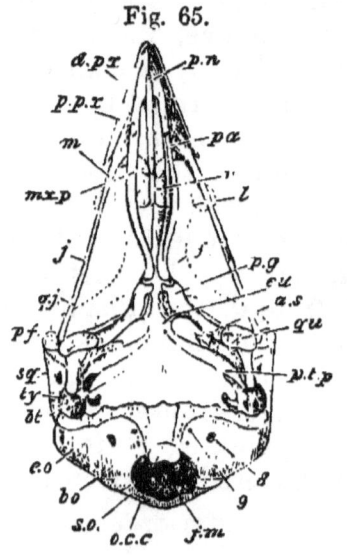

Fig. 65.

Chick two days old; under view of skull, with lower jaw removed; the upper bones are shown in outline on either side of the interorbital septum.

oc.c. occipital condyle; *f.m.* foramen magnum; *eu.* anterior opening of Eustachian tubes after traversing basitemporal bone; *pn.* prenasal cartilage; *pf.* sphenotic process; *ty.* tympanic cavity; 8, foramen for glossopharyngeal and vagus nerves; 9, foramen for hypoglossal nerve.

Bones: *b.o.* basioccipital; *e.o.* exoccipital; *s.o.* supraoccipital; *b.t.* basitemporal; *p.t.p.* pretemporal wing of basisphenoid; *qu.* quadrate; *sq.* squamosal; *a.s.* alisphenoid; *v.* vomer; *pa.* palatine; *pg.* pterygoid; *d.px.* dentary process of premaxillary; *p.p.x.* palatal process of premaxillary; *m.* maxillary; *mx.p.* maxillopalatine plate of maxillary; *j.* jugal; *q.j.* quadratojugal; *l.* lachrymal; *f.* frontal.

the nasal bone and the hinder edge of the dentary process of the premaxillary. The maxillopalatine plates (*mx.p.*

Fig. 65) are broader and reach nearly to the mid line, being separated partly by the nasal septum and partly by the small vomer, which is rounded in front, and split for a short distance behind. The forks of the vomer (*v.*) articulate with the inner and anterior points of the inner plates of the palatine bones, which lie side by side mesially, nearly concealing the rostrum. Behind these plates the palatine bones (*pa.* Fig. 65) are thickish angular rods, both together having an elongated lyriform shape. The hinder extremities are curved slightly outwards for articulation with the pterygoids. The latter (*pg.* Fig. 65) are short stout mallet-shaped bones, the handles being turned outwards and a little backwards to articulate by a cup with a little ball on the quadrate (§ 536), and also sending up above this a small epipterygoid plate. The head of the mallet is notched in front to receive the rounded posterior end of the palatine. The inner face of the head has a distinct oval plate of cartilage upon it, exactly corresponding to and articulating with the basipterygoid plate on the basisphenoid.

564. The quadrate (*qu.* Figs. 64, 65) has the apex of its orbital spur unossified; a similar condition is persistent in the Chelonia. The otic process has two cartilaginous surfaces close together; a rounded upper and outer head fitting in a cupped space on the squamosal, and a smaller one gliding on a cartilaginous facet of the prootic. There is a slight projection on the front of the otic process, but there is no postero-internal head as in most Birds, the relation to the tympanic wing of the exoccipital being lost (§ 521). The shaft of the quadrate is short, and bears a rounded condyle, almost transverse, and divided into a larger outer and a lesser inner portion. The quadrate is connected with the maxillary by the intervention of slender jugal and quadrato-jugal bones (*j., q.j.*), the latter being bent inwards to clasp the side of the quadrate above the condyle. The base and shaft of the stapes are ossified, and its outer end is a triradiate cartilage. The tympanic cavity will be described in the adult.

565. A new bone has appeared in the mandible—a rudiment of the *articular* (*art*. Figs. 63, 64); it projects inwards from the condylar region as an internal angular process, tipped with cartilage (*i.a.p.* Fig. 63). The posterior angular process (*p.a.p.*), and the shaft of the meckelian cartilage, which is three-fourths of the length of the jaw, are unossified. The dentaries of opposite sides have coalesced; they lie outside the cartilage and partly embrace it; their length is three-fourths of that of the jaw, and they are forked behind. On the inner face of the cartilage is a long-oblong splenial (*sp.*), leaving both the point and the hinder part of the cartilage bare. The angular below and the surangular above the posterior part of the cartilage leave its inner face considerably exposed, but they cover it externally, and meet.

566. Ossification has appeared in the two small cornua (minora) of the hyoid, and the ossification is continuous across the lingual cartilage. The median basal cartilage has two styliform bones, one in front of and the other behind the "thyrohyals" (=1st branchials). The only ossification in the latter cartilages in the last stage was a lower one on either side; there is now a second upper bone, but the curved tip is cartilaginous, as are the apices of all these osseous tracts.

567. The types suggested by a view of the chick's skull at this stage are the Hemipodiine, the Pterocline, and the Columbine outliers of the Gallinaceous group: subsequent development removes the Fowl greatly from its lower and more generalized relatives on one hand, and from its higher and more specialized congeners on the other. Every detail which has just been given will be found pregnant with meaning by the student who has an affinity for this subject. The coalescence of the various bones in the basisphenoidal region, the extension of the exoccipitals and the supraoccipitals into the ear-capsules, and the appearance of the epiotics and mesethmoid, are the chief events that have affected the brain-case. The

prenasal cartilage is relatively much reduced. The premaxillaries and dentaries respectively now form but one bone; and bony deposit has arisen in the stapes, the articular region of the mandible, and the small representatives of the hyoid and branchial arches.

Sixth Stage: Young Fowls up to nine months old.

568. In a chick three weeks old there are two features to be noted, (1) a distinct osseous centre in the *sphenotic* process (§ 542), subsequently uniting with the alisphenoid, and (2) the appearance of a small *opisthotic* plate on the outer occipital plane, in a notch of the superoexternal part of the exoccipital, soon becoming united with it.

569. In a chicken of two months old the structures have increased in massiveness, and the roofing bones of the skull are connected by delicately-toothed sutures; only narrow synchondrosial tracts are left in the cranium. An inner view of the occipito-otic region shows the three bones of the periotic capsule, as well as the alisphenoid, still remaining distinct; nor has the basisphenoid yet coalesced with the basioccipital or with the alisphenoid.

570. Above the optic foramen, wedged in between the alisphenoid and the interorbital septum, a four-sided bone has arisen in membrane on each side, and there is a smaller pair in front of and above the larger, helping to fill in the fenestræ left unoccupied by the orbital plates of the frontals. These are the anterior and posterior *orbitosphenoids* (o.s. Fig. 66). The anterior half of the interorbital septum is ossified, the mesethmoid encroaching on the anterior margin of the interorbital fenestra.

571. In chickens from three to four months old all the splint bones of the skull, with the exception of the early engrafted parasphenoidal elements, are separable by maceration; only the symmetrical premaxillaries and dentaries have united at the mid line, the former very early, the latter soon after hatching. Looking at the

skull below and behind, externally, the large supraoccipital still shows sutures with the exoccipitals, and they with the basioccipital—the latter being still distinct from the basisphenoid and its basitemporal plate. The posterior basicranial fontanelle is indicated by a deep median longitudinal chink in the posterior part of the basisphenoid. The complex basisphenoid is distinct from the alisphenoids, prootics, and basioccipital; but the small epiotic is uniting with the prootic and supraoccipital, and the opisthotic with the prootic and exoccipital. The sphenotic is confluent with the alisphenoid. The composition of the occipital condyle of three elements is still manifest.

572. In young fowls of the first winter, from seven to nine months old, a large number of embryonic characters persist (Fig. 66). The skull has become very thick and solid, with copious coarse diploë. The cranium proper is

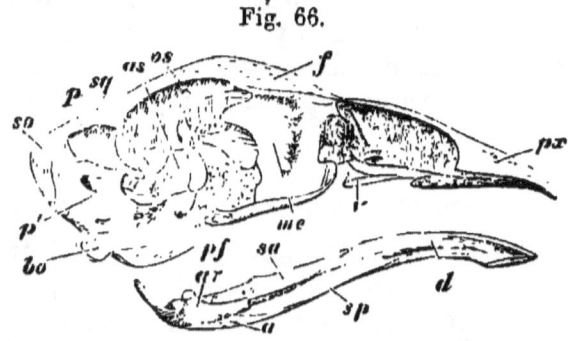

Fig. 66.

Fowl of first winter; median longitudinal section of skull.

b.o. basioccipital perfectly continuous with basitemporal, basisphenoid, and rostrum; *p'.* prootic; *s.o.* supraoccipital; *p.f.* pituitary fossa; *p.* parietal; *sq.* squamosal; *a.s.* alisphenoid; *o.s.* orbitosphenoid; *f.* frontal; *m.e.* mesethmoid; *v.* vomer; *px.* premaxillary; *d.* dentary; *sp* splenial; *s.a.* surangular; *a.* angular; *ar.* articular.

Cartilage is seen in the interorbital septum behind the mesethmoid; in the nasal septum in front of that bone, sending forward the small prenasal spur; and in the lower jaw slightly between the angular and surangular.

not half the length of the elongated skull, and is extremely narrow where it ends above the interorbital septum. The bones of the occipital ring, the periotic mass, and the basisphenoid are entirely confluent, as also are the anterior parts of the premaxillaries and dentaries. All the other lines of suture are still traceable.

573. In the more advanced specimens a small patch of bone can be seen between the orbitosphenoids in the cartilage of the septum, between the interorbital fenestra and the optic foramen; this is the first appearance of the *presphenoid*. Cartilage remains beneath this bone, in front of the interorbital fenestra, down to a vertical suture which divides the anterior part of the basisphenoid from the postero-inferior part of the mesethmoid: this connection is above the original rostrum, and is produced by extension of ossification. The mesethmoid ($m.e.$) has expanded superiorly into the aliethmoidal cartilage on either side, forming a flat plate, which can be seen between the frontals and nasals and behind the nasal processes of the premaxillaries (see Fig. 70). The mesethmoid ossifies the posterior part of the aliethmoidal roof, and extends behind into the spike mentioned in § 544: below this is a groove for the olfactory nerve, passing beneath the alar expansion on either side, in the angle between it and the vertical plate.

574. The rest of the nasal labyrinth is entirely unossified. The hinder part of the roof is flattened above, and its lateral edges grow inwards to form the upper turbinals ($u.tb.$ Figs. 69, 71), which near the middle line are again bent outwards, forming a bullate pouch opening outwardly; its edge is confluent behind with the antorbital wall. In this region the nasal septum is deficient by reason of the craniofacial notch, except for a slight cartilaginous ridge above, separating the nasal branches of the trigeminal nerve ($n.n.$) as they pass forwards to reach the septum.

575. No middle turbinal coil is developed on the face of the antorbital plate below the upper turbinal. In front of and below the latter, the inferior turbinal ($i.tb.$ Figs. 68, 71) forms a not very uniform scroll, coiled twice, the blind face of the coil being inwards, and its long axis

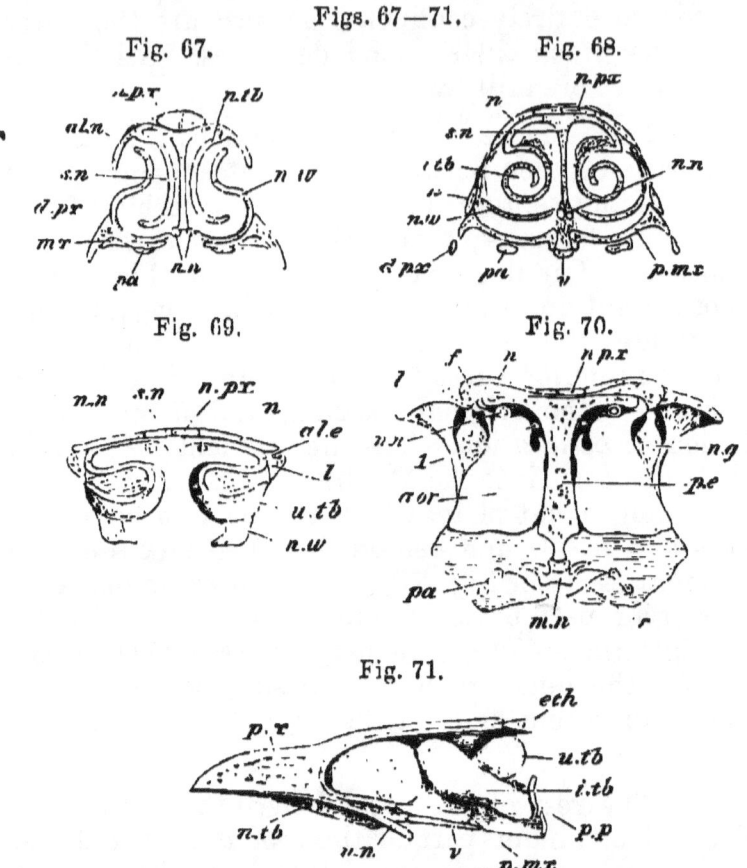

Views of nasal structures of Fowl of first winter.

Fig. 67. Transverse section through most expanded part of alinasal region; (*n.tb.* Fig. 71).

Fig. 68. Transverse section between aliethmoidal and aliseptal regions (through *i.tb.* Fig. 71).

Fig. 69. Transverse section in aliethmoidal region, with inferior turbinal cut away, in region of cranio-facial notch of septum (through *u.tb.* Fig. 71).

Fig. 70. Transverse section through most solid part of ethmoid, showing antorbital plate (*pp.* Fig. 71).

Fig. 71. View of nasal labyrinth from within, the septum being removed; less magnified than the preceding figures.

s.n. nasal septum; *al.n.* alinasal cartilage; *n.tb.* alinasal turbinal; *n.w.* nasal wall; *i.tb.* inferior turbinal; *u.tb.* upper turbinal; *al.e.* aliethmoid cartilage; *a.or.* (Fig. 70), *pp.* (Fig. 71), antorbital plate; *n.g.* nasal gland; 1, olfactory nerve; *n.n.* nasal nerve (from trigeminal), appearing as displaced from relation to the septum in Fig. 71; *m.n.* middle narial passage.

Bones: *n.px.* nasal process of premaxillary; *d.px.* dentary process of same; *n.* nasal (in Fig. 68 the lower fork of the nasal is cut through and seen separate from the upper); *f.* frontal; *l.* lachrymal; *p.e.* mesethmoid; *eth.* ethmoid; *v.* vomer; *pa.* palatine; *r.* rostrum; *mx.* maxillary; *p.mx.* palatal plate of maxillary.

directed obliquely. It grows out of the nasal roof and outer wall in front of the aliethmoidal region. Behind and below this, the nasal wall becomes imperfect for a space; and the hinder termination of the inferior turbinal is confluent with the lower outer angle of the antorbital plate.

576. Another turbinal occupies the fore part of the nasal cavity; it is the alinasal turbinal (*n. tb.* Figs. 67, 71), whose blind face is also inwards; at most it only forms half a coil. It arises from the outer edge of the nasal roof, but outside it the alinasal cartilage (*al. n.* Fig. 67) forms a complete lateral wall, continuous with the nasal floor except where it is slit obliquely for the anterior nasal opening. Posteriorly the alinasal turbinal, where it lies below the anterior part of the inferior turbinal, grows inwards, and joins a remarkable wing of cartilage continuous with the nasal septum, and pierced on either side by the nasal nerve (this is a development from the original bridge over this nerve, § 545). Anteriorly, the upper part of the alinasal cartilage terminates as a small ear-like process on either side, and between these the prenasal cartilage, which has not lessened since the time of hatching, lies in a groove of the premaxillary. The whole alinasal tract lies in an ovoid space formed between the processes of the premaxillary, and also partly in the notch between the processes of the nasals. In the Fowl the aliseptal region is very small in comparison with the aliethmoidal behind and the alinasal in front of it.

577. The bones of the palate have merely grown larger without altering their relations; those of the lower jaw are beginning to coalesce, and the posterior as well as the internal angular process is ossified by the articular.

The Skull in Fowls several years old.

578. By the time the fowl is a year old most of the cranial sutures are completely closed; and then, year by year, the skull becomes more and more dense, the periosteal layers filling in all spaces in the bony walls except those needed for the transmission of vessels and nerves. Thus in an aged bird (Figs. 72, 73) the occipital, otic, posterior and anterior sphenoidal, and ethmoidal regions have become one continuous bone, with the old landmarks all removed; and scarcely a sign is left of the once highly complex conditions of these parts. This state of things is quite normal for a Bird; and although carried to a very high degree in the Fowl, yet in some birds the obliteration of all signs of the once composite condition of the skull is still more perfect.

Fig. 72.

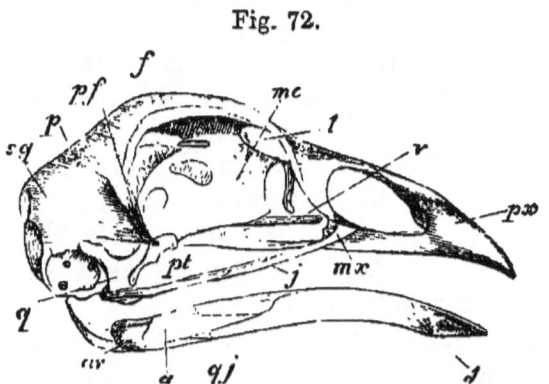

Fowl several years old; side view of skull.

Bones: *sq.* squamosal; *p.* parietal; *pf.* sphenotic process; *f.* frontal; *me.* mesethmoid; *l.* lachrymal; *px.* premaxillary; *mx.* maxillary; *v.* vomer; *j.* jugal; *q.j.* quadrato-jugal; between *v.* and *j.* palatine; *pt.* pterygoid; *q.* quadrate; *ar.* articular; *a.* angular; *d.* dentary.

579. In the side view of the skull of an old male (Fig. 72) it is seen that the face has gained considerably upon the cranium as compared with the condition found in the newly-hatched chick; although the maxillary bones (*mx.*) are relatively smaller than ever, being minute styles with a very small maxillopalatine ingrowth.

580. In an upper view the anterior part of the suture between the frontals is visible: the nasals have coalesced behind with the frontals, but are distinct from the forked nasal process of the premaxillary. Between these bones (frontals, nasals, and premaxillary) a considerable pentagonal tract of the top of the ethmoid is apparent. The following bones are permanently separate: lachrymals, maxillaries, jugals, quadrato-jugals, quadrates, pterygoids, palatines, vomer.

581. The whole interorbital plate ($m.\,e.$) has become one bone. In some old birds a trace of the interorbital fenestra can be seen; in others its outline is distinguishable, but it is filled up with periosteal bone. The alisphenoidal fenestra is also occupied by bone, and the alisphenoid has coalesced with the orbital plate of the frontal. Above this, the two pairs of orbitosphenoids, the presphenoid, and the anterior part of the orbital plates of the frontals, have coalesced together; and the fenestra which formerly existed on either side, between the edge of the orbital plate and the interorbital septum, is obliterated excepting for the antero-superior chink by which the olfactory nerve runs forwards on either side of the top of the mesethmoid.

582. In the cartilaginous isthmus above the craniofacial notch (§ 557) two small osseous centres have appeared, one behind the other; and there is a small nodule on either side between them. The rest of the nasal labyrinth is unossified save that in some cases a little bony matter passes from the mesethmoid to the antorbital plate, which otherwise hangs down from the ethmoidal roof free from the median parts.

583. An aponeurotic tract extending forwards from the middle of the outer surface of the squamosal, has become ossified, and meets and often coalesces with the sphenotic process, forming a temporal bridge. The occipital plane is much more vertical than in former stages, and the auditory capsule has recovered from its great obliquity, and is more erect than in the second stage. A

partial bony septum passes between the halves of the forebrain from the line of coalescence of the frontals, and there is a transverse septum corresponding to the coronal or parieto-frontal suture, to some extent separating the fore from the midbrain. A sharp crest passes from the projecting ridge of the prootic (in front of the meatus internus) inwards and forwards to the sides of the posterior clinoid wall of the basisphenoid. The arch of the anterior semi-circular canal, which is now nearly vertical, is connected by a ridge of bone with the fronto-parietal septum.

Fig. 73.

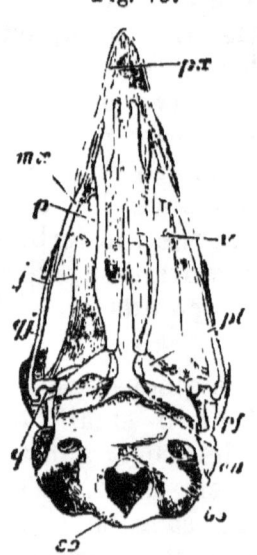

Fowl several years old; under view of skull with lower jaw removed.

so. supraoccipital; *b.o.* basioccipital; *eu.* orifice of Eustachian tubes; *pf.* sphenotic process; *pt.* pterygoid; *p.* palatine; *v.* vomer; *px.* premaxillary; *mx.* maxillary; *j.* jugal; *q.j.* quadrato-jugal; *q.* quadrate.

584. The basisphenoidal region, including the basitemporal plate, is of great thickness, and the thickening of the cranial floor reaches back nearly to the occipital condyle. The pituitary fossa is a deep well, elegantly rounded, turning backwards below, beneath the posterior clinoid wall; and it is perforated by the internal carotid arteries. One of these arteries enters the basis cranii outside

and in front of the exit of each vagus nerve, in a notch between the temporal wing of the exoccipital and the outer and posterior angle of the basitemporal, and these two bones coalesce together to form a bridge outside the artery. When the outer table of the bone, together with a portion of the diploë, is removed in the auditory region, the whole membranous labyrinth is seen to be enclosed in a bony case exactly corresponding to the membranous structures within, the cochleæ appearing in the floor of the skull at a moderate distance from each other, like little bon...

585. Besides the anterior tympanic recess which reaches forwards above the narrow Eustachian tubes (§ 541), there is at the summit of the tympanic cavity an upper tympanic recess[1], running into the diploë of the occiput on each side. The base of the stapes and the two fenestræ (ovalis and rotunda) lie at the bottom of a ... recess in the tympanic cavity.

6. The tympanic membrane is attached to a delicate f bone some distance within the margin of the bony tympanic hollow produced by the tympanic wing of the exoccipital: this margin is extended outwards by a thick fold of skin to form the proper meatus auditorius externus. The fine rim of attachment for the membrane can be traced just outside the three principal recesses of the tympanum, the anterior tympanic, the posterior tympanic, and the fenestral recess. The bony regions which furnish the attachment are the occipito-otic and basitemporal. The membrane is fixed in front to a strong aponeurosis passing down from the squamosal to the extremity of the pretemporal wing of the basisphenoid (§ 541), behind the head of the quadrate. This aponeurosis becomes ossified in old birds, thus excluding the quadrate from direct relation with the tympanic cavity in this type.

[1] In the Crocodile these recesses meet above the occipital arch between the outer and inner tables of bone.

In birds which have a second head on the otic process of the quadrate, the aponeurotic band here spoken of is attached to the quadrate and the hinder head is included in the cavity; the band itself frequently becomes the seat of tympanic ossifications, of which the number varies from one to seven in different types.

587. The columella (Fig. 74) is a very delicate styloid bone, having an ovoid stapedial plate in the fenestra ovalis. Externally it is cartilaginous, where it abuts against the middle of the tympanic membrane and distends it. The stapedial plate (*st.*) and the shaft of the columella or mediostapedial (*m. st.*) form a continuous bone. Distally there are three cartilaginous rays. Two outer ones, the

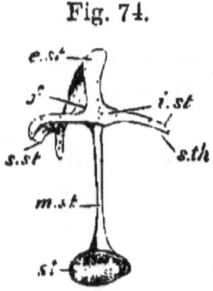

Fig. 74.

Adult Fowl, side view of columella auris.

St. stapedial (bony) plate; *m.st.* mediostapedial (bone); the remaining parts are cartilaginous; *e.st.* extrastapedial; *s.st.* suprastapedial; *f.* fenestra between these bars; *i.st.* interstapedial; *st.h.* stylohyal.

supra- and infrastapedial (*s. st.*, *i. st.*), pass at right angles from the shaft. The upper is bifurcate, and is attached by a membranous band to the posterior part of the tympanic roof. The third or median process (extrastapedial, *e. st.*) is broader than the rest, spatulate, and decurved. Extending from its outer edge to the extremity of the suprastapedial is a bar of cartilage, a fenestra (*f.*) being left in the angle between the extra- and suprastapedial. This bar lies against the middle and anterior part of the tympanic membrane towards the quadrate. The small infrastapedial is directed downwards and a little forwards, and is sometimes expanded below. This lower region was distinct in an early stage (§ 523), but has become confluent with the infrastapedial. It represents the top of the stylohyal (*st. h.*).

588. The mandible varies considerably in different individuals; in some specimens there is scarcely a trace of the great fenestra seen on either side in the Tetraonidæ between the forks of the dentary and the hinder elements. There is usually some remnant of the suture between the posterior wedge of the dentary and the angular and surangular; but all the other sutures close.

589. The cornua (minora) of the hyoid have coalesced to form a glossohyal: but the bone is tipped with cartilage anteriorly, and a chink-like foramen remains in the median part of the bone. This elegant sagittiform tongue-piece

Fig. 75.

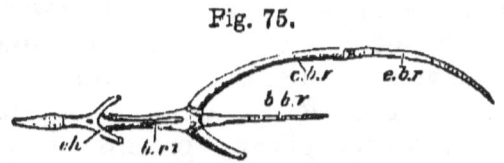

Fowl several years old; hyobranchial apparatus from above.
C.h. ceratohyal; *br.* 1, first basibranchial; *b.br.* second basibranchial; *c.br.* ceratobranchial; *e.br.* epibranchial.

(Fig. 75, *c. h.*) articulates by a synovial joint with the first basibranchial (*br.* 1), which itself has a similar articulation with the long slender proximally-ossified second basibranchial (*b. br.*). The branchial arch articulates on either side by a synovial joint with the posterior end of the first basibranchial, but the proximal (ventral) piece or ceratobranchial (*c. br.*) is united to the distal or epibranchial (*e. br.*) by a mass of fibrous tissue: the ceratobranchial is cartilaginous above, and the epibranchial at both ends.

590. SUMMARY. A striking feature in this history is the acceleration of the determining events; the characters which according to the theory of evolution are due

to very far-reaching inheritance, are assumed with great rapidity, the skull being definitely Sauropsidan at the end of the first week of incubation. Those details of structure, however, by which the Fowl is differentiated from other birds, are much later and slower in appearing, several of them being only gradually acquired after the first year of life. There is no fact in the early development of the Fowl's skull which does not illustrate or receive illustration from the previous histories. This book might be indefinitely enlarged if these correspondences were all pointed out: but the student who patiently meditates will perceive them for himself in due time.

591. Commencing with the simplest stage, we find the axial parts at the base of the skull developed continuously, as if the disjunction found in lower types were overpassed and their later condition of union at once attained. Conversely, distinct elements formed out of simple bars in other forms are here separate at first, without the necessity of any process of segmentation. That which was an event of development before has now become a primary condition. Again, the first appearance of the hyoid and branchial arches in the Fowl corresponds with the condition attained after metamorphosis in the Frog.

592. It is to be remarked how small a proportion of the brain-case itself is preformed in cartilage. The same elements are present as in previous cases, but their proportionate size in the structure is much less, and their total share in the completed skull is really small. Yet unity of plan is manifest in the occurrence of the same bony centres in cartilage, even if they form but insignificant nodules. Together with these facts in relation to the chondrified parts, we notice the increased size of the membrane-bones, the perfection of their adaptation in the adult organism, and their intricate interlocking with cartilage-bones. If the parasphenoidal elements in the Fowl are compared with the parasphenoid as previously

described, their essential likeness as well as their special diversities will be better understood.

593. In the first stage we see an elementary skull, already advanced to a condition reached in lower forms by coalescence and segmentation. In the second stage the principal outlines of the skull are visible: the occipital ring and the orbitonasal septum are in existence. There is a cleft palate like that of Struthious birds, and the angular processes of the mandibular cartilage indicate the type to which the Fowl belongs. In the next period numerous membrane-bones have appeared, and a few centres in cartilage; the nasal region is highly developed. In the fourth stage, still before hatching, the skull has attained many of its bones, the vomer being a notable addition; and the bony cranial floor has advanced in its remarkable process of composition. It is not till after hatching that some notable osseous centres appear, and the coalescence of bones becomes very marked: this goes on gradually until adult age, when most of the sutures are obliterated. The proportionate length of the precranial part of the skull increases, the bones become thick and solid, and mostly smooth externally. Much of the nasal labyrinth remains cartilaginous throughout life; some cartilage remains in the columella, the mandible, the palatopterygoid tracts, and the hyobranchial apparatus. Although bony deposit appears in the palatine and pterygoid tissues before true cartilage is formed in them, yet wherever in these tracts the ossification does not extend, cartilage is subsequently found.

APPENDIX ON THE SKULLS OF BIRDS[1].

594. Notwithstanding the great varieties of outward form to be seen in the skulls of Birds, there is extremely little modification in the general morphological relations of parts. The palatal structures present the largest amount of change, yet some of the features of importance in that respect are only to be discerned after very careful study. In Struthious birds there is a much greater persistence of sutures than in higher forms; the parietals are notably large. The rostrum is of greater size proportionally than in any other group of Birds. The cranio-facial notch is absent. The body of the basisphenoid intervenes (below) between the openings of the Eustachian tubes; in nearly all other birds the basitemporals meet below these tubes so that they open together close to the middle line. Only in some of the Struthionidæ are the Eustachian tubes at all enclosed in bone, and this occurs by the projection of a ridge of the basitemporal to meet another from the "pretympanic wing." The mesethmoid ossifies the whole interorbital and nasal septum. The otic process of the quadrate is not divided into two heads; its anterior fork remains tipped with cartilage in the adult. In this last particular, as well as in the structure of the palate, the genus *Tinamus* agrees with all Struthious birds, and this compound group may be spoken of as *dromæognathous* (*Dromæus*, the Emeu). The vomer is very broad, and unites in front with the maxillopalatine plates, while behind it receives the posterior extremities of the palatines and the anterior ends of the pterygoids, which are thus prevented from entering into any extensive articulation with the rostrum. There are strong basipterygoid (bony) processes from the basisphenoid, and these articulate with the pterygoids near their *outer* or hinder ends. The basitemporals are but feebly developed.

595. All other types have the posterior ends of the palatines and the anterior ends of the pterygoids articulating directly with the rostrum. The Fowl belongs to the *schizognathous* type (having a cleft palate). The vomer tapers to a point anteriorly; while posteriorly it embraces the rostrum. The maxillopalatine plates

---

[1] See Huxley "On the Classification of Birds," *Proc. Zool. Soc.* 1867, and "On the Alectoromorphæ," *ib.* 1868; also Parker, Art. "Birds," *Encycl. Brit.*, Vol. III.

leave a broader or narrower fissure in the palate between themselves and the vomer. The restricted Gallinaceous group (or Alectoromorphæ) excludes the Pigeons and Tinamous. The other principal schizognathous types are the Plovers, the Cranes and Rails, the Gulls, Pigeons, and Auks, and the Penguins. In Pigeons and Sandgrouse the vomer is absent. Some, as the Albatross and the Gull, possess an os uncinatum or antorbital bone ; others, as the Humming-Birds and Kagu, have septo-maxillaries. All except the Alectoromorphæ have mesopterygoids.

596. A group named *desmognathous* is distinguished by the following characters: vomer mostly small or absent; maxillopalatine plates united across the middle line, either directly or by the intermediation of ossifications in the nasal septum. Falcons and Geese are examples of what may be termed direct desmognathism; the maxillopalatine plates meet below at the middle line as in the mammal. In the Falcon the nasal septum is anchylosed to this hard palate; in the Goose it remains free. Indirect desmognathism is exemplified in Eagles, Vultures, and Owls; the maxillopalatine plates are anchylosed to the nasal septum, but are separated from one another by a chink. In *Dicholophus cristatus* the maxillopalatine plates are united by harmony suture and not by coalescence. In *Megalæma asiatica* they are closely articulated with, and separated by, a median ossification beneath the nasal septum, the "median septo-maxillary." The most advanced condition may be called double desmognathism, where the palatines as well as the maxillopalatine plates coalesce to form a continuous palate; this is seen in *Podargus*, and to a less extent in the large Hornbills. In Ducks and Swans a pair of ossicles is found between the palatines, stretching towards the maxillopalatine plate. In several *Ardeidæ* an additional bone, the postmaxillary, is formed behind the angle of the maxillary (this occurs also in the Emeu). In various desmognathous forms the palatines unite for a considerable distance behind the posterior nares, and send down a vertical crest at their junction. This is largest in the Pelican, where all the parts in front of the very mobile cranio-facial hinge are anchylosed into one mass, and the nasal labyrinth is in its most aborted state. In *Phœnicopterus* the median palatine keel is exquisitely thin in front, but where the palatines extend beneath the rostrum, a separate lamella is sent down from each bone and the two are bound together by fibrous

tissue, and between them the vomer is wedged: the ethmopalatine spurs are enormously long. In *Sula alba* the basitemporals are relatively as small as in *Dromæus;* the small Eustachian tubes, uncovered by bone, open at a little distance apart, in a wide shallow fossa where the three parasphenoidal elements meet. There is a postmaxillary; the cranio-facial hinge almost rivals that of the Parrot and Toucan; the columella auris is very long and bent, with a long attenuated cartilaginous infrastapedial process, terminated by a bony fusiform stylohyal. The hinge for the mandible is very far back, as also in the Cormorant.

597. In the (desmognathous) *Aëtomorphæ*, in addition to the hooking of the beak, there is another character; the maxillopalatine plates are generally united with an ossification beneath the nasal septum, or "median septo-maxillary." The vomer is azygous; the palatines often have a mediopalatine where they unite, or a pair of mesopterygoids. In some forms the supra-orbital process of the lachrymal is very large; in others (Hawks, &c.) there is a distinct supra-orbital at its extremity. In the Sparrow-Hawk distinct pterotic and sphenotic centres are developed; and the orbitosphenoids are preceded by cartilage.

598. The desmognathous *Parrots* are very uniform, having the most complete cranio-facial cleft, with a perfect hinge-joint between the frontal and nasal regions. There is no vomer; the palatines are vertically elongated posteriorly, while anteriorly they are horizontally flattened, and movably united with the rostrum. The lachrymal and postorbital (or sphenotic) bend towards one another, and unite by the intervention of a large os uncinatum or antorbital. In some also the temporal fossa is bridged over by the union of the zygomatic process of the squamosal with the os uncinatum. The nasal septum is a thick wall of bone; an annular ossicle is found in the alinasal cartilage of *Melopsittacus undulatus;* in *Palæornis torquata* this part is largely ossified and anchylosed to the upper jaw, and the alinasal turbinal is partly calcified. The Hornbills and Toucans have a median septo-maxillary in front of the azygous vomer. In *Podargus,* when the lower palatine floor is cut away, there are to be seen three small ossicles, the vomer (anterior), and two medio-palatines. In Kingfishers and Hoopoes there is no vomer.

599. The *ægithognathous* type (including the Passerines) is

distinguished by the union of the vomers with the alinasal wall and turbinal; and the embryo at least has a pair of "upper labial" cartilages, which the vomers ossify either partially or entirely. In most cases there are septo-maxillaries, attached to the angles of the vomers. In the fledgling Rook the basitemporal plate is an almost transverse band of bone; the rostrum bears no basipterygoid processes; the cranio-facial notch is nearly complete, but the nasals and nasal processes of the premaxillaries are set into the frontals as thin spurs. The palatines develope cartilage at their hinder (outer) angle, and this part ossifies separately and late, like a transpalatine. The mesopterygoid spur of the pterygoid coalesces with the palatine. The maxillopalatine processes are hooked and flattened. There is a chain of seven small tympanic bones, the principal of which becomes a perfect ring in the Crow, and surrounds the siphonium. In many Ægithognathæ the lachrymal cannot be seen at any stage; in many that have it, it soon anchyloses either with the nasal in front or with the ethmoid behind.

600. There is incomplete ægithognathism in the *Turnix* group; the large labials are imperfectly ossified by the vomers, and these bones are only strongly attached to the nasal labyrinth by fibrous tissue. The following forms of complete ægithognathism may be distinguished: (1) in *Pachyrhamphus*, *Pipra*, &c., the labials are often only imperfectly ossified by the vomers, and these centres are distinct from those ossifying the alinasal cartilages; but the union of the two tracts is complete; (2) in a great number of the Passerines, and in the Swifts, the labials are small and completely ossified by the vomers, but the bony deposit extends largely into the alinasal wall and turbinals; (3) a compound condition, namely, desmognathism in a perfectly ægithognathous type, the maxillopalatines uniting with a highly ossified alinasal wall and nasal septum (*Gymnorhina*, *Paradisœa*, *Artamus*); the transpalatine process is a long spike.

601. The *Saurognathous* group (Woodpeckers and Wrynecks) have their palatal structures arrested at a most simple and Lacertian stage, whilst in other respects they are metamorphosed and specialised beyond any other kind of birds. The principal characteristics are as follows: the retention and ossification of the trabecular cornua; the great number and distinctness of the vomerine series of bones (vomers, septo-maxillaries, median septo-

maxillary); absence of a distinct mesopterygoid; presence of a mediopalatine; no distinct transpalatine; abortive maxillopalatine plates; a distinct palatomaxillary on one side only. The nasal labyrinth is unusually simple; the inferior turbinal, which has three coils in *Rhea* and *Tinamus*, and two in most birds, in *Gecinus* is two-winged, and in *Yunx* makes less than a single turn. In *Gecinus*, which is very specialised, the columella has two suprastapedial spurs and two infrastapedial bands, which have united with the stylohyal. The small ceratohyals in the Woodpeckers early coalesce into one arrowhead-shaped bone, and behind it is a very long highly-ossified and elastic basihyal. Joined to this behind is a pair of ceratobranchials, half its length; the epibranchials are four times the length of the ceratobranchials, and, passing first down the sides of the upper part of the neck, they curve gently upwards and forwards, lying in a furrow on the top of the skull; bending slightly from the middle line, they end on the nasal roof.

CHAPTER VIII.

THE SKULL OF THE PIG.

First Stage: Embryos $7\frac{1}{2}$ to 8 lines long.

602. THE earliest embryos of the Pig in which the foundations of the skull are perceptible present strong resemblances in the cephalic region to those of lower vertebrates; nevertheless a few special features of morphological elevation are manifest. In a front view the second cerebral vesicle is seen as the uppermost of the neural bulbs, although it does not advance quite to the front; thus the forehead is sloping. The distinct cerebral hemispheres occupy the middle of the lateral aspect of the head, and are together rather broader than the second vesicle. Immediately beneath the hemispheres are the nasal sacs, reniform in shape; they are bounded inferiorly by the lateral angles of the nasofrontal process in which the trabecular cornua are contained. The eyeballs, not very prominent, are seen on either side of the front view, above and behind the nasal sacs. Between the latter is a nasofrontal tract of moderate width, forming with its horns the anterior and upper margin of the mouth. This opening is at present quite inferior and transverse, bounded laterally by maxillopalatine processes which grow forwards and downwards from the continuous superficial tissues of the sides of the head, and posteriorly by the mandibular arch, which constitutes a transverse bar below, and has its moieties convex outwardly at the sides. Behind the mouth the skin of the broadish throat is continuous and unmarked ventrally.

603. In a side view (Fig. 76), the great proportionate size of the second cerebral vesicle ($C\,2$), which occupies the whole crown of the head, is particularly marked. The third vesicle ($C3$) is much smaller, and on a lower level.

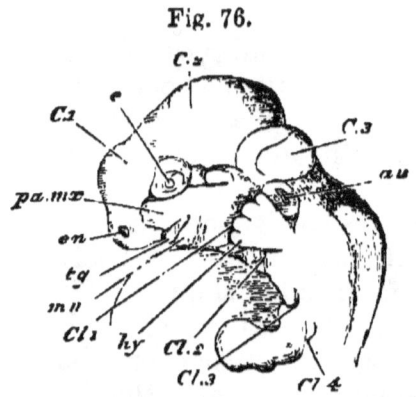

Fig. 76.

Embryo Pig, two-thirds of an inch long; side view of head and neck.

C 1, 2, 3, cerebral vesicles; *e.n.* external nasal opening; *e*, eyeball; *au.* auditory mass; *tg.* tongue-rudiment; *mn.* mandibular arch; *pa.mx.* maxillopalatine process; *cl.* 1, 2, 3, 4, visceral clefts; *hy.* hyoid arch, covered by the end of the fore limb.

There is a considerable depression between this vesicle and the spinal region, the cervical part of which is bent at a right angle with the dorsal, giving the embryo a strange humpbacked appearance. The olfactory sacs are seen at the antero-inferior extremity of the head; the eyeballs (*e*) are above and behind them in the angle between the first and second cerebral vesicles; and the pyriform ear-sacs (*au*), of approximately the same size as the other sense capsules, are below the hinder half of the third vesicle. The skin over them, as well as over the adjacent cerebral vesicle, is very thin as yet.

604. The side wall of the mouth and throat presents a continuous skin, interrupted below by four successive vertical clefts. The maxillopalatine process (*pa. mx.*) is perfectly continuous with the wall of the mouth and throat above, and projects downwards and forwards in the subocular

region, a cleft remaining between it and the eyeball. The hinder visceral clefts are overlapped by the palmate rudiment of the forwardly-directed fore limb.

605. In a median longitudinal and vertical section the brain vesicles are seen to be hollow, having only a film of soft neural substance lining the membranous cranium. The mesocephalic flexure is even more intense than in other types. The notochord, underlying the hinder vesicle, bends suddenly upwards so as to be situated in front of it. It does not extend so far as the summit of the third vesicle or the base of the second, but ends in a free blunted point, just opposite to that superior and posterior infolding of the membranous cranium which partially severs the second and third vesicles. Above the apex of the notochord is a mass of gelatinous stroma (the "middle trabecula") forming the hilus of the long kidney-shaped second vesicle. In front of the notochord at its upward bend is the pituitary body, and immediately anterior to that, the small first cerebral vesicle giving off its large hemispheres. The internal carotid artery is distinct on either side of the pituitary body.

606. Parachordal hyaline cartilages (Fig. 77, *pa. ch.*) already exist, being rather below the level of the notochord (*nc.*) on either side; they follow its curve, stopping somewhat short of its apex. The ear-capsules lie external to the parachordal cartilages, with walls commencing to chondrify. They are tuberous bodies, having a nearly straight inner margin, and a sublobular outer edge, and are broadest behind. The aqueductus vestibuli is left as an opening in the cartilage on the upper and inner edge: the facial nerve enters the capsule a little behind this, its passage being the aqueductus Fallopii. The semicircular canals are but just beginning to bud out from the postero-superior region of the ear-cavity, and the cochlea is indicated by a styliform anterior projection of the same cavity. There is an external deficiency in the capsule, filled with a plug of gelatinous stroma. The glossopharyn-

geal (8*a*), vagus (8*b*), and hypoglossal (9) nerves pass through soft stroma in the posterior angle between the auditory sac and the parachordal cartilage. The trigeminal is distinguishable in front of the ear-sac.

Fig. 77.

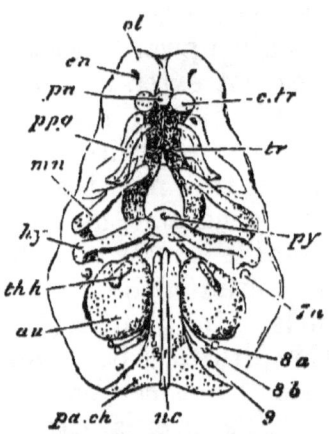

Embryo Pig, two-thirds of an inch long; elements of the skull seen somewhat diagrammatically from below.

pa.ch. parachordal cartilage; *nc.* notochord; *au.* auditory capsule; *py.* pituitary body; *tr.* trabeculæ; *c.tr.* trabecular cornu; *pn.* prenasal cartilage; *e.n.* external nasal opening; *ol.* nasal capsule; *p.pg.* palato-pterygoid tract enclosed in the maxillopalatine process; *mn.* mandibular arch; *hy.* hyoid arch; *th.h.* first branchial arch; 7*a*, facial nerve; 8*a*, glossopharyngeal; 8*b*, vagus; 9, hypoglossal.

607. The trabeculæ (*tr.*) are well marked out in the fore part of the cranial floor, lying beneath the olfactory sacs and the first cerebral vesicle, and extending upwards and backwards to embrace the pituitary body. They are shaped like callipers, and their pointed (notochordal) apices are some distance apart. They reach nearly as far back as the fore end of the parachordals, but are considerably below them in position. The trabecular curve is not moulded on the sides of the pituitary body, but is very distinct from it; and the intertrabecular space is much more extensive than that body. In their fore part they very early coalesce and thicken: the commissure

gradually extends backwards, until the uncoalesced part merely corresponds to the pituitary body. The fore end of each trabecula does not take part in the fusion, but becomes first clubbed and then bent somewhat outwards and inferiorly, forming the trabecular cornua (*c. tr.*); and these cause the skin to project on either side in the palate, mesiad of the internal nostrils. Between these prominences of the cornua a median backwardly directed process soon arises, which includes an azygous growth from the trabecular commissure, the prenasal or rostral process. (Figs. 77, 78, *pn*.)

608. The trabeculæ, as well as the other skeletal elements, are not yet truly cartilaginous, although unmistakeable; they are embedded in a gelatinous tissue rich with young cells, whose protoplasmic substance takes up carmine very freely. The differentiation of the rods is at present a matter of degree, that part of the blastema which will become hyaline cartilage being the most compact and crowded with young cells; next to this the nascent perichondrium; and the most gelatinous part outside is the rudimentary condition of the areolar connective tissue.

609. The nasal sacs are intimately related to the front part of the trabeculæ; and are already complicated. There is a squarish median internasal region, of considerable thickness, beneath the first vesicle, and in front of the at present down-turned trabecular extremity. The inner walls of the nasal cavities are simple, but the outer walls exhibit various processes projecting inwards. There is a lower bilobate process, the anterior lobe of which becomes the alinasal turbinal (see Fig. 79, p. 281), and the hinder lobe the inferior turbinal. Another swelling below the fore part of the nasal roof is the rudiment of the nasal turbinal; while a mass in the hinder angle beneath the olfactory crus gives rise to the upper and middle turbinals and the true olfactory region. By reason of these processes the nasal meatus is already tortuous, first ascending, proceeding backwards, and then descending; but the external and internal nares are on the same level, partitioned from

one another by the forward part of the maxillopalatine process and the tissue containing the trabecular cornu.

610. The internasal mass contains a pair of cartilaginous septal laminæ, each being continuous below with the corresponding trabecular moiety at its inner and upper edge, very near the middle line, and forming a concave inner wall to each nasal sac. The arch extends considerably into the upper nasal boundary; it may be called aliseptal cartilage.

611. It has already been stated (§ 602) that the nasal sacs are *beneath* the cerebral hemispheres. A transverse vertical section through the head at the region of the internal nares cuts through the olfactory crura partially

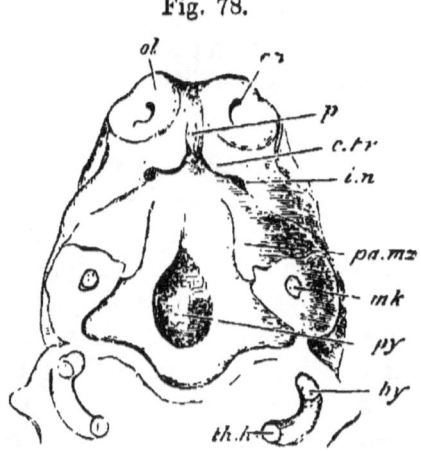

Fig. 78.

Embryo Pig, two-thirds of an inch long; palatal view, the lower jaw and lower side of throat having been removed.

p. prominence enclosing prenasal cartilage; *c.tr.* that covering the trabecular cornu; *ol.* nasal sac; *e.n.* external nostril; *i.n.* primary internal nostril; *pa.mx.* maxillopalatine process; *mk.* section of meckelian cartilage; *hy.* section of hyoid arch; *th.h.* section of branchial arch, applied to lower extremity of hyoid; *py.* tissue beneath the pituitary fossa.

separated by a median ethmoidal partition of mesoblastic tissue; the trabeculæ lying as rounded rods somewhat below this partition; and the irregularly shaped nasal sacs beneath the cranial cavity and external to the trabeculæ. The thick maxillopalatine process forms the outer and lower margin of the section (see Fig. 78), enclosing, in that part which is close to the internal nostril, the small rudiment of the palatopterygoid bar. In the upper and outer tract of the maxillopalatine process the cleft between it and the upper facial region (§ 604) is cut through; it passes downwards into the posterior portion of the nasal passage, and upwards to the base of the fore part of the orbital region (the future "inner canthus"); it becomes the lachrymal passage.

612. The palatopterygoid bars ($p.pg.$, Fig. 77) are less definitely developed at present than the other skeletal elements; they are small sigmoid granular rods, in the inner edge of the maxillopalatine processes, which extend inwards to form a considerable part of the palatal roof. They are continued forwards beneath the trabecular region, approximating as they advance; but they do not meet, so that there is a primary cleft palate or median palatal hollow. The posterior ends of the palatopterygoid bars diverge outwards, instead of turning inwards like the postoral arches.

613. The mandibular bars ($mn.$ Fig. 77) are much stouter and more perfect than the palatopterygoids; they occupy most of the lower jaw, but do not meet in the middle line. Each bar is sigmoid, and strongly inhooked proximally, where it extends towards the auditory sac. It comes into close relation with the first visceral or mandibulo-hyoid cleft, which at its upper part is of considerable extent owing to the thickness of the throat-wall. The cleft becomes constricted at the point where the mandibular bar is in contact with its anterior wall; this is the position of the future tympanic membrane.

614. The hyoid arch (Fig. 77, $hy.$) is much like the mandibular; but it is flatter, and the right and left bars approach one another more closely below, at an obtuse angle; its apex also is more sharply inturned, making the

shoulder stand out like the tubercle of a rib. The apex of the bar grows into relation with the mandibulo-hyoid cleft behind, in a similar manner to the bar in front. The main part of the facial nerve (7*a*) curves backwards round the fore part of the auditory capsule, and behind the angle of the hyoid bar. At some distance behind the hyoid, in the lower part of the third visceral arch (*th.h*), is a much smaller cartilage on each side, attached to the fore part of the sides of the larynx. This is homologous with the branchial rods of lower vertebrates.

615. The primordial skeleton of this most highly-specialised Mammal is as simple as that of the lowest Fish we have examined; the head, with its intense mesocephalic flexure, compares very well in shape with that of the Skate or the Frog. The elements are more distinct from one another than in the Fowl. There is everything to show the persistency of morphological impress; the visceral clefts which appear, all of them transitory except one, are more in number than the arches which contribute anything to the adult skeleton. The embryo of the air-breathing Pig gives evidence of its relationship with forms breathing through water. The prenasal nodule and the little cornua are further manifestations of unity with other types, cunningly adapted and hidden in subsequent history.

616. But the Pig even at this stage shows evidence of its high grade. The notable posterior clinoid ridge is prepared for by the high upward curving of the parachordals, above the trabecular level. The early internasal fusion of the trabeculæ and the development of the rudiments of turbinal folds while as yet the sense capsules are sub-equal in size, are indications of the future predominance of the olfactory organs. A distinctive palatopterygoid tract exists in the maxillopalatine process of the mandibular arch, but the palatopterygoid itself shows no appearance of being a process from the mandibular cartilage. The principal arches of the skull at their proximal extremity

already begin to be specialised; the first visceral cleft is becoming tortuous, and the apex of the mandibular arch (future manubrium of the malleus), which is plainly comparable to the apex of the otic process of the quadrate in Birds, grows inwards into the cleft in a way not occurring in the Fowl, so as to become involved in the tympanic membrane.

617. Here then in the last of our types we find the same parachordal and trabecular elements in the cranial floor and the same relations of the notochord as in previous cases; with two principal arches, mandibular and hyoid, and smaller rudiments, the palatopterygoid and first branchial. All these parts are quite distinct at first. But there are already two openings to each nasal cavity, and these are approximately on the same level; the "posterior nares" are in the anterior part of the oral cavity, as in Dipnoi and Amphibia. The roof of the mouth, or primary palate, is the floor of the cranium. A free growth of cartilage, binding the various bars together, would produce a very good parallel to the cartilaginous skull of Lepidosiren.

Second Stage: *Embryos one inch long.*

618. In this stage chondrification has fairly set in, although the cells of the hyaline cartilage are still close together. Ossification has also commenced in fibrous tracts related to the mouth. Differentiation of parts has gone on very rapidly in the olfactory and auditory regions, whilst the embryo has merely become longer by one half. The mesocephalic flexure is partially lost, but the apex of the notochord and the direction of the parachordals still make a considerable angle with the trabeculæ.

619. The parachordals have grown round the notochord to a large extent, but do not unite in front of it. They are narrow in the interauditory region, and much broader behind where they coalesce with the inner and hinder edges of the periotic masses; they extend back-

wards considerably behind the latter, ending pointedly on either side. The occipital ring of cartilage is developing.

620. The trabeculæ have coalesced in their whole extent except around the pituitary space, which is itself becoming floored with cartilage; and this cartilage unites with the parachordal below the level of the clinoid ridge. The cranial floor is being extended laterally by trabecular outgrowths, after the manner in which the occipital ring is being formed; thus alisphenoidal and orbitosphenoidal laminæ are arising.

621. The alinasal cartilage has become completely chondrified up to the fore end of the nasal region, and the primary moieties are more or less confluent with the trabeculæ and their cornua into a single cartilaginous septum. This is thick and low in front, but much thinner and more elevated behind, in the mesethmoid region. The outer nasal wall is fully chondrified, and sends cartilaginous processes into the various turbinal folds. There is no trace of separate trabeculæ in the base of the mesethmoid; that lamina merely gradually thickens towards its base, and just outside the mucous membrane covering it below is the internal nostril on either side. The base of the mesethmoid tract is underlaid by an elongated granular mass which is undergoing endostosis to form the azygous *vomer*. A dense premaxillary stroma is found under the trabecular cornua in front, ready to ossify; while far on either side of the ethmoidal region there is a faint beginning of the *maxillary*, above and quite distinct from a rudimentary tooth-pulp. The compressed granular palatopterygoid rod is considerably mesiad of this tooth-pulp, bounding the median cleft of the palate.

622. In the ear-sac the semicircular canals and the coils of the cochlea have now become differentiated. The median lobular protuberance of the outer wall of the capsule seen in the first stage is separate to a great extent from the rest of the auditory wall, but remains as a plug,

thus forming the stapes, lying in the fenestra ovalis. It has two papular elevations on its external surface, and is covered externally by tissue commencing to chondrify.

623. By the thickening of the side-wall of the throat the first visceral cleft is so much lengthened that it may now be called tympano-eustachian canal; its Eustachian part is directed forwards into the mouth; its tympanic portion is becoming closed by the formation of the tympanic membrane as the upper end of the mandibular arch grows inwards towards the ear-capsule. An external tube outside the tympanic chamber is being formed by the growth of an opercular skin-fold round the primary orifice of the cleft; cartilage soon arises in this fold. The external edge of the periotic mass is produced as a pterotic ridge or tegmen tympani, covering and partially surrounding the upper ends of both the mandibular and the hyoid arches.

624. The mandibular bar becomes more and more incurved at its proximal extremity, so that it is very hook-shaped (see Figs. 80, 81, representing a later stage). It has a small bulbous end, and a thickened portion at the bend or shoulder, which articulates with the upper part of the next bar. The incurved apex becomes the manubrium, or handle of the malleus, and the remainder is the meckelian cartilage. It is the ingrowth of the manubrium into and over the tympanic canal that, carrying the skin with it, gives rise to the tympanic membrane. The head of the malleus further grows out into a boss at its junction with the main bar, for the attachment of the tensor tympani muscle.

625. The proximal region of the hyoid bar is more differentiated than the preceding arch, and also undergoes distinct segmentation. The incurved and backwardly-directed hook has been cut off from the shaft, and becomes shaped into the incus with its processes; while the remaining bar shifts its position backwards along the incus; its proximal end becomes somewhat two-headed, to articulate externally with the tegmen tympani, and to

unite internally with the ear-sac close in front of the exit of the facial nerve.

626. It was the tubercular part or shoulder of the unsegmented arch which came into contact with the mandibular arch growing backwards and forming the tympanic membrane: while the apex or head was bent inwards and backwards towards the ear-sac. Consequently, when segmentation takes place, the tubercular part of the hyoid is left as the front and outer end of the incus, articulating with the head of the malleus (Fig. 81). A small boss arises on the outer edge of the incus, to form the short crus; the part of the incus growing towards the ear-sac remains as the long crus, and its small rounded head is the orbicular process, coming into close contact with the stapes.

627. The distal part of the mandibular arch extends forwards in the lower jaw; and half encircling it on the outer side is a mass of granular tissue or nascent cartilage, reniform in section, in the axis of which the *dentary* bone is being formed. The remainder of the hyoid arch (which may be called stylo- and ceratohyal) extends downwards to the fore part of the larynx, and is fastened to a median cartilage, which is the first basibranchial (basihyal of human anatomy); this lies between a pair of small cartilages which represent the first branchial arch.

628. In this minute embryo some of the features most characteristic of the adult as belonging to the mammalian group are already distinguishable. Chondrification has advanced greatly, and is rapidly completing its work. The notochord is still visible, but the basilar plate is well formed, though incompletely united as yet with the auditory capsules; and the exoccipital region is arising. The trabeculæ have fused together almost completely, but are not yet united to the basilar plate; lateral cartilages, however, are growing in the ali- and orbitosphenoidal regions. The cochlea and semicircular canals are well formed. The nasal labyrinth has increased in size and completeness; and a median septum is found instead

of two distinct inner nasal walls. The trabecular cornua appear as recurrent cartilages.

629. The palate is still distinctively cleft; there are rudiments of the vomer and the maxillaries. The palatopterygoid bar does not completely chondrify, but has already begun to be ossified. As a contrast to this, the stroma of the dentary (very distinct from the meckelian cartilage) chondrifies very largely before it ossifies. The outer ear is also becoming cartilaginous, while the proximal parts of the two principal arches are rapidly assuming their adult relations. The mandibular apex becomes shaped into the parts of the malleus, without any segmentation taking place; while the hyoid, becoming bent so as to articulate with the malleus, gives rise to the incus, which is also in relation to the commencing stapes formed out of the tissue of the ear-capsule. The incus being segmented from it, the main hyoid arch is withdrawn backwards, approaching the stapes and becoming connected with it and the ear-capsule by an interhyal ligament.

Third Stage: Embryos an inch and a third long.

630. There is now continuous cartilage occupying the whole median basicranial and facial regions; the lateral cranial walls are partially chondrified, the occipital ring being complete. The parachordals have united in a basilar plate, entirely enclosing the diminished notochord (Fig. 80). The anterior end of the basilar plate forms the hinder boundary of the pituitary space (posterior clinoid wall, *p. cl.* Fig. 79); but does not rise so high relatively, nor form so great an upward curve as did the parachordals in the first stage. No remnant of the notochord is found in this posterior clinoid wall—a sharp contrast to the first stage, when the notochord projected beyond the parachordals.

631. The basilar plate (*b. o.*) is narrow between the ear-capsules, which are continuous with it; internally

and especially posteriorly they are on the same level with it, but externally they are elevated so as to constitute the lower half of the lateral cranial wall. Behind the auditory masses the basilar plate widens, and curves gradually upwards on both sides to form the occipital arch. The supraoccipital region (*s. o.*) rests on and is continuous with the hinder half of the ear-capsules. The foramen magnum (*f. m.*) is large and transversely elliptical, with a supero-median notch: the crescentic condyles are external to the foramen, extending towards the middle line below.

632. The foramen for the vagus and glossopharyngeal nerves (Fig. 79, 8) is immediately behind the ear-mass internally; the hypoglossal foramen is somewhat behind this. The auditory cartilage, protruding into the cranial cavity, presents several depressions and openings. There is a depression beneath the arch of the anterior canal to receive the cerebellar process (flocculus); antero-inferiorly the facial and auditory nerves perforate the cartilage, close together, but separated by a distinct bridge. The condition of the auditory cavity will be described later.

633. In front of the basilar plate, the floor of the cranial cavity presents a deep pituitary cup (*py.*), floored with cartilage; a slightly elevated anterior clinoid wall, with a depression in front of it on which the optic chiasma lies (presphenoidal region, *p. s.*): and a more elevated ethmoidal region, the level of which is approximately the same as that of the basilar plate. The ethmoidal cartilage projects slightly upwards at the extreme fore part of the brain case, forming a rudimentary anterior wall.

634. The lateral portions of the cranial floor and the lowest regions of its side walls contain cartilaginous orbitosphenoidal and alisphenoidal growths, continuous with the median floor. The alisphenoidal cartilage (*al. s.*) is short and thick, and is attached to the anterior and

posterior clinoid tracts. Its most noteworthy part extends downwards to the palatopterygoid bar, to form the external pterygoid cartilage. Between the alisphenoid and the ear-sac is a large shallow fossa for the Gasserian ganglion, and the largest division of the trigeminal nerve

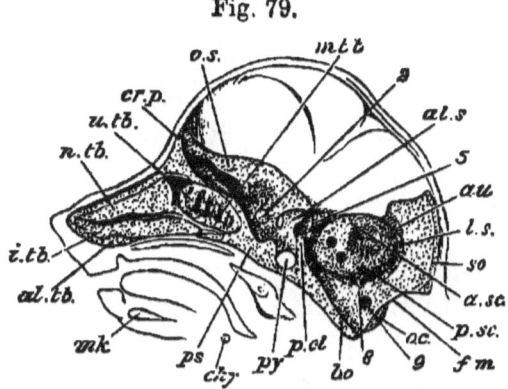

Fig. 79.

Embryo Pig, an inch and a third long; median longitudinal section of head, with nasal septum removed.

al.tb. alinasal turbinal; *i.tb.* inferior turbinal; *n.tb.* nasal turbinal; *m.tb.* middle turbinal; *u.tb.* upper turbinal; *cr.p.* cribriform (ethmoidal) plate (the line should be prolonged to a point over the turbinal folds); *p.s.* presphenoid region; *o.s.* orbitosphenoid; *al.s.* alisphenoid; *py.* pituitary fossa; *p.cl.* posterior clinoid ridge; *au.* ear-capsule; *a.s.c.* anterior semicircular canal; *p.s.c.* posterior semicircular canal; *s.o.* supraoccipital cartilage; *l.s.* position of the lateral sinus in dura mater; *f.m.* foramen magnum; *o.c.* occipital condyle; *b.o.* basilar plate; 2, optic foramen; 5, foramen ovale; 8, foramen for glossopharyngeal and vagus nerves; 9, foramen for hypoglossal; *mk.* meckelian cartilage; above it, the tongue; *c.hy.* ceratohyal.

(the inferior maxillary) passes out between these two cartilaginous regions (5). The superior maxillary division of the trigeminal passes out between the orbito- and alisphenoids. The orbitosphenoids (*o.s.*) are far the larger, spreading from their presphenoidal base into a sickle-shaped lamina extending from the fore part of the cranium almost to the auditory masses. The optic nerve (2) passes out by a foramen in the base of the orbitosphenoid. The thin *frontals* have arisen in the stroma outside and above the outer edge of the orbitosphenoids.

635. The nasal cavities extend backwards beneath the cranial cavity in its ethmoidal portion, and on each side of the median ethmoidal plate the nasal roof is soft, and the olfactory lobes of the brain lie upon it. The nerve filaments perforate the roof to reach the turbinal growths of the nasal sacs; and very soon cartilage appears between the groups of nerve fibres, and gradually becomes perfected into a cribriform plate (*cr. p.*).

636. The nasal cavities already occupy more than half the length of the head. Rather more than half their extent belongs to the snout, which is a rostrum-like projection from the bulbous head. The nasal septum is a complete vertical lamina of cartilage, of considerable and almost uniform height, being most elevated at the junction of the snout with the cranial box. The septum bears no conspicuous projections on its lateral surfaces; but superiorly it is continuous with the upper and outer cartilaginous nasal walls, which are distinguishable from before backwards into alinasal, aliseptal, and aliethmoidal regions. Anteriorly the septum is broader at the base, owing to its continuity with the primary trabecular cornua; but a little posteriorly these cornua have become recurrent cartilages distinct from the septum; and they lie for some distance close to its base, diminishing to small styliform rods. The greater portion of the septum is floored by condensed tissue very like nascent cartilage, separating it from the roof of the mouth, or from the palatal plates where they are uniting to form the true palate. This tissue early ossifies as the vomer.

637. The outer nasal wall becomes more complete and more complex in passing from before backwards; each of the main turbinal outgrowths from the upper and outer walls contains a small cartilaginous blade continuous with the capsular wall. The nasal sacs are widest where they underlie the fore part of the cranial cavity. They diminish in size as they approach the presphenoidal region, under which they die out: this is the part of the nasal cavity known as the sphenoidal sinus. The *premaxillaries*

have appeared in the roof of the mouth, beneath each nasal mass, short of its extreme anterior extremity.

638. The trihedral mass of stroma forming the outer part of the palate roof and the upper lip, sends inwards in its middle region a thin tongue of tissue to meet its fellow and form the secondary palatal roof. This meeting takes place at present only for a short distance; elsewhere the palate is still cleft. The palatopterygoid rod lies just at the junction of this inner palatal region with the outer stroma; it is kidney-shaped in transverse section. Beneath the hinder part of the nasal capsule this imperfect cartilage is ossifying to form the *palatine*; further back a small rudiment of the *pterygoid* appears (the future internal pterygoid plate), adjacent to the downgrowth from the alisphenoid which becomes the external pterygoid plate. The maxillary bones first arise beneath the nasal sacs under the anterior cranial region, in the palatal roof. It is evident that it is the palatal part of the maxillary which is the first formed; outside and behind it, a row of tooth-pulps is developing, close to the edge of the lip, corresponding to a similar set in the lower jaw.

639. The floor of the mouth is occupied by the prominent tongue, and below its level on either side is the stroma in which the meckelian cartilage is embedded. This primary rod is flanked externally and partly embraced by a dentary stroma, which ossifies rapidly, becoming however distinctly cartilaginous in the interim. The dentary bone is vertically elevated behind; a row of tooth-pulps lies above it in the greater part of its extent. The meckelian cartilages have fused together anteriorly.

640. The arrangement of the parts of the ear-capsule is as follows. The coils of the cochlea (*cl.* Fig. 80) are fully developed, and are entirely anterior, and near the middle line and base of the skull. The facial nerve passes into the capsule above and externally to the cochlea, and curves backwards and outwards to reach the roof of the tympanic cavity. The anterior semicircular canal

arches high up on the antero-external part of the auditory mass: the horizontal canal is almost directly below it, and

Fig. 80.

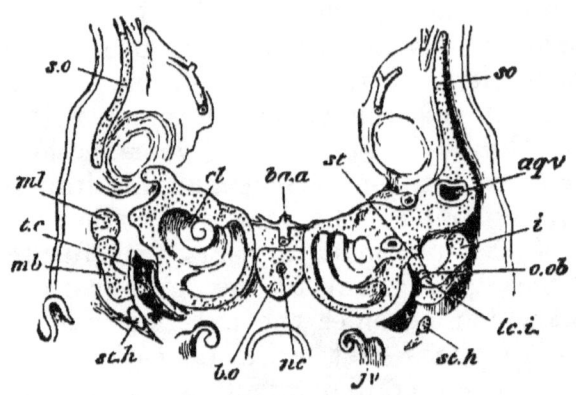

Embryo Pig, an inch and a third long; posterior view of a section through the basal region of the skull. The section being not perfectly transverse shows a different condition on each side.

The left-hand view shows parts in front of those displayed on the right; the manubrium of the malleus *mb.* is cut through at its thickest (posterior) part, and the main part or shoulder *ml.* is seen, at the part which articulates with the incus; within *mb.* is the tympanic cavity *t.c.* On the right side the incus *i.* is seen, covering the malleus, and having its short crus cut away; its orbicular head *o.ob.* is applied to the stapes *st.* The stylohyal cartilage *st.h.* is seen in oblique section as it passes downwards and forwards. The supraoccipital cartilage *s.o.* is seen as if discontinuous with the auditory tract on the left side, in consequence of the section passing through the notch in the front end of the cartilage. *b.o.* basilar cartilage, narrow between the ear-capsules; a little of its upper surface is seen, as well as the section enclosing the notochord *nc.*; *ba.a.* basilar artery; *j.v.* jugular vein, with a coiled branch; *l.c.i.* long crus of incus; *aq.v.* aqueductus vestibuli; *cl.* cochlea. Under the basilar cartilage the throat is partially represented.

the main part of the labyrinth is internal, behind the cochlea. The posterior canal, convex backwards, touches the horizontal canal externally, and ends internally in the labyrinth. The middle and hinder part of the external edge of the ear-mass, containing part of the posterior canal, projects as a roof or tegmen tympani (pterotic ridge) over the tympanic cavity; so that the fenestra in which the stapes lies places the labyrinth in direct contiguity with the tympanic cavity. The short or external crus of the incus articulates with the tegmen.

641. The tympano-eustachian canal has increased in length and become more definite; it is completely occluded by the tympanic membrane, in which the handle of the malleus (Figs. 80, 81, *mb.*) is embedded. The latter piece has become flattened, but is not segmented from the remainder of the mandibular arch. The incus (*i.*) has

Fig. 81.

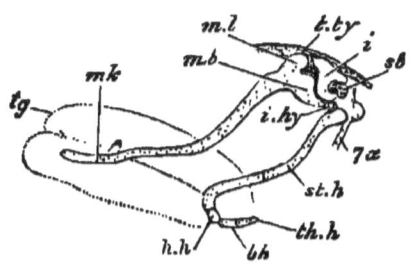

Embryo Pig, an inch and a third long; side view of mandibular and hyoid arches. The main hyoid arch is seen as displaced backwards after segmentation from the incus.

tg. tongue; *mk.* meckelian cartilage; *ml.* body of malleus; *mb.* manubrium or handle of the malleus; *t.ty.* tegmen tympani; *i.* incus; *st.* stapes; *i.hy.* interhyal ligament; *st.h.* stylohyal cartilage; *h.h.* hypohyal; *b.h.* basibranchial; *th.h.* rudiment of first branchial arch; *7a,* facial nerve.

similarly progressed, applying the orbicular head (*o.ob.*) of its long crus to the stapes (Figs. 80, 81, *st.*) The latter has begun to assume its distinctive shape, the two tubercles which it bore at first being now connected by a bridge of cartilage in the fenestra ovalis. The main hyoid rod (*st.h.*) is attached by ligament to the orbicular end of the incus, and is in contact with the auditory capsule behind the stapes. In this interhyal ligament (*i.hy.*) a small cartilage has arisen. The facial nerve (*7a.*) passes (above the tympanic membrane) to get behind the hyoid arch, parallel with which it proceeds to its destination.

642. The main hyoid bar passes directly downwards from the auditory region for some distance, and then

curves forwards and inwards, being connected with the small hypohyal segment (*h.h.*). The latter is articulated to the median azygous rudiment of the next arch (basibranchial), and to its lateral or branchial cartilages (so-called basi- and thyrohyal).

"The proper territories of each investing bone in the Pig evidently only want *time* that they might all become true cartilage; ossification sets in too soon for the formation of the intercellular substance, but each tract, before ossification, is a true morphological element or organ."

(*Phil. Trans.* 1874, p. 306.)

643. We have now considered the structure of the skull in an embryo Pig only an inch and a third long, and have found it completely constituted as to cartilage, with all the normal elements continuous, viz. basilar plate, trabeculæ, occipital ring, auditory capsules, ali- and orbitosphenoids, nasal septum, walls, and turbinals. The principal arches have almost perfected their permanent relations. The predominance of the orbitosphenoid cartilages, the large size of the nasal capsules, their extension backwards beneath the brain-case, and the formation of the cribriform plate are to be noticed. The true cartilage-bones have not yet appeared; but there are present the vomer, frontals, premaxillaries, maxillaries, palatines, pterygoids, and dentaries.

Fourth Stage: Embryo two and a half inches long.

644. Development has proceeded very rapidly; the snout is proportionately longer, and the forehead has become sloping instead of vertical. The well-marked granular territories that at first invested the primordial skull are now largely ossified, and these ossifications are massive in relation to so small a skull. The cartilaginous parts also are undergoing endostosis at many points.

645. The cartilaginous cranium has not changed much in its main outlines: the floor has thickened considerably.

The cribriform ethmoidal plate has now four slender bars passing between the olfactory filaments, and a common outer band which does not extend backwards to the orbito-sphenoidal cartilage. The latter has grown forwards to unite with the aliethmoid cartilage, and backwards to overlap the antero-external angle of the auditory cartilage, quite overshadowing the alisphenoid. But the external pterygoid plate (Fig. 82, *e.pg.*) arising from both the basisphenoidal and the alisphenoidal regions, is well developed, and articulates with the palatopterygoid bar.

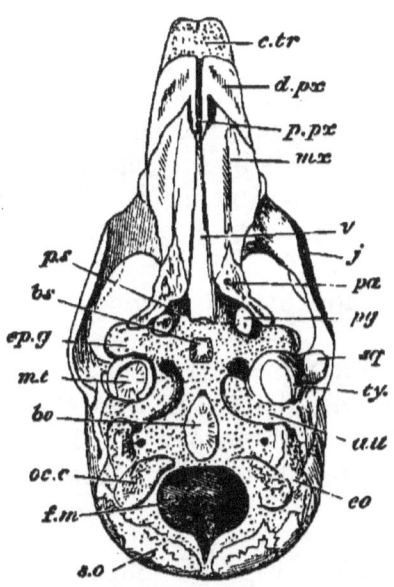

Embryo Pig, 2½ inches long; under view of skull with lower jaw removed.

f.m. foramen magnum; *oc.c.* occipital condyle; *au.* auditory capsule; *m.t.* tympanic membrane; *e.pg.* external pterygoid process; *p.s.* presphenoid region; *c.tr.* trabecular cornua.

Bones: *b.o.* basioccipital; *e.o.* exoccipital; *s.o.* supraoccipital; *ty.* tympanic; *b.s.* basisphenoid; *sq.* squamosal; *pg.* pterygoid; *pa.* palatine; *v.* vomer; *d.px.* dentary plate of premaxillary; *p.px.* palatal process of same; *mx.* maxillary; *j.* jugal.

646. The cartilage bones are principally endosteal at present, although they rapidly gain the surface and affect

the perichondrium. There is a thick *basioccipital* (*b.o*) in the basilar plate, seen both above and below; it is shaped like a spear-head, and is fast obliterating the notochord. In the *exoccipital* region there is a large growth of bone (*e.o.*) extending inwards towards the upper edge of the foramen magnum, and outwards and downwards in the external or paroccipital process. Besides this, a distinct ossification arises in the massive condyle, which coalesces with the exoccipital before long. The *supraoccipital* (*s.o.*) is double at first, but the two patches run into one another in a day or two.

647. The cartilage in the pituitary floor is ossifying to form the *basisphenoid* (*b.s.*); a little later, a distinct centre arises on either side for the *alisphenoid*. There is no presphenoid, but a small *orbitosphenoid* ossification is found outside and behind each optic foramen. The auditory capsule is unossified at present.

648. The hinder nasal region has become much broader than the anterior. The upper and middle turbinal folds with their included cartilaginous laminæ occupy the back part of the nasal cavity; and the inferior turbinal is further forward. Anteriorly the recurrent trabecular cornua (*c.tr.*) still persist, on either side of the base of the septum.

649. The vomer (Figs. 82, 83, *v.*) embraces the base of the septum along the hinder two-thirds of the nasal region. The premaxillaries are found on the under surface of the fore part of the snout, not reaching to its extremity. Each has an expanded dentary portion (*d.px.*) grooved by tooth-sacs, and a small medio-palatal spur (*p.px.*) extending towards the vomer. The end of the snout around and above the external nostrils is unclothed by bone: but behind this region the premaxillary of each side sends up a facial lamina to meet the outer edge of the nasal, also coming in contact behind with the facial plate of the maxillary. The *nasals*, *frontals*, and *parietals* form a

regular double series of thick bones. The parietals do not extend so far back as to the supraoccipital; they occupy the main part of the vertex of the skull. The frontals are the largest pair. They stretch considerably over the face, touching the lachrymal and maxillary bones; they arch over the orbit, and send inwards from the eave an orbital plate which is in the side wall of the skull, and reaches the orbitosphenoid cartilage. The anterior and posterior median fontanelles (fronto-parietal and parieto-occipital) are still widely open.

650. The maxillary (*mx.*) has its facial and palatal plates largely developed. The facial lamina extends from the premaxillary along the lower part of the face to underlap the *jugal*. Above, the maxillary is in contact with the frontal and lachrymal, and does not form any part of the margin of the orbit. The *lachrymal* is a small bone enclosing the upper end of the orbitonasal (lachrymal) canal, and occupying some space on the face between the frontal and jugal bones. The latter (*j.*) bounds the lower margin of the orbit, and passes backwards underneath the zygomatic process of the squamosal, nearly to the ear-sac.

651. The *squamosal* (*sq.*) is a large triradiate membrane bone, with a squamous part lying on the infero-lateral wall of the cranium, between and below the frontal and parietal; it has also a long sickle-shaped process directed backwardly by the side of the ear-capsule above and outside the tympanum; and a zygomatic process growing outwards and forwards to the back part and external edge of the orbit, resting on the jugal. At the point of junction of these three parts the squamosal has an orbicular concave surface, external to the tympanic membrane; with this surface the osseous mandible articulates.

652. Returning to the palate, the greater portion of it is occupied by the large oblong palatal plates of the maxillaries, which do not meet in the middle line but leave the vomer exposed between them. The maxillaries

are grooved by vessels in the middle of their palatal plates, whilst the dentary portions are hollow and shell-like, containing large growing tooth-germs. The palatines (*pa.*), pointed in front, rod-like behind, extend outwards and backwards on the palate from the maxillaries to the external pterygoid cartilaginous plate. Their ascending lamina reaches to the hind part of the nasal septum. Internally to the hinder portion of each palatine, is a small osseous nodule, the pterygoid (*pg.*).

Fig. 83.

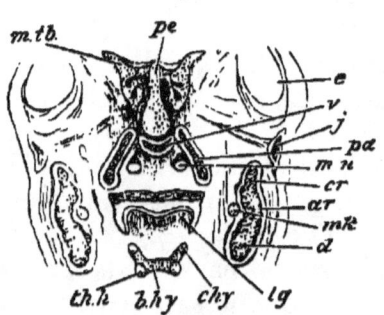

Embryo Pig, 2½ inches long; vertical section of head, showing structures between and beneath the orbits.

p.e. mesthmoid plate; *m.tb.* middle turbinal coil; *e.* eyeball; *m.n.* nasal passage above secondary palate; *mk.* meckelian cartilage; *b.hy.* basibranchial (body of hyoid); *c.hy.* ceratohyal; *th.h.* first branchial arch; *tg.* tongue.

Bones: *v.* vomer; *pa.* palatine; *cr.* coronoid process; *ar.* articular tract; *d.* dentary; *j.* jugal.

653. The proper mandible is now an extensive ossification of considerable vertical extent, ensheathing its own cartilage, sending up a backwardly curved coronoid process behind (Fig. 83, *cr.*), and having a cartilaginous condyle beneath the curve of the coronoid, articulating with the squamosal. There is a protuberant angular region of cartilage below the condyle. The osseous symphysis is not formed at present: but the symphysis of the meckelian rod is very extensive. This cartilage lies within the bony mandible unchanged; the *malleus* and its manubrium have become ossified, but are still perfectly continuous

with the rest of the primary arch, just within the articular condyle.

654. The tympanum is becoming rapidly perfected; the Eustachian tube opens into the back of the mouth beneath the basisphenoid. The tympanic membrane is partially enringed by a *tympanic* parostosis (Fig. 82, *ty.*) deficient behind. The *incus* is ossifying: behind it the stylohyal is continuous with the ear-capsule below the horizontal semicircular canal. At its lower end the stylohyal region passes into the hypobyal. The hypohyals, basibranchial (*b.hy.*), and the pair of branchials (*th.h.*) are articulated together; and they support the root of the tongue.

655. This stage is an excellent one for comparison with the adult Fish, Amphibian, or Reptile, or the ripe chick of the common Fowl. It also corresponds very closely with an early stage of the skull of *Balæna japonica*, figured by Eschricht. The non-segmentation of the cartilage of the mandibular arch, and the articulation of the dentary with the squamosal, at once distinguish the mammalian type; yet if the palate be compared with that of the young Ostrich (*Phil. Trans.* 1866, Plate VII. Fig. 4), the conformity is more remarkable than the difference. The new bones which have appeared are the four occipitals; the basi-, ali-, and orbitosphenoids; the nasals and parietals; the lachrymals, jugals, squamosals, and tympanics; the incus and the malleus.

Fifth Stage: Embryo six inches long from snout to ischium.

656. The general form of the head has not altered much; the snout is longer and its upper surface slopes more gradually from the frontal region forwards. The bones have become very dense; on the roof of the skull they are applied to each other edge to edge by sutures, in

certain places even overlapping. The anterior fontanelle (fronto-parietal) is still open, but is much lessened; the parietal and occipital bones form a good lambdoidal suture. The sickle-shaped process of the squamosal which lies alongside of the ear-capsule, has an increased surface, projecting backwards and upwards to the supraoccipital (Fig. 84, *s.o.*), and downwards in apposition with the paroccipital process (*p.oc.*) of the exoccipital. The end of the snout, around the small external nostrils, is unclothed by bone. The nasals project forwards mesially. The orbital plate of the frontal has developed very largely, so as to form the whole bony roof of the orbit. The upper and lateral bones of the skull require but little relative change, with increase of size, to bring them to their adult condition.

657. The previously existing cartilage-bones have increased in size. The bones forming the occipital ring are not yet in contact with one another. The exoccipitals (Fig. 84, *e.o.*) ossify the condyles and the paroccipital processes. The supraoccipital (*s.o.*) is a large concave shell of bone, separated by a considerable tract of cartilage from the exoccipitals.

658. The basisphenoid is a thick bone, extending partially into the posterior clinoid ridge, and continuous with the alisphenoids, which are of considerable size, and extend downwards into the external pterygoid plates. The presphenoid cartilage is well ossified, and continuous with the orbitosphenoids, which have ossified the inner half of the corresponding cartilages. The latter have not grown with the growth of the skull, and consequently appear relatively contracted; they are separated from the aliethmoid cartilage anteriorly, and are at some distance from the auditory capsule behind. The remainder of the facial axis and nasal septum is one sheet of solid cartilage, perfectly continuous on either side with the nasal labyrinth and its now highly complex turbinal growths and cribriform plate.

659. The proper periotic bones are now distinct. The *prootic* (*pr.o.*) is an endosteal patch surrounding the meatus internus, lying under the fore part of the cochlea and extending supero-posteriorly to the junction of the anterior and posterior canals: externally it is seen above and in front of the fenestra ovalis (*f.ov.*). The *opisthotic*

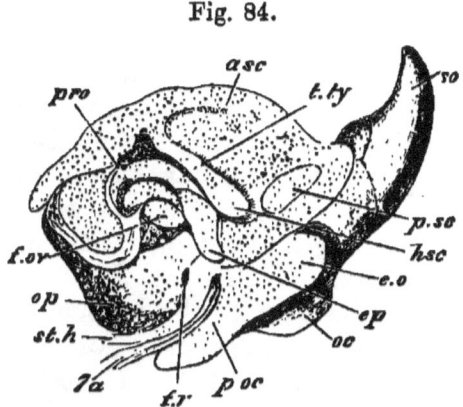

Fig. 84.

Embryo Pig, six inches long; outer view of occipital and auditory regions.

t.ty. tegmen tympani; *f.ov.* fenestra ovalis; *a.s.c.* anterior, *h.s.c.* horizontal, *p.s.c.* posterior semicircular canals; *f.r.* fenestra rotunda, the line should be carried upwards to the angle behind *f.ov.*; *7a.* facial nerve; *p.oc.* paroccipital process; *o.c.* occipital condyle; *st.h.* stylohyal cartilage.

Bones, in some cases not well indicated: *e.o.* exoccipital; *s.o.* supraoccipital; *pr.o.* prootic; *op.* opisthotic; *ep.* epiotic.

(*op.*) is on the under surface of the capsule behind, and covers the most bulbous part of the cochlea below: one of its processes lies between the fenestra ovalis and the fenestra rotunda (*f.r.*) close in front of the head of the stylohyal cartilage. The *epiotic* centre (*ep.*) is at present but a small scute *above* the head of the latter cartilage and below the hinder end of the tegmen tympani or pterotic ridge (*t.ty.*).

660. The structures of the middle ear have acquired almost their full development. The stapes is not yet ossified. The tympanic bone is still an imperfect ring, but is thicker and broader, although no meatus externus

is as yet formed. There is an additional small ossicle clinging to the inner edge of the tympanic: it arises in the connective tissue forming the floor of the tympanum; it is a small *os bullæ*.

661. The palate has not become much broader: it is now regularly oblong. The maxillaries meet the premaxillaries, leaving the incisive foramen near the middle line on each side. The palatal plates of the maxillaries meet all along the middle line, hiding the vomer. The palatal plates of the palatine bones continue the general surface of the palate, being triangular with the bases adjacent and the apices directed outwards and backwards. The ascending laminæ of the palatines curve round the posterior nares and ascend to the base of the nasal labyrinth in its hinder region. The pterygoids are quite small. The ossification of the lower jaw is almost complete; the coronoid process is narrow and curved backwards; the condyle is still cartilaginous.

Sixth Stage: *New-born Pigs*.

662. Since the last stage the head has almost doubled in length, and the process of ossification has gone on very rapidly: the form of the skull has become much more specialized. The whole of the occipital and sphenoidal cartilages are ossified. The spheno-occipital synchondrosis is of small extent, and a scarcely thicker tract of cartilage remains between the basi- and presphenoid.

663. The perpendicular ethmoid and nasal septum are still unossified; but the inferior turbinals are almost completely, and the middle and upper turbinals (so-called lateral masses of the ethmoid) are partially converted into endosteal bone. The cribriform plate is soft, and so is the snout; but this latter is everywhere burrowed with vessels prior to ossification. The external nostrils are here inferior, the alinasal cartilage being continued forwards and

downwards, and having coalesced with the recurved trabecular horns and their continuation backwards along the lower edge of the septum. The external nostril is bounded anteriorly and partially externally by the alinasal cartilage, externally by a cartilage of the "labial" category, distinguished as appendix alæ nasi, and within and behind by the trabecular cornu. The prenasal cartilage is now undistinguishable.

664. The bony palate is now most compact, and its postero-external angles are continued by the pterygoids and the external pterygoid plates. The pterygoid is fixed posteriorly to the solid nut-like tympanic, indenting it —a temporary state of things only. The lower jaw is well ossified, while the adjacent meckelian bar has degenerated into a band of fibrous tissue.

665. The three periotic centres have completely ossified the capsular cartilage, and coalesced with one another. The exoccipital (Fig. 85, *e.o.*) and its paroccipital process (*p.oc.*) are apposed to the periotic bone behind, the process being separated from the stylohyal bar (*st.h.*) by the facial nerve (*7a*), which lies in the styloid canal, and gives off at its upper part its anterior branch (the chorda tympani, *7a′*) running upwards and forwards to curve over the tympanic membrane and enter the Glaserian fissure.

666. The processus gracilis of the malleus (the continuation of the primary mandibular arch) is reduced to a style ending in fibrous tissue (Fig. 86, *p.gr.*); the manubrium (*mb.*) is flat and slightly arcuate; the head (*ml.*) articulates with the incus by a synovial joint, the miniature of the tibio-astragalar joint in the same animal. The head of the malleus sends inwards a rounded process (*i.p.m.*), and the manubrium at its base has a posterior process to which is attached the tensor tympani muscle.

667. The hollow of the tegmen tympani (*t.ty.*) has the head of the incus in its hinder recess, which pos-

teriorly has a round cup-like facet for the short crus of the incus (*s.c.i.*); the head is also partially roofed by the ingrowing squamosal. The hooked long crus of the incus (*l.c.i.*) is tipped by a distinct orbicular ossification (*o.o.*).

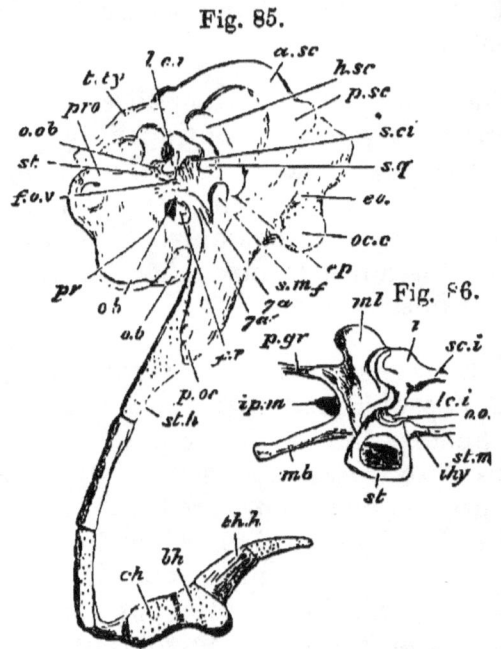

Fig. 85.

Fig. 86.

Fig. 85. The Pig at birth; outer view of auditory capsule, &c., the squamosal and tympanic having been removed.

o.b. os bullæ; *pr.* promontory; *f.ov.* fenestra ovalis; *st.* stapes; *o ob.* os orbiculare; *pr.o.* prootic; *t.ty.* tegmen tympani; *l.c.i.* long crus of incus; *a.sc. h.sc. p.sc.* semicircular canals; *s.c.i.* short crus of incus; *sq.* a small piece of the squamosal in the hinder part of the tympanum; *e.o.* exoccipital; *oc.c.* occipital condyle; *ep.* epiotic; *s.m.f.* stylo-mastoid foramen; *7a,* facial nerve; *7a',* chorda tympani branch; *f.r.* fenestra rotunda; *p oc.* end of paroccipital process; *st.h.* stylohyal; *ch.* ceratohyal; *b.h.* basihyal; *th.h.* "thyrohyal."

Fig. 86. Auditory chain of bones.

ml. body of malleus; *mb.* manubrium; *p.gr.* processus gracilis; *i.p.m.* internal process; *i.* incus; *s.c.i.* short crus; *l.c.i.* long crus; *i.hy.* interhyal; *o.o.* os orbiculare; *st.m.* stapedius muscle; *st.* stapes.

applied to the cup on the summit of the arch of the stapes. The base of the latter fits into the fenestra ovalis.

The little nucleus of cartilage which was early formed in the interhyal ligament (*i.hy.*) is now attached by its base below the head of the stapes, while its pointed distal end is buried in the fibres of the stapedial muscle (*st.m.*).

668. In the head of the stylohyal cartilage, confluent below and behind the preceding parts with the auditory capsule, a bony centre has appeared, forming the *tympanohyal*. The middle of the bar, outside the skull, ossifies as the *stylohyal*. The distal end of the bar is tied by ligament to the small hyoid cornu (*c.h.*); the larger cornua (*th.h.*), belonging to the first branchial arch, are ossified.

Seventh Stage: The Skull of a Pig six months old.

669. In this stage the greater number of the sutural landmarks, largely obliterated in the adult, are still in existence. The long angular skull is an irregular pyramid, with two equal and two unequal sides and an oblique base. A complete contrast in outward form to the human skull, that of the Pig is the straightest of all the types; it is very strongly built, but its bone-tissue is inferior in density to that of the Sheep, being intermediate in this respect between the bone of a Ruminant and that of a Cetacean. The flat top of the skull, with its orbits flush with the surface, indicates the semi-aquatic habits of its owner; and the depth and squareness of the base of the pyramid is correlative with the high neck and strong shoulders.

670. The long straight nasals overlap the snout in front, and are articulated by suture along their outer margin with the upper edge of the long premaxillaries, and for a less extent with the maxillaries; they terminate in a transverse line of suture with the frontals. The latter together form a somewhat pentagonal plate, divided along the middle line by the sagittal suture. The anterior third is deeply grooved, the grooves issuing from the

supraorbital foramina; the posterior half of the outer margin of the bones bears the thick and somewhat prominent supraorbital ridge. The frontals have a large orbital plate, bounded behind and above by the short postorbital process, and lower down and within by the orbitosphenoid. The upper surface of the parietals is narrow, and divided by the continuation of the sagittal suture; laterally they are much compressed, forming the inner wall of the large temporal fossa; behind, the upper part of the high supraoccipital abuts against them.

671. The premaxillaries have a large facial and a lesser palatal region, the palatal spurs being slender and compressed, lying together in the middle line, with an anterior palatine foramen on either side. The huge maxillary forms most of the side of the face and the anterior root of the zygoma; and the part containing the hindermost tooth-socket is bound to the external pterygoid plate and the descending part of the palatine. The horizontal plates of the maxillaries form three-fourths of the grooved and ribbed hard palate. The palatine bones complete the palate, being much longer at the middle line than externally. Altogether, the median suture of the hard palate is two-thirds of the length of the skull.

672. The inner ascending plate of the palatine articulates with the vomer, and sends forward a long scoop-like process beneath the lateral ethmoidal masses. Externally and behind they form a thick boss, which articulates with the maxillary and the external pterygoid process on the outside, and with the proper pterygoid on the inside. The latter bone has a very thin ascending part, and an upper squamous plate underlying the presphenoid—a separate piece (mesopterygoid) in some examples.

673. The thin dentate posterior end of the vomer extends to the same transverse line as the mesopterygoid plate; in front it reaches almost to the fore end of the premaxillaries. Its lower edge lies, behind, on the up-

turned edge of the palatal plate of the palatine bones, and anteriorly to this, on the long harmony-suture of the palatal plates of the maxillaries. Its upper edge is grooved by the nasal septum. The complex nasal labyrinth is well ossified, but the septum, between the inferior turbinals, is still soft. The common trabecular and alinasal cartilage projects forward beyond the bony structures of the skull, and in the adult becomes ossified endosteally as the median snout-bone.

674. The lachrymal is nearly equally developed within and without the orbit; it articulates with the frontal above and behind, with the maxillary in front, and with the jugal below. It has both an upper and a lower canal, immediately in front of the rim of the orbit. The jugal or malar bounds the lower and anterior portion of the orbit: it is massive in front, while its hinder half shelves away under the zygomatic process of the squamosal. About an inch of space intervenes between the highest part of the jugal and the postorbital spur of the frontal.

675. The squamosal by its squamous lamina forms the lower posterior boundary of the side of the skull, overlapping the hinder edge of the frontal and the lower part of the parietal. It is turned outwards at its middle region below, to form the transversely-extended glenoid cavity for the mandibular condyle. The zygomatic process passes forwards from the outer side of this articular surface to rest upon the jugal. The line of the zygoma is continued backwards and upwards by an acute ridge rising towards the upper part of the supraoccipital, bounding the deep temporal fossa. The squamosal also coalesces with the upturned mouth of the external auditory meatus, and thence gives two processes, the post-tympanic, fixed on the side of the tympanic bone, and in front of this, a postglenoid process. In front of the glenoid facet the squamosal is strongly sutured to the alisphenoid and to its external pterygoid plate.

676. The main part of the tympanic bone is somewhat

like a large filbert, ridged below; it principally consists of a mass of square-chambered diploë. It has a small cavity, and across its mouth the tympanic membrane is stretched; from its outer angle behind, a tubular meatus has grown upwards and backwards. There is a large foramen lacerum for nerves and vessels along the inner and hinder edge of the tympanic bone, between it and the basisphenoid and basioccipital. Part of the periotic mass is visible through this foramen; it is very largely unanchylosed to the surrounding parts. The tympanic abuts on the alisphenoid in front, the squamosal externally, and the exoccipital behind. The post-tympanic spur of the squamosal on the outside, the tympanic bone on the inside, and the paroccipital process behind, encircle a canal through which the styloid cartilage and the facial nerve descend, and in which the tympanohyal bone is impacted.

677. The great occipital plane is scooped above, beneath a strong transverse backwardly-turned ridge. The supraoccipital sends a wedge forwards between the parietals, expands laterally above, and descends so as to occupy a small part of the margin of the foramen magnum. The exoccipitals spread widely outwards, meeting the hinder edge of the squamosals and their post-tympanic processes, and lying behind the epiotic or "mastoid" portions of the periotic bones. From this region the exoccipitals run downwards to form the paroccipital processes; while their middle region projects and constitutes the diverging semioval occipital condyles. Within the paroccipital process, on the base of the skull, is a considerable foramen for the hypoglossal nerve.

678. The basioccipital is a pentagonal bone joining the exoccipitals by suture, and separated from the basisphenoid by a narrow synchondrosis. The basisphenoid, half the length of the basioccipital, is continuous with the ascending alisphenoids. A narrow cartilaginous tract separates the basi- from the presphenoid, which is hidden below by the vomer. The large orbitosphenoids appear

in the posterior part of the orbit around and above the optic foramen. No change of importance has taken place in the auditory ossicles: the preceding description (p. 295) is sufficient.

679. The mandibular rami are quite distinct from each other. They are large deep bones, with a small notch between the coronoid and articular regions, the coronoid process scarcely rising higher than the condyle.

680. In the three preceding stages the enlargement of the face as compared with the brain-case, and the development of the tracts which are specially strengthened for muscular attachment, have given the skull approximately its adult conformation. We may note that the ossifications of the auditory capsule were found in the fifth stage, as well as the small os bullæ; the turbinals, the os orbiculare, the tympano- and stylohyals. and the hyoid cornua in the sixth; while it is only in the latest stage that the mesethmoid and cribriform plates are ossified, and the tubular part forming the meatus is added to the tympanic. Further description of the adult skull of the Pig may be found in Prof. Flower's *Osteology of the Mammalia*.

APPENDIX ON THE SKULLS OF MAMMALIA.

681. The Mammalian skull presents great uniformity in its broad morphological features, and the extraordinary varieties of external contour which are found depend mainly upon the relative size and prominence of bones, their distinctness or anchylosis, or their overlapping one another. Remarkable cases of asymmetrical skulls occur among the Cetacea. The most important forms of the Mammalian skull are admirably elucidated in Prof. Flower's *Osteology of the Mammalia*, in Prof. Huxley's *Lectures on Comparative Anatomy*, and his *Anatomy of Vertebrated Animals*. We are yet in great ignorance of the history of the skull in most Mammalian groups; consequently only a few facts of importance in general morphology can be noticed here.

682. Reptilian affinities in the skulls of *Echidna* and *Ornithorhynchus* are found in the absence of an ascending ramus in the mandible, in the slight bending of the cochlea, in the stapes being imperforate and columelliform, in the malleus being very large and the incus small. In the *Marsupials* many of the cranial sutures persist throughout life; the squamosal, the periotic mass, and the tympanic remain separate, not uniting to form a "temporal" bone. Very frequently the jugal extends far backwards and furnishes part of the articular surface for the mandible. The internal carotid arteries pierce the basisphenoid and enter the bottom of the pituitary fossa, as in Birds, instead of coming in at the sides of the basisphenoid. There is an anterior tympanic recess comparable to that of Birds (called alisphenoid bulla). The palatal plates of the palatines have large deficiencies in Marsupials, and in the Hedgehog. Palatal plates of the pterygoid exist in *Myrmecophaga*, in *Manis*, and many Cetacea. The pterygoid is hollowed by cells in Cetacea. In *Manis* several bones of the skull have air-passages, and by this means the two tympanic cavities are in communication with one another.

683. In *Ruminants* the external pterygoid process comes distinctly from the basisphenoid. The whole sphenoid mass is developed from a pair of alisphenoid and a pair of orbitosphenoid centres; the basi- and presphenoid centres do not appear. The tympanohyal has two centres in the Sheep. In *Rhinoceros* the squamosal sends down a large postglenoid process, which unites with the post-tympanic process of the same bone to form a sort of external auditory meatus, the tympanic bone not possessing a tubular part. The "mastoid" portion of the periotic bone is hidden by the combined post-tympanic process of the squamosal and paroccipital process of the exoccipital. The horns of *Ungulata* are well known and need not be described here. In *Sirenia* the frontals are prolonged into broad supraorbital processes, and the nasals are abortive. The anchylosed tympano-periotic mass can be readily separated from the rest of the skull.

684. In *Cetacea* the supraoccipital and interparietal separate the small parietals and unite with the frontals, each of which has a great supraorbital plate. The very large maxillary extends far backwards and outwards, often largely overlapping the frontal, as well as far forwards, bounding nearly the whole of the gape. The premaxillaries have very long nasal processes extending to the an-

terior nostrils, which are superior in position. The nasal passages are nearly vertical, the nasals and turbinals small or rudimentary. The periotic mass is but loosely connected with the squamosal and tympanic, and is readily isolated. The lower jaw has no ascending ramus, and the condyle is posterior. The hyoid is broad and well ossified, with two pairs of bony cornua. In the *Carnivora* there is a remarkable tympanic bulla partly due to the periotic cartilage and partly to an extension of the tympanic bone. The periotic portion is traversed by the internal carotid artery. For the structure and variations of the bulla see Prof. Flower's *Osteology* and his important paper in *Proc. Zool. Soc.* 1869.

685. In *Rodentia* the presphenoid is mostly very distinct as a median bone; it is seldom so in the other Mammalian groups, the median tract being ossified by the orbitosphenoids. There is frequently a large distinct posterior ossification of the alisphenoid. In the Rabbit the presphenoid is high and much compressed, so as to form an interorbital septum; and the optic foramina run into one, as in some Seals. The tympanic and periotic bones are anchylosed together but not with other bones. The pituitary fossa is permanently unfloored by bone, so that there is a perforation in the middle line in the adult skull. The premaxillary is very large, with great palatal plates, and is much perforated. In the Guinea-pig (*Cavia aperea*), the pterygoids do not reach the base of the cranium, but are fixed on basipterygoid processes like those of Ostriches, which are really homologous with the external pterygoid plates of the Pig. From the sides of the basis cranii above and behind these basipterygoid processes a long continuous bony lingula passes backwards on either side, extending to the front of the auditory bulla; at the end of each is a delicate sigmoid bone, separate at first, anchylosed afterwards. The first answers to the pretympanic wing of the Bird, the second to the basitemporal. In old specimens the vomer is still found in two separate moieties. These facts indicate the high antiquity of this type. The vomer appears on the palate in the Cat and the Pangolin. In the Capybara and the Guinea-pig the alveolar border of the maxillary is prolonged far backwards beneath the orbit so as to unite with the squamosal at a level with the anterior border of the glenoid fossa. In some Rodents, as *Dipus*, *Chinchilla*, there is in addition to the tympanic bulla another expansion above the tympanic cavity, forming a rounded prominence on the postero-

external angle of the skull. In the Beaver it forms a definite angular process.

686. The Hedgehog has large pretympanic wings whose poster or margin nearly reaches the foramen magnum. In Moles and Shrews much of the skull is primarily cartilaginous: a large periotic crest projects from the upper edge of the ear-capsule, distinct from the supraoccipital, and becomes ossified, occupying a large part of the cranial wall. The squamosal is partly outside it, also appearing in front and below.

The Human Skull.

687. The circumstances which contribute most to modify the form of the human skull and the condition of its component bon. as compared with that of animals, are—1st, the proportionally large size of the brain and the corresponding expansion of the cranial bones which enclose it; 2nd, the smaller development of the face as a whole, and especially of the jaws, which brings the facial bones almost entirely under the fore part of the brain-case, instead of in front of it, as occurs in all animals, with the partial exception of the anthropoid apes; and 3rd, the adaptation of the human skeleton to the erect posture, which, as regards the head, is attended with the sudden bend of the basicranial axis at a considerable angle upon the line of the erect vertebral column; and along with this the advance of the occipito-vertebral articulation to such an extent as to make the head nearly balanced on the upper extremity of the spine. The downward opening of the nostrils, the forward aspect of the orbits and eyes, the nearly vertical forehead, and more or less oval-shaped face, are accompaniments of these human peculiarities in the form of the head, which, together with those already mentioned, strongly contrast with the smaller cranium and its strong crests of bone, the larger projecting face and jaws, and the other characteristic features of the skull in most animals[1].

688. A general acquaintance with the topography of the human skull will be assumed, in order that the references to its structure

---

[1] This paragraph is quoted from Quain and Sharpey's *Anatomy*, 8th edition, Vol. I. p. 73; this work also contains much of the information given below, but differently arranged. See also Huxley and Flower in the works before mentioned.

and development may be abbreviated. There is on the whole a more complete consolidation or anchylosis of the osseous elements than in most other mammals. Features of contrast between man and many lower forms are the union of the frontals and the moieties of the mandible; of the premaxillary with the maxillary; the distinctness of the presphenoid; the fusion of the interparietal with the supraoccipital; the large proportionate size of the squamosal and its union with the periotic mass; the comparatively small size of the tympanic, and the absence of an auditory bulla; the occurrence of fewer and less extensive elements in the hyoid arch.

689. Briefly catalogued, the following are the elements distinctly developed in the human skull, in comparison with the bones as named works on human anatomy:

Occipital = basioccipital, exoccipitals, supraoccipital, interparietal.

Sphenoid = basisphenoid (including sella turcica), alisphenoids with external pterygoid plates, pterygoids, presphenoid, orbitosphenoids.

Ethmoid = mesethmoid, ectethmoids with proper superior and inferior turbinals. Inferior turbinals or maxillo-turbinals; vomer; sphenoidal turbinal.

Parietals, frontals, nasals, lachrymals; premaxillaries, maxillaries, palatines; jugals (malars).

Temporal, composed of several principal parts:—squamosal (including zygomatic process), tympanic (including tympanic ring, external auditory meatus, and postglenoid process), and petromastoid (including prootic, epiotic, opisthotic); tympanohyal, stylohyal.

Malleus, incus, os orbiculare, stapes.

Inferior maxillary = dentary, coronary, splenial, mento-meckelian.

Hyoid = basibranchial (body), ceratohyals (lesser cornua), first branchials (greater cornua).

690. The occipital bone is almost horizontally placed, and the condyles are not far from the middle of the floor of the cranium. The supraoccipital originates by four centres, in pairs above and below: these speedily unite, but four fissures long remain between them. The upper pair, developed in membrane, represent the interparietal. There are in early life anterior and posterior fontanelles in the middle line, at the anterior and posterior angles of the parietals; there is also a fontanelle at the postero-inferior angle of each parietal.

The basioccipital and basisphenoid are separated by synchondros's up to the twentieth year. In the adult large air-cells are found in the conjoined basi- and presphenoid. A considerable portion of the nasal septum remains unossified, in front of the mesethmoid. There is a persistent aliseptal cartilage on either side, antero-inferior to the nasal bone, and united with the top of the septum; also a pair of alinasal cartilages, each bent on itself, and the two co-applied over the termination of the septum. Thus the alinasal cartilages bound the first part of the nasal cavities above, mesially, and externally, but do not reach to the base of the nostril in either position.

691. The alisphenoid nuclei appear about the eighth week of fœtal life between the foramina rotunda and ovalia, and spread thence outwards, and also downwards into the external pterygoid processes. The basisphenoid arises from two granules, lying side by side in the sella turcica, and uniting about the fourth month. After their union two other centres appear, forming the lingulæ (basitemporals), just outside the carotid grooves. The internal pterygoid plates (proper pterygoids) arise by separate nuclei in the fourth month. The alisphenoids are united to the basisphenoid in the first year after birth. The orbitosphenoids appear as a pair of nuclei, one outside each optic foramen; these extend by growth into the orbitosphenoid cartilages (lesser wings of sphenoid). Another pair of nuclei is found on the inner side of the optic foramina, and the presphenoid is formed by their union, or there may be an independent centre. At birth the alisphenoids are only suturally united with the lingulæ, which are still large in comparison with the basisphenoid. The orbitosphenoids and presphenoid are anchylosed together above the optic foramina, but the part of the orbitosphenoid beneath the foramen abuts against the basisphenoid. Later still, the basisphenoid becomes larger in proportion to the lingulæ, and the posterior clinoid processes are ossified. The sphenoidal turbinal appears after birth, applied to the front of the body of the sphenoid (pre- and basisphenoid). Each is in early life a hollow pyramid formed of three laminæ; an inferior, constituting the adult turbinal; an external or orbital portion, situated in the adult between the orbital plates of the lateral ethmoid, alisphenoid, frontal, and palatine; and a superior, forming the inner wall and roof of the original sphenoidal sinus, becoming partially absorbed and partially united to the presphenoid, which by the enlargement of the nasal cavities and sphenoidal sinuses

is ultimately reduced to the thin sphenoidal septum and rostrum. The presphenoid is for a year or two broad, and rounded inferiorly; it gradually becomes narrower and more prominent.

692. Ossification arises in the ectethmoids (orbital plates) in the fourth or fifth month, and gradually extends into the turbinal coils; during the first year the mesethmoid with the cribriform plate is ossified by a single nucleus, uniting with the lateral masses about the beginning of the second year. The ethmoidal cells are developed in the fourth or fifth year. It is only after birth, and with the gradual advance to adult years, that the spheno-occipital and sphenoethmoidal synchondroses are obliterated; the occipital mass is completely united into one bone; the vomer may become anchylosed with the mesethmoid. The epiotic fenestra existing for a long time beneath the arch of the superior semicircular canal is finally occupied by bone.

693. The maxillary, arising in the maxillopalatine process, begins to ossify immediately after the mandible and the clavicle, from several nuclei. Béclard (Meckel's *Archiv*, VI. 432) describes centres for the alveolar arch, the palatal plate, the orbito-malar tract, the nasal and facial, and the incisor. These are united at the end of the third month of fœtal life. The antrum begins as a shallow depression seen before birth on the inner surface of the bone; it deepens, extends outwards, and gradually separates the orbital and palatal portions, which at birth are close together. The premaxillary, occupying the nasofrontal process, is distinct, but is covered on its facial aspect by a process of the maxillary, which unites with it completely before birth. In all young skulls an incisor fissure is traceable on the palate between the premaxillary and the maxillary, passing outwards from the incisor foramen to the front of the canine socket. The palatine arises by one centre; the vomer by one, which developes two laminæ embracing the septal cartilage. These two plates undergo increased union from behind forwards up to puberty, forming a single median plate, and ultimately leaving only a groove on the anterior and superior surface, in which the mesethmoid fits. The nasal and lachrymal have one centre each; the jugal (malar) has one or two; the inferior turbinal one.

694. The squamo-zygomatic portion of the temporal arises by one nucleus, the tympanic by another around the tympanic mem-

brane, and these centres soon unite. The periotic capsule ossifies late, by three centres; (1) the prootic, forming the roof of the cochlea, the superior and part of the posterior semicircular canals, the internal auditory meatus, the tegmen tympani, and the upper part of the girdle of the fenestra ovalis; thus the prootic gives rise to the greater part of the petrous portion of the temporal bone, as well as the upper portion of the mastoid; (2) the opisthotic, constituting all of the petrous portion visible on the base of the skull, the floor of the cochlea, the investment of the fenestra rotunda, and half the girdle of the fenestra ovalis; it also developes the lamina which completes the carotid canal, and furnishes the inner part of the floor of the tympanum; (3) the epiotic, forming the lower and main part of the mastoid process.

695. At birth the petromastoid or periotic bone is separated from the squamosal by cartilage; bony union takes place in the first year. The "mastoid" tract is then flat, the glenoid fossa shallow, the articular eminence of the zygoma scarcely perceptible, the styloid process cartilaginous. The external auditory meatus is only developed after birth, by the arching outwards of the united bones, and the growth of the special tubular part from the external surface of the tympanic ring. The mastoid process gets prominent in the second year, but its cells only arise after puberty. A tympanohyal is found at birth and is distinct for a few years after, as a little cylindrical plug, in a depression in the hinder border of the tympanic, just antero-internally to the stylomastoid foramen. It is soon anchylosed with the periotic mass, and becomes ensheathed by the vaginal process of the tympanic. The styloid cartilaginous process is continuous with the tympanohyal, begins to ossify before birth, and may remain unanchylosed with surrounding parts even to middle life.

696. The moieties of the mandible arise at first in fibrous tissue around the meckelian cartilage. Mr Callender[1] has discovered a mento-meckelian ossification of the cartilage at the symphysis; and there appear to be splenial and coronary elements on the inner side, in addition to the dentary on the outside. The upper end of the meckelian cartilage becomes ossified as the malleus, and the cartilage extends downwards and forwards between the tympanic bone and the periotic capsule; the processus gracilis of the malleus, lying

[1] See *Phil. Trans.* 1869, p. 163.

in the Glaserian fissure between the squamosal and the tympanic, is the limit of ossification, the rest of the cartilage becoming gradually obliterated. The hyoid arch is represented above by the incus, the tympanohyal, and the styloid process or stylohyal. The auditory ossicles are at first altogether outside the tympanic cavity; and as the latter enlarges, its mucous membrane is reflected around the ossicles. The stapes is described as having three ossific centres. The hyoid bone is developed by five centres, for the two pairs of cornua and the body.

697. The dates of appearance of the different osseous centres are as follows ; sixth or seventh week of gestation, dentary, premaxillary, maxillary, frontal ; seventh to eighth week, basioccipital, exoccipital, supraoccipital, basisphenoid, alisphenoid, presphenoid, squamosal, parietal, palatine, vomer, nasal, lachrymal, jugal or malar; fourth month, pterygoid, lateral ethmoid (middle and upper turbinal); fifth month, inferior turbinal; fifth and sixth months, periotic bones; eighth month, hyoid (great cornua, body, small cornua), tympanic, tympanohyal; stylohyal; first year, sphenoidal turbinal, mesethmoid and cribriform plate.

698. The union of distinct bones takes place in the following order: the two basisphenoid centres unite in the fourth month of fœtal life; the pterygoid and the external pterygoid plate in the sixth month; the presphenoid and basisphenoid in the eighth month. During the first year after birth the periotic mass unites with the squamosal, the alisphenoid with the basisphenoid, and the moieties of the lower jaw coalesce at the symphysis. The second year witnesses the union of the frontals with one another, and of the mesethmoid with the lateral masses. The exoccipitals coalesce with the supra-occipital between the second and fourth years, with the basioccipital from the fifth to the sixth year. The basioccipital and basisphenoid are anchylosed only after the twentieth year. The styloid ossification unites with the "temporal" in adult life; the osseous union of the hyoid cornua with the body of the bone does not occur till after middle age.

CHAPTER IX.

THE MORPHOLOGY OF THE SKULL.

699. HAVING now described in a condensed fashion the more important facts which have been ascertained respecting the history and structure of the skull in various forms, we proceed to summarise those facts, and to indicate some conclusions which appear reasonable. A larger measure of consideration is asked for these conclusions than if they merely arose from a verification of the facts which have been detailed; for they have become developed in the course of many years of exploring work, in concomitance with a transition from the darkness of archetypal fancies to the clear light of actual verifiable history. In many cases the views expressed are based upon a much more extensive acquaintance with skulls in all stages of development than has been indicated; a mass of detailed information remains for future publication.

The Cartilaginous Skull.

700. In all cases the primary elements of the cartilaginous brain-case consist of two pairs, arising beneath the matrix from which the dura mater developes; (1) the parachordals posteriorly, under the greater part of the hinder division of the brain; and (2) the trabeculæ beneath the forebrain. Whatever cartilage appears subsequently in the brain-case grows nearly always in direct continuity with these elements, by gradual chondrification of mesoblast. In this way a more or less complete cartilaginous box is ultimately formed.

701. The notochord is related to the whole length of the parachordals and to the hinder extremity of the trabeculæ. The stratum in which it originates is immediately below the nervous axis, and somewhat above that in which the cartilaginous elements arise. The anterior end of the notochord is often curved upwards towards the second brain vesicle; but in some cases it is turned downwards again between the trabeculæ; and it has moreover been detected in Elasmobranchs actually bent double, and running backwards some distance before terminating. In other forms the flexure of the notochord is very slight; it simply bends downwards a little. In embryo Sharks and Skates from half an inch to an inch in length, the notochord is seen to present a number of beadings in its extreme anterior portion; this feature is evanescent. In the Fowl at an early stage the cranial notochord has two constrictions dividing it imperfectly into three spindle-shaped regions; and this continues observable for a considerable time. So far as we can make out, there is nothing in the structure of the notochord itself which gives us very definite information about its segmental relations or about the segmentation of the skull. The point to be remembered is, that the notochord has a relation to the trabeculæ as well as to the parachordals.

702. The parachordals become identified sooner or later with the tube of tissue surrounding the notochord, which in several cases (Bombinator, Dogfish, &c.) is definitely chondrified previously to this confluence, and which is continuous with the cylinder of cartilage surrounding the rest of the notochord. Each parachordal becomes coalesced with the trabecula and the ear-capsule of its own side, and developes more or less of the lateral occipital wall, before uniting with its fellow. In fact, in the majority of cases the parachordals during development recede from one another in their anterior half or two-thirds; and concurrently with this, the notochord becomes, by the growth of surrounding parts, apparently retracted

in the cranial floor, so as to appear only between the hinder part of the parachordals. Thus for a time there is constituted a posterior basicranial fontanelle, or membranous space in the hinder part of the floor of the chondrocranium.

703. When the parachordals unite in the region where the notochord still persists, it is by the growth of cartilaginous bridges both over and under it. The bridge beneath the notochord is very marked and becomes thick; the cartilage is thinner above, and often non-existent for a long time, so that the notochord lies in a groove on the basilar plate constituted by the union of the parachordals. In many cases where a basicranial fontanelle exists, the cartilages do not approach one another again, and the fontanelle is only closed by bony growth; but in other types, especially in Birds, the space is nearly obliterated by the growth of cartilage, a later bony deposit completing the work. The whole of the cranial notochord is gradually aborted in most instances, and its place is occupied by cartilage; but in various forms a remnant is left as a slender string, embedded in the basioccipital bone or cartilage.

704. The length of the parachordal cartilages compared with other parts varies greatly. In some cases (Salmon, Fowl) the parachordals stretch anteriorly almost to the extreme notochordal apex, and posteriorly to a considerable distance behind the ear-capsules, being at first much more bulky than the trabeculæ. It is noteworthy that the parachordals extend in the Elasmobranchs into a very considerable region of the neck. A segmentation takes place afterwards, separating the cranial from the cervical cartilages at the usual place, and forming two basal condyles convex backwards. In all other known forms the hinder extremity of each parachordal primarily forms a condyle: and thus in every cartilaginous condition of the vertebrate skull where condyles are formed, there are two occipital condyles.

705. The parachordal is always bent outwards more

or less, behind and in front of the ear-capsule, and partially undergirds it. In Elasmobranchs it grows quite underneath the otic mass, and is even carried outwards somewhat beyond it. The paroccipital process in higher vertebrates is an extension of the parachordal around, outside, and beyond the ear-capsule. The preauditory growth of the parachordal is more or less in confluence with the hinder end of the trabecula.

706. The trabeculæ are at first simply solidifications of tissue, afterwards chondrified, in the sides of the floor of the first cerebral vesicle, which almost bulges down between them. The hinder termination of the trabeculæ (Salmon, Frog) is often very slender at first. In others, as in Urodeles, they are largest behind, and early embrace the anterior half of the cranial notochord; in Sharks they are about of equal breadth throughout. In cases where they are early in contact with the parachordals, they lie over them at an angle. The two trabeculæ have always a more or less lyriform appearance, being approximated in front, beneath the fore part of the brain-case, and also behind. In the posterior portion of the intertrabecular space the infundibulum and pituitary body are found; but the primary interval between the trabeculæ is related to the whole of the first cerebral vesicle, and not merely to these bodies.

707. The primitive trabeculæ are flat or more or less rounded in section. The position they occupy in the head undergoes change in consequence of the mesocephalic flexure; at first, where early distinct, they are on the whole on the same level as the parachordals; but the mesocephalic flexure may bend the former down to an angle of $120°$ or more with the latter, and in other cases they occupy this position when first distinguishable. As growth advances this flexure is entirely lost, and the level of the trabeculæ and the parachordals may again become approximately identical.

708. It is on the whole in the lower types that the

trabeculæ are perceptible as primitively distinct from the parachordals; and this condition would appear to be due to the mesocephalic flexure. In higher forms (Bird, Pig) they can at no time be found clearly separated, although a demarcation may appear to exist as the rudiment of the posterior clinoid ridge. Where the dimensions of the trabeculæ are least bulky behind, they may barely embrace the apex of the notochord: in every case where they are distinct they do this. Where they are of greater size posteriorly, they are related to a varying length of the notochord; in the extreme case (Axolotl), to more than half the cranial notochord, tapering backwards in close contact with it, and attaining their greatest width opposite the notochordal apex.

709. There is a most interesting variation in the relative size and date of appearance of parachordals and trabeculæ. In some types the parachordals are large and definite from the first, and the trabeculæ disproportionately small and little solidified (Salmon); in others the trabeculæ are clearly manifest and solid, and have taken up their definite relations considerably before the parachordals have appeared at all (Axolotl, Frog). In Elasmobranchs the two are very similar, but the trabeculæ chondrify first.

710. If not existing at the earliest time of solidification, a prechordal (postpituitary) bridge is formed as a meeting-place of trabeculæ and parachordals, except where the parachordals do not extend so far forwards. In most cases this bridge appears to be wholly or in great part formed by the trabeculæ. Furthermore the posterior clinoid ridge appears usually to be clearly a trabecular product, either arising as a transverse ridge on the united basicranial cartilage (Fowl, Pig), or being produced at first by the trabeculæ overlapping the parachordals.

711. At the broadest region of the trabecula, in front of the notochordal apex, a small curved tongue or lingula may be given off, remaining free from other parts, and directed backwards (Fowl); or such a process may be

large and become united with an antero-external process of the parachordal (*Carcharias glaucus*).

712. The anterior extension of the trabeculæ at their first appearance varies very greatly with the extent of the early development of the snout. They always from the beginning underlie the whole of the anterior part of the brain-case; but in some cases (Frogs and certain Newts) they pass between the nasal sacs to the prenasal region very early, and throw out cornua. In others they very soon extend to the internasal region, where the septum subsequently forms. The trabecular modifications in the region of the cranial cavity will be first dealt with.

713. The trabeculæ underneath the fore-brain, which are at first separated by an interval, but are closest together anteriorly, in many forms very early become approximated all along their length, except where the pituitary body and the carotid arteries lie between them. They may be placed quite flat together (Elasmobranchs) or obliquely at an obtuse or right angle looking downwards (Salmon). Coalescence quickly takes place in all the region of apposition, and a trabecular plate occupies the whole floor of the brain-case in front, adapting itself to the shape of the brain, being flat or gently curved, often rising more or less towards the anterior or ethmoidal extremity of the brain-case. The prepituitary part of the trabecular cartilage may develope a distinct anterior clinoid ridge, and thus a squarish or circular pituitary fossa may become limited. At a later period in many cases a thin floor of cartilage, continuous with the trabeculæ, arises in the bottom of the pituitary fossa, interrupted only by the perforations for the internal carotid arteries. The trabecular plate sometimes extends outwards beyond the cranial cavity, so as to partially support the eyeballs.

714. In Amphibia generally the primary trabeculæ remain apart where they underlie the forebrain, but in the Anura a thin cartilaginous floor unites the bars and fills up the primordial fontanelle. In Snakes the trabeculæ

are permanently separate as far as the nasal region, and retain a simple form. In Eels they are separate anteriorly.

715. The most remarkable modification of the cranial trabecular region is that which gives rise to an interorbital septum in Osseous Fishes, Lizards, and Birds. The roof of the mouth gradually becomes separated by a considerable distance from the floor of the cranium, and the deeper portion of the eyeballs comes to lie between the palate and the cranial floor. A median membrano-cartilaginous vertical wall (interorbital septum) separates the orbits from one another; and the cartilage which it acquires is usually due to the growth of a median trabecular crest. Thus the forebrain becomes supported mesially on a wall which rises high in front of the pituitary region; and lateral outgrowths from the top of this septum underlie the sides of the brain. The interorbital septum may be thick and solid, or thin with some thickening at the base; and a greater or less extent of it may never chondrify, or may become membranous after having once been cartilaginous.

716. When the trabeculæ primarily extend as two distinct rods into the internasal region, they are early coalesced into an internasal plate, which forms the base of the nasal septum where that is definitely developed. The same term of internasal plate is also applied when the internasal chondrification, continuous with the trabeculæ, is single from the first. This plate may be a simple bar at the base of the internasal region, very like that produced by the early coalescence of the two trabeculæ as above described, appearing as a continuation from the level of the cranial floor and sometimes permanently retaining this form; or it may first arise as a prolongation of the vertical interorbital septum just described. In any case, anteriorly to the internasal tract, an azygous prenasal structure is very often developed; either directed straight forwards (Skate), or curved downwards as the axis of the beak (Bird). In the Skate it attains a very large development forwards as the axis of the rostrum,

and has a grooved upper surface continuous with the internasal floor. In the Chick at an early stage, and in the Tortoise, it is bent underneath the axial parts and becomes recurrent. Where the adult has a very blunt snout, the prenasal element is short or almost absent (Salmon, Axolotl, Frog).

717. The main part of the nasal septum may at first arise double, as in the Pig, the two laminæ very early coalescing. When the nasal organs are widely separated there are wings of cartilage (nasoseptal laminæ), which arch widely outwards from the basal internasal region and unite with the olfactory capsules. They may become so outspread in adult life as to appear continuous with the internasal plate, only rising gently on each side.

718. Lateral outgrowths always proceed from the axial cartilages in the nasal region. The olfactory organs become supported by anterior and posterior growths, the trabecular cornua, and the antorbital or lateral ethmoidal cartilages. The cornua in some types are visible at almost the earliest stage of the growth of the skull, as outward curvings of the trabeculæ, jutting forwards in front of the nasal sacs: and they may soon show traces of division into two lobes, of which one, external, broadens and partly underlies the nasal sac and defends it anteriorly, while the other bends inwards and downwards, becoming more or less recurrent underneath the internasal floor; and finally these recurrent cartilages, in some Birds, coalesce into one, quite independently of the prenasal part, which exists at the same time.

719. The antorbital cartilages in some respects agree with the cornua, in having a similar relation to the nasal capsule, in arising early from the trabeculæ at their anterior junction: and they may be almost completely recurrent (Skate). In Urodeles they arise quite separately from the ethmoidal angles of the skull. They frequently acquire a greater vertical depth than the cornua, in accordance with the height of the corresponding parts;

and they may become the points of attachment or coalescence of a region of the mandibular arch, or cartilaginous parts developed from it. The antorbital frequently coalesces with the back of the nasal wall, and in that circumstance is sometimes carried far outwards by the growth of the capsule, being disjoined from the axial part from which it at first arose.

720. We have now to trace the formation of a cartilaginous brain-case in the tissue surrounding the primary membranous cranium, in continuity more or less perfect with the basal structures which have been described. Separate cartilages are seldom developed in any part of the cranial boundary, but in many cases it is quite impossible to determine whether the chondrification has been by a conversion of the cells at every point of the tissue, or by a proliferation of the cells in cartilage already existing: probably both processes occur. The formation of lateral walls in cartilage is frequently not so complete as the furnishing of basal parts: and their extent and style seem to be very much influenced by the impaction of the three pairs of sense-capsules. The roof is still less perfectly chondrified in many types, and has its own special regions of fenestration.

721. In all cases a cartilaginous occipital ring, bounding the foramen magnum, is formed by the upward growth of the parachordals on either side, meeting above and coalescing. This ring is perfectly definite and continuous behind the periotic masses: but is complicated with them in their hinder region. The occipital ring in most cases is not limited to the region behind the ear-capsule, but the cartilage extends above and supero-laterally between the ear-capsules in the cranial roof, sometimes stretching forwards for two-thirds of their length. The occipital investment may be left somewhat imperfect below in the notochordal tract; but in these cases there is a remnant of the notochord.

722. A cranial wall distinct from the ear-capsule cannot be clearly made out in the auditory region.

Further forwards there is a cartilaginous wall, interrupted by nerve foramina and by fenestræ. This wall may be approximately vertical in lower types; but in ascending the scale, it becomes more and more tilted outwards and overhanging, so that ultimately (Pig) parts which were constituents of the wall in lower types enter into the composition of the floor of the large cranium, and the greater part of the wall and roof is unchondrified, and forms a vast fontanelle in the cartilaginous skull. It is especially to be noted that the types which possess an interorbital cartilaginous septum are as a rule deficient in their cranial walls. The regions of lateral cartilage will be defined when we come to speak of nerves; but it must be mentioned here that in some cases (*e.g.* Snakes) a small cartilage called alisphenoid arises in the tract immediately in front of the ear-capsule, in the cranial wall or external part of the floor; and this may also appear as a *process* of basicranial cartilage, not cut off from it. In front of this a larger wing of cartilage, the orbitosphenoid, may be developed, also having definite relations.

723. The lateral cranial wall originates in the simpler types by the growth of a longitudinal crest upon the trabeculæ; this is most prominent behind, and often early confluent with the auditory capsule behind and the nasal in front. Where the cranial region is definitely marked off (in all but Elasmobranchs), this crest is continued inwards in the ethmoidal region, the two crests meeting so as to form with the floor a trough or barge-like hollow in which the brain lies: and at the same time the ethmoidal tract may be confluent with the nasal sacs.

724. The lateral chondrification may become continuous with a superior (supraorbital) ridge of cartilage connecting the ethmoidal and nasal regions with the auditory, and finally the chondrification may proceed inwards towards the centre of the cranial roof. The tegmen cranii (Frogs, Sharks, Rays, &c.) is of varied size, usually leaving one large fontanelle in front, or two hinder

small ones in addition (Frog). The growth may take place chiefly from before backwards, as in the Salmon and Frog; in the Salmon it almost entirely covers the brain-case with a thick roof or tegmen, excepting only two small postero-lateral or parietal fontanelles in which cartilage is permanently deficient. In Urodeles the growth of the cranial floor may be arrested early and the trabeculæ may remain comparatively widely separated, and finally appear as totally absorbed into the side walls, forming part of them; so that the cranium has two complete side walls but no cartilaginous roof or floor in the trabecular region.

The Sense Capsules.

725. The organs of special sense influence the architecture of the skull in a very marked degree, the eyeballs being no less potent than the nasal and the auditory capsules in affecting the external form of the skull and the style of its intimate structure. The sense-capsules have in common a more or less complete cartilaginous investment, and an intimate association with and protection by the cranial cartilage. They are situated in pairs at the sides of the skull, and are related to a more or less extensive fenestration of the cranial cartilage where they abut upon it. Each capsule receives a special nerve of sense, besides being supplied with other nerves. The cartilaginous investment is not perfectly complete in any case, or if it is complete at any period, does not remain so, orifices occurring through which the efficient agencies in exciting sensation may work, or in some cases as the relic of a primary involution. But the olfactory investment differs from the other two in being usually like a cap or dome, more or less widely open below, while they are more complete, and rather resemble balls.

726. The auditory masses are on the whole of larger size relatively to the cranium in lower types, and in ascending we find them more and more subordinated to

the enlarging brain-case, and inwrought into its general contour. In the lower groups the height of the ear-capsule is nearly equal to that of the cranium, and its breadth may be as great ; and the close impaction of the organ upon the brain-case does not allow of the formation of a distinct cranial wall of cartilage at that region, and leads to a coalescence of cranium and capsule at all the borders, with marks of distinctness at various points. Thus we may speak of the lateral tract where there is no distinct cranial cartilage as constituting a large auditory fenestra. In higher forms we are not able to perceive the formation of cranial and auditory cartilages separately from one another; the growth is continuous from the first: but the real fenestration may become obvious at a later stage, after ossification has taken place, by the lack of anchylosis with other parts throughout the whole or great part of the borders of the periotic mass.

727. In many instances a small postero-superior region of the capsule never chondrifies, or remains unchondrified till a late period of development, forming an epiotic fenestra. It marks the primary connection of the auditory mass with the epiblast. Another fenestra is formed in the Salmon infero-mesially, appearing at first as a space between the parachordal cartilage and the capsule; but the former travelling outwards and the capsule lying over it, and becoming identified with it, the fenestra travels also outwards, and becomes really an opening in the lateral aspect and floor of the vestibule, afterwards closed. Its ultimate position is similar to that in which the stapes arises in Amphibia. A deficiency occurs in the cranial surface of the ear-mass in some Teleostean fishes, so that there is no cartilage at that part between the membranous labyrinth and the cranial cavity.

728. A remarkable feature about the ear-capsule is the formation at an early period in nearly all types of an external more or less horizontal projection, called the pterotic ridge: it is related to and often contains part of

the horizontal semicircular canal. Where there is no tympanum this ridge overhangs the articulation of the appendicular elements of the skull. When a tympanic cavity is developed, this ridge is specialised into a tegmen tympani, or tympanic roof; and then there is a definite scooping of the ear-mass beneath the ridge. This cavity may be still more defined by a posterior boundary and partial floor derived from the exoccipital cartilage, and by a cartilaginous floor derived from the ear-cartilage itself (auditory bulla).

729. The surfaces of articulation and regions of coalescence of the periotic capsule with appendicular parts will be mentioned later; but one point must be referred to here: namely, that a fenestra (ovalis), which has an intimate relation to the function of hearing, arises in the part of the capsule adjacent to the tympanum, either by dehiscence of the cartilage, or by the cutting out of a small segment, the stapes, which remains in the hole out of which it had been cut. The stapes is formed in other cases by the chondrification of the tissue in the membranous fenestra. In the Urodeles the origin of the stapes may be described as the cutting off of an opercular fold over the fenestral slit; and here the stapes is only connected by ligament with parts at some distance from it. But in higher forms the stapes is always closely tied to, or coalesced with, or originates in continuity with extraneous elements. Another fenestra (rotunda) originates in the wall of the cochlear rudiment in many Reptiles, and occurs in a similar situation in all Birds and Mammals.

730. The mobile eyeballs are not amalgamated with the cranium like the auditory masses; but the cavities in which they are lodged show most admirably the pliability of the axial parts, and the manifold shades of adaptation which they can assume. Whether the eyes occupy the sides of the whole fore part of the brain-case, or are embedded beneath it, or elevated above it (Chimæra), or carried outwards (Zygæna), the orbits are bounded by the same cartilaginous structures little modified. In the

simple lateral position of the eyes (Sharks), the outward extension of the trabecular base of the cranium provides a partial floor for the orbit, the projecting supraorbital tract protects it above, and the antorbital and postorbital or sphenotic terminations of that tract wall it in before and behind. The palatal structures and aponeurosis form the rest of the orbital floor. In the Amphibians, which have simple lateral orbits, there is little furnishing of protecting laminæ; but the history of the suborbital cartilages in the Anura is of great interest, the subocular bar being first the mandibular suspensorium and afterwards the palatopterygoid cartilage. The eyeballs in an early stage rest almost entirely on cartilage; but as the subocular fenestra is gradually widened with the growth of the head, the eyes come to rest almost exclusively on membrane. In these types the optic foramen is always larger, often much larger, than the nerve which passes through it, the rest of the space being filled by membrane.

731. When the eyeball is more deeply embedded and more thoroughly protected, we have a modification of the cranial cartilage which is as if the lower regions of its side-walls were forced together and converted into a mere vertical septum. The fore part of the cranial cavity is elevated upon this septum, the portion of the brain therein contained becoming small proportionately, and lying between the upper part of the eyes, or even totally above them. In this condition, as growth proceeds, a more or less extensive fenestration of the septum very frequently occurs, and there may also be (Lizard) fenestration in the slanting cranial roof of the orbit (orbitosphenoidal tract), as well as an optic foramen larger than the nerve.

732. In Birds the orbit is protected by the cranial wall slanting outwards above and behind, by the prominent supraorbital ridge above and behind, and by the antorbital or postnasal plate of the nasal capsule, which may be prolonged into an external and inferior process (os uncinatum of Finches and Parrots). No very definite floor is developed, the palatine tracts being little specialised

in relation to the orbit. In the Mammal an additional circumstance influences the position and relation of the orbits, namely, the extension backwards of the nasal capsules beneath the fore part of the cranium and between the eyes. In its cartilaginous condition the orbit is roofed by the orbitosphenoidal tract, and partly floored and walled by the lateral ethmoidal cartilage. It is subsequently completed by bones, but may continue to be very largely supported by membrane beneath. A very considerable portion of the orbit may in these cases be below the level of the cranial floor.

733. The sclerotic investment of the eyeball is cartilaginous in most types except the mammalian. In many Elasmobranchs and some Osseous Fishes the eyeball is supported near the entrance of the optic nerve by a movable cartilaginous rod or pedicle, articulated with the sclerotic and with the cranial cartilage.

734. In its simple condition in the Elasmobranchs the nasal capsule is a dome of cartilage widely open below. Its sensory membrane is produced into many folds, but these are not supported by cartilage. The capsular cartilage does not grow out from the cranium, but early becomes continuous with it behind. More or fewer labial cartilages become involved in the valvular inferior opening, and may coalesce with the proper nasal investment. A lateral union of the capsules with the internasal cartilage occurs by the intervention of nasoseptal laminae. A fenestra may arise in the inner superior aspect of the capsule, in addition to the large nerve fenestra more posteriorly, through which the olfactory fibres pass downwards.

735. In the Salmon separate capsular cartilages cannot be said to exist: there is at an early period a median internasal tract of cartilage with anterior and posterior horns, and a floor formed by the conjoined trabeculæ. In the adult the membranous nasal capsule lines a scooped hollow on either side of the massive snout cartilage, and is protected by bones. In most Urodeles there is the

same massive internasal cartilage and anterior growth from the cornua; but there is more or less of distinct roofing cartilage belonging to the capsule; no labial cartilages here become related to the nasal openings.

736. In the Frog a special vertical nasal septum is formed. The olfactory sacs rest at first on the anterior prolongations of the trabeculæ. Later, they lie between them and the distal end of each mandibular suspensorium —a remarkable condition. Finally, the nasal septum being formed, a supero-external proper nasal roof arises, and the cornua and labials are dovetailed into the boundaries of the capsules, so as to leave two openings, a superior and an inferior. The roof bends downwards behind so as to form a hollow conchoidal recess. In other Anura which have a nasal septum, the true capsular cartilage may be very imperfect or almost non-existent, together with a most extensive adaptation of true labials to the defence of the sacs (Pipa, Dactylethra). In the Frog we see the first appearance of cartilaginous laminæ within the capsule at its fore part. Similar laminæ occur in many Reptiles.

737. In Birds and Mammals the olfactory capsules rise to a very high grade of complication. They may increase in size so as to occupy far the greater part of the length of the skull; but some considerable part of the complication connected with them is due to bones which are not preformed in cartilage; these need not enter into our consideration of the capsules proper. Essentially we may consider the capsules in these classes as consisting of the median vertical septum, and of outgrowths from it or arising continuously with it, forming roofs, side-walls, posterior and partial anterior walls, and more or less of a floor on the external aspect, by folding in of the side-wall. In a few forms this floor becomes complete (Falcons). From the inner surfaces of these capsular cartilages laminæ of cartilage proceed inwards towards the septum and become coiled in very various styles. Three principal regions are distinguished, aliethmoidal (posterior), aliseptal (middle), and alinasal (anterior), the latter being related

to the external or anterior nostril. Turbinal laminæ may arise in all these regions, having special characters of their own: those of the aliseptal region (inferior turbinal, nasal turbinal) are often preponderant in size; those of the aliethmoidal region (middle and upper turbinals) are especially related to the sense of smell. In Mammals all these growths are seen at their highest perfection, and in many cases the nasal sacs burrow backwards beneath the cranial cavity into the sphenoidal region. Both in the Bird and in the Mammal there is an adaptation of primordial labial cartilages to the purposes of the anterior nostril: in the Pig this orifice is inferior. The posterior nostril is also provided with special labial cartilages, appearing in the vomerine region, in Snakes and many Birds.

738. In the Mammalia the space in which the olfactory fibres pass down into the capsules is at first a large membranous fenestra comparable to that for the optic nerve of the Frog; the fibres pass in a scattered fashion through this membrane. During growth cartilage appears in the interspaces between the fibres, and forms the cribriform plate continuous with the top of the hinder part of the nasal septum. Another fenestration, corresponding to that which occurs in the interorbital septum, arises in the nasal septum of many Birds, and this may become very large, aborting the previously existing cartilage so much as to produce a large notch, opening below: this is principally in the aliseptal region. In some cases there are several fenestræ in the septum.

The Arches.

739. The arches appended to the cranium have certain resemblances to those of the trunk. They enclose a cavity, are liable to segmentation, and, have very similar nerve-relations. The non-attachment of the gill-arches to the cranium or to vertebræ contrasts with the adherence of ribs to the vertebral column.

740. All the postoral arches may bear gills in some vertebrates or at some period of development. The mandibular and hyoid are highly specialised in regard to various functions; the last branchials also may be considerably modified. A simple idea of the series may first be considered, viz. the average branchial arch in most Fishes. A continuous rod of cartilage originates in each visceral fold at the sides of the throat, bending more or less backwards in the pharyngeal roof, free from the cranium and vertebræ, and passing forwards and inwards below. The inferior ends of each pair of arches usually become connected by the intervention of a median keystone piece or basibranchial, very comparable with the azygous elements of the sternum. By a simple segmentation these branchial arches are perfected. Four segments arise, and a more or less perfect jointing allows of mobility of the parts. The upper and the lower segments are usually more or less horizontal in position; the two others are the more important, and are chiefly lateral. The number of the branchial arches in Fishes varies from five to seven; the hindermost of them is modified, attached to the one before it, and deficient in one or more of its segments. The basibranchials become attached to one another in the median ventral line.

741. The hyoid arch in Fishes becomes divided by segmentation into the same number of parts as the branchials, and there is a basihyal: but the mandibular arch on each side may be only segmented into two main pieces, a suspensorial and a meckelian. The suspensorial cartilage in all oviparous vertebrates has a forward prolongation which enters into the maxillopalatine process of the cheek, and may be so large as to constitute the entire upper jaw and enter into no definite union with other parts (Shark); or again, it may unite with a retral antorbital or palatine growth to form the main palato-quadrate bar (Frog, Salmon). In this case we have a conjunction of evidently distinct parts. The meckelian cartilage does not undergo segmentation, nor does it form a basal piece. In some

cases the right and left cartilages fuse at the symphysis. The joint between the lower jaw and the suspensorium appears to correspond in position to those between the epibranchials and ceratobranchials.

742. The relations of the upper end of the mandibular arch to the cranium, and the methods of suspension of the lower jaw, offer numerous points for consideration. In certain forms the mandibular arch is directly and strongly fixed to the cranial floor or wall, or to the ear-capsule; and concurrently with these conditions there is a complete distinction between the hyoid arch and the mandibular; or again, the former is so tied to the latter as to take a considerable part in the suspension of the lower jaw. When the mandibular suspensorium entirely or principally supports the lower jaw, the skull has been called *autostylic*[1]; when it largely shares the work with the hyoid, the term *amphistylic* is applied. But in certain cases the mandibular arch does not retain its primary relation to the cranium, but becomes distant from it, and not directly fastened to it; it is firmly tied to the upper portion of the hyoid arch, which thus becomes the principal suspensor of the jaw; these skulls are *hyostylic*.

743. The primary relation of the mandibular arch to the cranium is to its side-wall and base. In the Axolotl it is at first applied to the trabeculæ far forwards, but soon becomes placed in the usual situation, namely, just in front of the ear-capsule and near the primary apex of the trabecula. An ascending process in some forms brings the arch into relation with the side-wall of the cranium (various Urodeles); and either of these regions may coalesce with the cranial cartilage. As a result of transformation, or as a primary relation in higher types, the mandibular arch may be brought into connection with the capsular cartilage of the ear, may develope a surface for articulation with it (Frog), or may unite with it. A backward prolongation of the upper region of the arch (otic process) may lie

[1] Huxley, *Ceratodus*, l.c.

along the side of the capsule or coalesce with it. Furthermore, a hinder and superficial leaf of the suspensorium is cut off in many Anura, and becomes adapted to the tympanic membrane as the annulus.

744. The suspension of the lower jaw in hyostylic skulls is effected by a smaller or greater modification of the primitive form of the hyoid arch. The upper part of the arch either arises separately, or is cut off from the rest of the arch; being carried forwards at its lower end, it is firmly attached to the lower part of the suspensorium and to the jaw. In a primitive condition (Shark) the upper end of this hyomandibular element is articulated both with the basis cranii and the auditory capsule, underneath the pterotic ridge.

745. The segmentation of the hyoid arch to form this hyomandibular may either be transverse or oblique, in a position comparable to that of the joint between the epi- and ceratobranchials; or it may be longitudinal (Salmon), affecting nearly the whole length of the arch, the anterior (hyomandibular) piece becoming superior, and the posterior inferior. In amphi- and autostylic skulls the hyomandibular piece may be comparatively small, or may not exist.

746. In many Anura and Sauropsida, when the hyoid arch chondrifies, it appears to be only the lower portion of an arch. Sometimes at an early, sometimes at a comparatively late period of metamorphosis, a series of structures originates near the auditory capsule and behind the mandibular suspensorium, in precisely the same situation as the hyomandibular, the nerve and cleft relations being identical. They failed formerly to be recognized as portions of the hyoid arch, because they did not originate continuously or contemporaneously with its lower segments, or because in the adult they may not be continuous, their directions being different. These structures form the columella auris, and always enter into connection, sometimes coalesce, with the stapes belonging

to the auditory capsule. One or more segments may arise in this upper hyoid tract; and its outer end is usually in relation with the tympanic membrane, and helps to support it.

747. In Mammalia, while the primitive condition is more complete, the metamorphosis is in some respects more extraordinary. Both the mandibular and the hyoid arches are of full length and development at first; and their upper ends become related to each other and to the tympanic cavity. The mandibular arch is never segmented; but its meckelian region does not form part of the functional lower jaw. Its upper tract, remaining small, is specially moulded for purposes connected with hearing, lies in the tympanic cavity, and supports by one of its processes the tympanic membrane. The upper part of the hyoid arch is bent inwards to be fastened to the stapes, and then the hook, apposed by its elbow to the top of the mandibular arch, is cut off, becoming the incus; the main part of the arch is carried backwards and coalesces with the auditory capsule. The processes of the incus are exactly paralleled by the processes of the columella in Frogs.

748. Retrogression in the cartilaginous cranium of particular forms can hardly be said to exist, except in the Urodeles, where some of the axial basicranial cartilage is absorbed. But as we pass from branchiate to higher types the proportionate amount of cartilage that occurs in the cranial investment is diminished.

749. The mandibular arch can scarcely be said to undergo retrogression: at a varying period of development it ceases to grow, and becomes a less important factor in the lower jaw in each ascent of type. The hyoid and branchial arches may however be said to retrograde, but the change is related to new functional adaptations. These phenomena occur especially in the caducibranchiate Amphibia. The main (lower) part of the hyoid arch, from being short and massive, and attached to the suspensorium,

becomes long and slender and attached to the ear-capsule; in some it is entirely absorbed. In Frogs the branchial arches are originally quite distinct from one another, but become united into a basket-work by transverse bars both above and below; and then gradually during the loss of the water-breathing function, the whole branchial structure is almost completely absorbed, leaving rudiments which are related to the larynx, and which unite with a broad basihyobranchial plate very much resembling a sternum. In other cases the branchial arches in becoming modified are variously tied to one another; and the remnants are of very different patterns. Frequently what looks like a process of transformation is merely a cessation of growth combined with adherence to a basal attachment.

750. In the abranchiate forms the early hyoid arch may elongate, remaining slender, and there may be continuous cartilage between the basihyal region and the ear-capsule; or a portion of the arch becomes merely ligamentous, leaving cartilage proximally and distally. One pair of branchial cartilages is found, ceasing to grow at an early date; they mostly do not disappear, but become applied to, and support the larynx: a relation to the orifice of the breathing apparatus in air-breathers parallel with that of the branchial arches to the respiratory orifices in water-breathers. In the Frog and in Mammals paired branchial rudiments partially embrace the larynx. In Birds it is the posterior end of the basibranchial bar that is applied to the larynx: in Urodeles it is the same element which is thus adapted, but it is completely segmented off from the remaining parts. In Lizards there is no application of the branchial arches to the larynx; and there is a considerable first branchial arch. In Woodpeckers there is no basibranchial, but an immensely long first branchial arch is extended completely, in other birds partially, round the head, and lies over the nasal roof.

The Cranial Nerves.

751. The relations of the brain-case to the foramina of exit of the cranial nerves are of much importance in the study of the skull. The positions of these nerves assuredly serve as landmarks; but their precise significance cannot yet be definitely expressed. A certain order is always observed in their arrangement; but the distances of the foramina from one another vary greatly, and two of the nerves (trigeminal and vagus) often perforate the cranium as two or more trunks. The order in which the nerves pass out from before backwards is the following: olfactory, optic, oculo-motor (several), trigeminal, facial, auditory, glosso-pharyngeal, vagus, hypoglossal. These nerves are certainly not of equal morphological value. How far they correspond, in relation to segments of the body, with the ordinary nerves, *e.g.* of the dorsal region, is not yet ascertained[1]. Consequently deductions from them with regard to the composition of the cranium are premature. Those who hold that the olfactory and optic nerves are of so special a nature as to be quite unconnected with the ordinary segmental arrangement of nerves, are certainly not entitled, from the position of these foramina, to make any dogmatic conclusions respecting the cranial segments and their homology with vertebral segments. We believe that these nerves will ultimately be found to have some definite connection with primordial segments and their nerves.

752. The position in which the cranial nerves emerge may be generally defined to be the lateral region of the basis cranii or the lower part of the side-walls. The emergence is never in the middle line, the nearest approach to it being in the case of the olfactory nerves. These mostly pass out of the primordial cranium through a large fenestra in the back part of the nasal walls, the fibres

---

[1] See Mr Balfour's very able discussion of this question in his *Development of Elasmobranch Fishes*, *Journ. Anat.* Vol. xi., Part 3, April, 1877, and also Mr Marshall's valuable paper on the cranial nerves of the Fowl in the same journal.

being not closely packed, but more or less widely distributed in the fenestra. The olfactory fenestra is at or very near the anterior end of the cranium.

753. The optic nerves enter the hinder or middle region of the orbits by a foramen frequently larger than the nerve. The motor nerves of the eyeball may enter it by separate small foramina behind the optic foramen, or in common with the anterior division of the trigeminal. The latter is always situated near the junction of the cranial wall with the ear-capsule. Several branches proceed from it, becoming distinct before emergence. They may all pass out of the brain-case by a common foramen, or by as many as three distinct foramina, corresponding to the three main divisions of the nerve. The facial nerve is more or less closely connected with the auditory, and in higher forms one foramen may be common to the two nerves In many Anura the facial emerges in front of the ear-capsule, in the same foramen with the trigeminal. In other forms it bores into the front part of the ear-capsule and has a winding course within its anterior and lower region. In Sharks and Axolotls the facial foramen is in the line of junction between the true basilar cartilage and the ear-capsule. The auditory nerve enters its capsule directly. The glossopharyngeal and vagus foramina are always found in or near the line of junction of the occipital and auditory cartilage; being more in the base or side-walls according to the position of this junction. They may emerge by a common foramen; or by two foramina not far from one another; or the vagus, separated into a number of strands, may have a foramen for each, as in Notidanus. The glossopharyngeal appears frequently as if included in the auditory cartilage: the vagus not so. The hypoglossal nerve (where it exists) comes out more internally than these nerves, just in front of or near the condyle.

754. The greater part of the tracts in front of and above the mouth are supplied by branches of the trigeminal nerve. The main anterior branch (ophthalmic

or orbitonasal) has a general distribution throughout much of this territory, especially around and in the orbits and nasal cavities. In Birds its nasal branch becomes specially related to the prenasal cartilage: and it may pierce through the anterior facial tissue and appear on the extreme fore part of the face in them and in mammals. There are several indications that this nerve is a dorsal branch, whose distribution is extended far forwards[1]. The co-extension of the orbitonasal and the trabecular growths is very remarkable, whatever may be its interpretation.

755. Of the oculomotor nerves it may be said that they supply a series of muscles which are developed, at least in lower forms (up to the Frog), from the first muscular segment of the body, appearing primarily on either side of the anterior part of the notochord; and permanently taking origin from the basicranial canal in the Salmon. The oculomotor nerves are more or less clearly divisible into an anterior and a posterior division.

756. The orbitonasal nerve runs along the side-wall of the skull in its orbital path; frequently it ascends high up under the supraorbital ridge. In every case where an ascending process of the mandibular suspensorium is developed, this lies above the orbitonasal at its divergence from the remaining trigeminal trunks. The two other divisions of the trigeminal are the superior and inferior maxillary. They diverge almost directly outwards in most cases over the suspensorium in autostylic skulls. The distribution of the superior maxillary corresponds tolerably closely with the extent of the maxillopalatine process: that of the inferior maxillary with the lower jaw.

757. The facial nerve in almost all cases sends forwards a notable branch, soon after its exit from the cranial cavity, outside and at a little lower level than the orbitonasal. This is the palatine or vidian nerve, and it is supplied to preoral parts like the last, but does not extend

[1] Mr Balfour has brought forward this view prominently in his paper above cited.

so far forwards morphologically: it has nothing to do with the prenasal region. The facial trunk further divides sooner or later into two branches, one of which (known as the chorda tympani, in mammals), turns forwards and unites more or less closely with the inferior maxillary nerve. The forking of the facial is always above the first visceral cleft, and is often very closely related to it. Consequently also this branching always takes place in or in close proximity to the ear-capsule. The course of the facial in the periotic cartilage is notable for its great concavity looking backwards—which may be matched by the concavity looking forwards, frequently seen in the course of the glossopharyngeal. The posterior branch of the facial is primarily related to the fore edge of the hyoid arch. The distribution of the facial is very complicated and little explained in higher animals; it spreads over a large territory on the face, and is the motor nerve connected with the development of facial expression.

758. The glossopharyngeal nerve is not so much implicated in the ear-capsule as the facial, but yet may be united by branches both with the facial and the vagus. It supplies the hinder part of the hyoid and the fore part of the first branchial arch, the branches being separated by the second visceral cleft where that exists. The vagus is distributed to the branchial arches, its branches (like so many distinct nerves) forking above the arches after the manner of the glossopharyngeal. Besides this the vagus is distributed to many tracts of the body, and its whole significance cannot be entered upon here. In abranchiate forms it supplies those parts which are in any way recognizable as corresponding to branchial arches: but this is but a very small fraction of the whole distribution of the nerve. The hypoglossal is the nerve of several muscles behind the hyoid arch as well as in front of it.

759. It should be noticed how remarkably several of the cranial nerves pass forwards, and appear to go beyond any simple segmental distribution. The orbitonasal, the vidian, the glossopharyngeal, and the hypoglossal are all of

this nature, while we see the vagus distributed far backwards, even in the simplest forms: thus, supposing evolution to have occurred, the lowest types referred to in this book have passed very far beyond any primordial arrangement of segments and nerves that can be conceived.

Plan and Segmentation of the Cartilaginous Skull.

760. We are thus led naturally to consider what conception can be formed of the real structure of the cartilaginous skull. This must necessarily involve reference to the cartilaginous vertebral column, which shares with the brain-case the function of surrounding and protecting the axial nervous system and emitting nerves for distribution over the body.

761. We do not conceive of the skull as being composed of a number of coalesced vertebræ; not having perceived indications of any process of coalescence in the embryo, and being unaware of any evidence of the past occurrence of such a transformation in ancient times. A large portion of each vertebral centrum is due to chondrification of a tube of tissue around the notochord. Such chondrification takes place distinctly around the cranial part of the notochord in two types examined; and this cartilage is no doubt homologous with the spinal cartilage just mentioned. As evidence relating to its segmental value in the chondrocranium we have merely the constriction of the cranial notochord in two places in the Fowl, giving the appearance of three notochordal segments; and one constriction in the Urodeles. The notochord and its enveloping cartilage are never found in the anterior part of the cranium, and consequently. judging by ordinary standards of homology, the prepituitary structures contain nothing resembling the main part of the vertebral centra.

762. The brain becomes enveloped in lower vertebrates by almost continuous cartilage, except where nervous

structures pass out. The origin of this investment from two paired bars, the trabeculæ and parachordals, which have different histories, do not originate contemporaneously in some types, and lie in the base of the anterior and posterior regions of the cranium, is apparently unlike anything in the history of the vertebral column. What then is the meaning of these bars? Are they similar to one another?

763. Viewing embryos at the stage when the mesocephalic flexure is strongest, gives rise to the idea that the trabeculæ and parachordals are not of the same order. The early anterior coalescence of the trabeculæ presents some resemblances to the coalescence of the ventral extremities of a visceral arch. The curved outline of the trabeculæ in various forms further gives an appearance of possibility to the view that the trabeculæ might be forwardly turned visceral arches supporting the brain; but the same fact has made it conceivable by some that the trabeculæ are a pair of down-turned neural arches.

764. Yet when we consider that the trabeculæ, like the parachordals, arise in the mesoblast in the floor of the cranium; that the occurrence of the mesocephalic flexure does not change these relations, and would appear to be due to the great expansion of the brain at an early stage, producing a discontinuity of the basal bars at the region of the curvature; that the flexure becomes obliterated, and the basal cartilages coalesce into one on either side, and then by junction of the right and left bars a continuous floor arises; that the side walls of the fore part of the cranium are formed by growth from the trabeculæ, just as, posteriorly, the walls are formed from the parachordals; that nerves are similarly emitted through the trabecular and the occipital walls; when it is seen, in short, that the trabeculæ are neural in all their relations, as completely as and in similar fashion to the parachordals; it seems impossible to resist the conclusion that the trabeculæ and the parachordals must be placed in one and the same category.

765. If we further consider in what way the structure of a wide neural tube will differ from that of a simple cylindrical one, it will appear that the formation of a broad floor is a feature of unlikeness, but only in degree. If then the trabeculæ and parachordals are looked upon as representing in a continuous form the basal parts of a series of cartilages forming the neural arches of vertebræ, flattened and widened, we think some advance is made in comprehending the skull; and we are then prepared to view the lateral parts of the cranium as continuous neural arches protecting the brain, extending into the roof and uniting more or less extensively with each other. And just as in certain forms of vertebræ the bases of the neural arches nearly or quite meet above the centrum surrounding the notochord; similarly in the part of the cranium destitute of notochord, the bases of the neural investment meet and form a basal tract, but without any proper centrum beneath.

766. The most complete brain-case of this kind is found in forms which do not develope bones; in other types the cranium is mostly never perfected in cartilage; often the roof is very incomplete, and even the side walls are considerably deficient. The occipital region always however remains more or less perfect; sometimes it is imperfect basally in the notochordal tract.

767. There is no definite evidence of segmentation in the history of the highly-perfected chondrocranium of the Elasmobranchs, unless the separate origin of the trabeculæ and parachordals is regarded as such, which we cannot think. It has been said that it is still very doubtful what is the real significance of the position of exit of nerves from the cranium as regards segmentation. Until the doubts as to the homology or equivalence of the cranial nerves are cleared up, it would be rash to frame a theory of skull structure upon their disposition. If we trusted to them, we should be led to consider that at fewest six segments are represented in the simplest skull,

without reckoning more than one as related to the vagus[1].

768. In Elasmobranchs there is a precranial floor perfectly continuous with the cranium, but it is destitute of a roof, and its side-walls unite with the nasal capsules. Where shall we mark off the termination of the cranium and the commencement of that which is not cranium? We cannot tell. We see no logical ground for separating into two categories the cartilage between the nasal sacs of a Skate and the cranial floor: the scooping of the rostrum in the same skull as well as its continuity of origin from the trabecula appears to indicate that it is not to be dissociated from the cranial floor. These parts then must have originally had some relation to body-segments. The long precranial valley found in an early stage of Dactylethra may be considered as throwing light on this matter.

769. There appears to be no reason to doubt that the Elasmobranch skull is in many ways the lowest and simplest type we have examined; and it leads us to speculate on the former existence of a still simpler form of skull, related to a large number of segments of the body, in which the brain was less concentrated and little expanded, and extended along a lengthy brain-case. Ascent of type appears to correspond to expansion of the capacity of the cranium together with concentration of the brain, its retraction from between the nasal capsules, and in higher forms its very great diminution in the interorbital region, or almost total retraction behind the orbits.

770. Thus we view the chondrocranium as having had a long history; and it may be impossible to discover and

---

[1] This mode of computation is adopted provisionally, in preference to adding the full number of branchial arches to the cranial segments. We do not see that it follows, because a nerve is supplied to several segments, that the portion of the skull to which it is related must be itself a condensation of several segments. The emergence of the vagus by several cords in Notidanus may be an indication that a series of cranial segments corresponding to the branchial arches has disappeared.

formulate accurately the number of segments of the primitive vertebral body to which the present chondrocranium corresponds. It appears to have grown beyond the primordial simple construction, and to have arisen into a higher and more complicated form, more perfected and specialised in relation to conditions of life, combining all kinds of potentiality, of adaptability to different uses. If these views are read in connection with those of Professor Huxley on Amphioxus (Proc. Roy. Soc. 1874), their great agreement will be perceived.

771. Then in those higher types in which bone occurs, we get a more or less distinct origination of alisphenoid and orbitosphenoid pieces in the cranial wall; but we do not find any warrant for saying that each of these is homologous with a neural arch. It does not appear to us that the cranium is a structure made up of such pieces soldered together; their distinctness is like that of disconnected portions of a continuous structure; they belong to the higher type into which the brain-case grows as an osteocranium.

772. The intimate relation of the three sense-capsules to the cranium has already been noticed at length. Each pair is built into the cranium in its own style, being variously and very perfectly protected or supported by it. A partial floor, inner wall, and roof may be provided for each organ by cranial cartilage, in addition to its own chondrifications. In ascending from the lower to the higher types we are struck by the fact that each organ appears to press continually closer to the middle line, and either to compress or get beneath the cranial cavity or cartilage. In the Salmon we find a very simple state of the inter- and prenasal cartilage, which is expanded into a thick mass not directly comparable with the internasal region of a Dogfish. There is a simplification of this condition in the Urodeles, the inner walls of the proper nasal capsules and the median tissue coalescing. But in higher forms a process occurs which appears to be intelligible by imagining the lateral parts of the internasal plate of

the Dogfish, and the inner nasal wall, to be elevated and constituted into vertical walls, which being approximated unite to form the nasal septum. In these types also it becomes increasingly difficult to distinguish what is internasal (trabecular) cartilage and what belongs to the proper capsular wall. The floor of the nasal sac in the Frog and other forms is made out of the primary cornua, and appears very comparable to the basilar cartilage underlying the ear-capsule in the Elasmobranchs and in the Axolotl, and the trabecular shelf supporting the eyeballs in the former group.

773. Approach of the orbits to the median line of the head occurs in many forms, concurrently with development of an interorbital septum: and the brain is either elevated above them or much retracted. The approach of the ear-capsules to the middle line in Birds and Mammals appears to be a phenomenon of the same order: but in them the organs are largely covered by the brain in its region of highest development.

774. The fenestration of the cranial cartilage by the sense-capsules has also been alluded to: the capsules appear to be related to three tracts of cranial cartilage, and at any rate in some forms cartilage is aborted where they abut. This is exemplified in the fenestration of internasal, interorbital, and orbitosphenoidal tracts, and the frequently greater size of the olfactory and optic foramina than the nerves which pass through them; as well as by the non-formation of lateral cranial cartilage internal to the ear-capsules in Elasmobranchs and others, a fact less perceptible in higher forms in consequence of the continuity of cranial and capsular cartilage in them from the first.

775. It does not appear that the sense-capsules bear any sure testimony to the segmental structure of the skull. They were probably originally situated in definite segments, but their large size may have extended them over the tracts due to several primordial segments. See-

ing what is our uncertainty about the composition of the brain-case, we cannot tell what segments of it especially correspond to the capsules: only the regions can be roughly indicated.

776. Of what value are the visceral arches as indications of the composition of the skull? The mandibular arch is constantly related to the basis cranii just in front of the trigeminal foramen; the hyoid to the basilar plate where it extends outwards beneath the ear-capsule, and in other cases to the median region of the same capsule; the branchial arches are behind these, and not directly connected with the cranium. Do all the branchial arches belong to the segments which the cranium represents? They are certainly all supplied with nerves from the cranium; but to relate all the branchial arches to cranial segments requires the assumption of the disappearance or condensation of a series of body segments so far as their axial and neural parts are concerned; and our reckoning of the number of segments that have thus disappeared will vary according to the view taken of the perfect quota of branchial arches. If we view the first two branchial arches as specially related to the cranium (namely, those supplied by the glossopharyngeal and the first branch of the vagus), and the hinder branchial arches as belonging really to the neck, we have four arches related to the cranium at the side of and behind the pituitary body[1]. The bearing of this view on the structure of the osteocranium must be considered later.

777. We believe that it is justifiable to regard the cornua trabeculæ of the Dogfish and Frog, (recurrent cartilages of Pig, Passerine Bird,) and the antorbital plate or bar, as preoral representatives of visceral arches; and the method of segmentation of certain palatine tracts in Urodeles, Lizards, and Birds gives some colour to the imagination that another segmental piece is there to be

---

[1] We have no wish to cast any discredit on the opposed view, of the disappearance or condensation of segments in the hinder part of the skull. The question cannot yet be decided.

traced; but the light is very uncertain at present. There is no certainty that all the body segments concerned in the cranium are provided with representatives of arches. But on this view, derived from arches or their representatives, not fewer than seven segments are concerned in the formation of the cartilaginous skull.

778. The meaning of the labials and other superficial cartilages is very obscure. No definite relations between the labials and the arches can be made out at present: yet their almost constant occurrence, their predominance in the Lampreys and Elasmobranchs, in the early stages of Frogs, and the importance of the changes in which they take part, lead to the idea that they must have some intelligible position in reference to the skull. The extra-branchials of the Dogfish are at any rate superficial cartilages related to the branchial arches, and they appear to be homologous with the scapulo-coracoid cartilages.

The Osseous Skull.

779. Calcareous deposit occurs in vertebrates in the following tracts; epidermis or epithelium (enamel of teeth, outer layer of Ganoid scales): dermis (dentine of teeth, Ganoid and Teleostean scales): subcutaneous fibrous mesh; immediately outside the perichondrium of a cartilaginous tract (parostosis); immediately within perichondrium, and eating into cartilage, (ectostosis): a little way beneath the surface of cartilage (superficial endostosis): and deep in its substance (true endostosis, central or subcentral). In most of these tracts the calcification may be such as not to gain the title of bone; but in all except the first, true bones may result from the process.

780. In various embryos it has been noticed that the calcification of teeth occurs before the bony plate that bears them and unites a series has appeared. The principal membrane-bones arise previous to any ossification of cartilage. On the other hand, some types very deficient

or entirely wanting in ectostoses have superficial calcifications of cartilage; and in others, deep endosteal deposit takes place before ectostosis, but it is soon either absorbed or superseded by the latter.

781. Any of the calcified tracts mentioned above may unite with the next so as to be undistinguishable from it. Dermosteal and parosteal tracts are frequently conjoined: again and again do bones developed parosteally graft themselves on cartilage as ectostoses, or ectostoses extend into outer tracts of fibrous tissue: and ectostoses similarly unite with endostoses.

782. These facts, combined with our present knowledge of fossil forms, seem to furnish a clue to the processes by which the bony skull has become what it is in various types. We conceive of primordial vertebrates which possessed a calcified exoskeleton and a cartilaginous brain-case; these calcifications becoming true bones, and continually ossifying deep tracts; the chondrocranium gradually acquiring its proper bony centres; the parostoses extending more deeply and being applied to and moulded upon the cartilaginous parts, in many cases uniting with their proper bones, or aborting the cartilage.

783. Scales and dermal bones in vertebrates have probably a relation to the segmentation of the body. The bony armour of Callichthys, the lateral line series of scales each with its mucous duct, of Teleosteans, the bony scutes of the Sturgeon and the Crocodile, are but a few out of many facts suggestive of this idea. The dermal bones and parostoses of the head are representatives of the same thing, occurring serially and in rows which can be traced more or less clearly. This relation has been proved with regard to the parostoses related to the cartilaginous parts of the sternum and shoulder-girdle[1]; and if a student desired to be prepared in the best way for understanding the difficulties and complicated relations of tracts of bone and cartilage in the skull, he could not do better than

[1] *On the Shoulder-Girdle and Breast Bone.* Ray Society, 1866.

master the simpler problem of the sternum and shoulder-girdle.

784. The body of Callichthys is enveloped in a right and left series of elongated supero-lateral and infero-lateral (ossified) dermal plates. The supero-lateral bones are directed downwards and forwards, the infero-lateral downwards and backwards. The upper and lower series are very much alike; only the upper plates are pierced at their base in front and notched behind, for mucous glands. If this portion were cut off by a suture in each upper plate we should have a series comparable to the mucous scales of the lateral line of typical Teleostei. This does in fact take place in the first cincture behind the head, forming a post-temporal bone. The cinctures are not perfect above or below; space is left for small dermo-spinal plates and spines in the fins; and for dermo-ventrals and ventral spines.

785. In the work referred to it is shown, in the case of Callichthys, that the two supraoccipital derm-bones, the single parietal, the symmetrical frontals, the single dermo-ethmoid are all serial homologues, whether azygous or symmetrical, of the upper three-fourths of the supero-lateral dermal plates clothing the body. The lower part of the first body-plate (containing the mucous duct) is cut off as a supratemporal. Then this series becomes double around the eye, and we have the dermal post-frontal, and postorbital, and supraorbital, the suborbital and lachrymal, and then the nasal. The infero-lateral plate is subdivided where it is in relation with the shoulder-girdle, into three; supraclavicle, clavicle, and interclavicle. The opercular repeats the supraclavicle, the subopercular (not found in Callichthys) the clavicle; but the hyoid dermo-cincture is completed by the many branchiostegals (three in Callichthys); the basibranchiostegal is a lower spine-bone, and seems to correspond with the interclavicular region. The squamosal, in relation to the mandibular pier, carries on the series of supraclavicle and opercular; the clavicle and subopercular are represented by the inter-

opercular; the dermal bones of the lower hyoid and interclavicular regions are represented by the jugulars, and the splints of the mandible. The maxillary, jugal, and quadrato-jugal, homologous with part or the whole of the squamosal, belong to the infero-lateral series in the fore part of the head.

786. The ossifications which occur in the cartilaginous brain-case and its anterior continuations, independently of the proper capsular bones, are comparatively few, and on the whole simple in character so long as we have to do merely with cartilage. The most constant ectostoses in the cranium are the exoccipitals, appearing on either side of the foramen magnum, enclosing the vagus and often the glossopharyngeal and hypoglossal nerves, bearing the occipital condyles where there are two, or contributing the lateral portions when the condyle is single. In every skull which is truly ossified these bones occur; and when neither basi- nor supraoccipital exists, the exoccipitals extend largely into the floor and roof of the cranium, very much after the manner of the neural arches of vertebræ in various animals.

787. The orbitosphenoid is, next to the exoccipital, the most constant bone of the same category. It is the ossification of the lateral cranial wall anteriorly, or of so much of it as developes cartilage; it is perforated by, or is anterior to the exit of the optic nerve. Very frequently it is conjoined in development with the similar tract in front (ethmoidal), beyond the brain-case: and the lateral ossifications may become united by median bone, forming the complicated sphenethmoid. It is noteworthy that the orbitosphenoid may arise by two centres on each side. The alisphenoid is the only other bone which can be classed with these: it is in the lateral cranial wall, or so much of it as is cartilaginous, between the exit of the optic and the trigeminal nerve, or partially or entirely enclosing the latter. It also in some cases arises by two centres on each side. Both the orbito- and alisphenoids

may lie in the lateral part of the floor of the cranium in higher types with large brains.

788. The basioccipital, rarely found in Amphibia, but occurring in all higher types, is an expression of ossification arising immediately around the notochord, after the manner of the main part of a vertebral centrum, and extending on either side into the proper parachordal cartilage. Developing at first in the Chick around the hinder portion of the cranial notochord, it subsequently grows so as to enclose the whole of its three segments, when the notochord has relatively retired and the posterior basicranial fontanelle is established. In the Axolotl, where no basioccipital is formed, there is an apical ossification around the notochord (cephalostyle), which afterwards loses its identity in the parasphenoid, or is absorbed. We thus see that the basioccipital corresponds very closely with a vertebral centrum, but its relation to the whole of the cranial notochord makes it doubtful whether it is not equivalent to two or three vertebral centra.

789. The next bone, the basisphenoid, corresponds to a certain extent with the basioccipital, in being more or less laid down in cartilage which is the result of the junction of lateral masses, the trabeculæ, as the basioccipital is largely due to the parachordals. But the basisphenoid has no relation to the notochord, or to anything like a vertebral centrum. A basisphenoid formed from cartilage occurs less frequently than a basioccipital; in most cases it arises from a pair of centres on the anterolateral margin of the pituitary body, which afterwards join. It is not till we reach the Birds and Mammals that a basisphenoid arising from a single median centre is found, but even in the highest form, in Man, we find a pair of centres. The basisphenoid may be increased in bulk very largely by ossification in adjacent membrane.

790. A rudiment of an anterior median centre, the presphenoid, is found in Lizards, and more markedly in Birds: it is in the upper part of the interorbital septum,

the expression of conjoined and compressed trabeculœ. A distinct presphenoid is marked in various Mammals, most definitely in rodents; but in many cases the region is ossified by the inward extension and union of the orbitosphenoids. Anteriorly to this, the median cartilage is ossified by a vertical mesethmoid where a true nasal septum is found: it occupies the hinder part of this region and the fore part of the interorbital. The cribriform plate in Mammals is simply a lateral ossification continuous with the top of the back part of the mesethmoid, and extending through the interspaces between the olfactory fibres where they pass out of the cranium. The mesethmoid has a definite extent forwards, not passing beyond the cranio-facial fenestra (Birds), or into the aliseptal region as formerly defined: but in various birds two or three centres may ossify part of the anterior region of the septum.

791. Other forms of ossification of precranial cartilage are found in the Cod tribe, where there is an anterior median ethmoid, and in the Pike, where small paired ossifications invade the cartilage. One lateral ectostosis has to be mentioned, the ectethmoid (prefrontal) in the antorbital wall, occurring in the Salmon and again in Mammals. In the adult Pig a prenasal bone is found in the extreme anterior cartilage of the head (prenasal and alinasal). Ossifications in the turbinal ingrowths of the nasal capsules arise only in Birds and Mammals. They present no important phenomena beyond those of the cartilaginous structures in which they take their origin.

792. In the ear-capsule we have a series of bones very definitely related, and very persistent. The prootic is about the anterior semicircular canal, and is related to the antero-inferior region of the capsule; the epiotic is over the junction of the anterior and posterior canals; the opisthotic is postero-inferior, and frequently furnishes the hinder part of the margin of the fenestra ovalis. These are the most constant ossific centres, remaining distinct from one another either throughout life or till an advanced

stage, in Fishes, Amphibians, and Reptiles. In Fishes there is a pterotic over the external canal, frequently bearing the horizontal articular facet for the hyomandibular: this centre reappears in the Axolotl, but is not found in the higher types. There may be another centre in front of this, the sphenotic, occupying the antero-external region of the capsule; but the cartilage in which it arises is always of a composite character, being due to a confluence of proper cranial cartilage with that of the capsule.

793. There is a strong tendency for the epiotic centre to unite with the supraoccipital, and the opisthotic with the exoccipital; and in some forms the periotic bones do not arise separately, but the supraoccipital and exoccipitals extend into the epiotic and opisthotic regions respectively. Where there is no supraoccipital, the exoccipital may ossify the whole posterior portion of the capsule. In Birds the epiotic and opisthotic are quite small, and the adjacent bones grow into the auditory cartilage and then annex these centres to themselves. The prootic on the other hand has much more independence and persistence; it is the most constant periotic bone, and is never aborted by adjacent bones. It may grow far into the floor of the cranium, and may even occupy a considerable portion of its roof in Amphibians; sometimes it further developes a process which lies in the side wall of the skull anteriorly to the capsule, and is comparable to an alisphenoid.

794. The cartilage-bones of the outworks of the skull have in their way, as curious a persistence as those of the cranium proper. But some regions comparatively early in the ascent of types become affected by membrane ossification, and cease to develope cartilage, or develope it only when the ossifying force is exhausted, or develope it in the embryo and abort it under the influence of the growth of neighbouring parostoses. The parts which persist as cartilage-bones are selected, as we ascend the scale, for specially important functions, being usually

reduced in relative size and becoming more elegantly shaped.

795. It appears that the palatine of the osseous Fish represents the antorbital or palatine of Amphibia, and partially the os uncinatum of a Bird. It cannot be definitely stated whether the pterygoid and mesopterygoid of an osseous Fish belong to a proper subocular arch or to the forward growth of the upper mandibular segment: probably the latter is the case. There is no doubt that the metapterygoid and quadrate of Fishes belong to the upper mandibular segment; and they correspond regionally with the epi- and pharyngobranchials. In the classes above Fishes the palatine and pterygoid are very largely formed out of membrane, the cartilage being either ossified in addition, or more or less aborted as in Amphibia; or only coming into existence at a late period, in various tracts, which remain permanently unossified, as in Birds. A specialised portion of the pterygoid tract may however exist in cartilage within and above the membrane-pterygoid, and ossification of this developes the epipterygoid of Lizard and Turtles.

796. In Reptiles and Birds the whole upper mandibular segment has but one bone, the quadrate; and this may be said to answer in a general way to both quadrate and metapterygoid of the lower forms; but it is more highly specialised, and itself becomes, like the palatine and pterygoid tracts, increasingly unified with the cranium. The mandible usually has but one centre, the articular, corresponding to a ceratobranchial; in the Frog and in Man there is an inferior or anterior (mentomeckelian) centre, apparently like a hypobranchial. But the whole meckelian cartilage never becomes proportionally so much ossified as the branchial cartilages. In ascending the scale the articular is very persistent, but becomes small in the Birds (largest in Struthionidæ) and specialised in form. In Mammals no articular is formed, and the representative of the quadrate (the malleus) is very small,

highly specialised and intimately involved in the cranial structure.

797. The upper segment of the hyoid arch is ossified by two centres, the hyomandibular and the symplectic, in osseous Fishes. Where it exists in higher forms it is from the first specialised for the service of the ear, and in some is ossified by two centres corresponding to the larger ichthyic centres. Of course the part that ossifies the proper stapedial plate, due to the auditory capsule, is not to be considered as representing any portion of the hyoid arch: but there is often no distinction between the stapedial and the hyoid centres. In the mammalian type, however, the stapes is quite distinct. In most columellas there is but one bone, which may be considered as a specialised hyomandibular. The incus is its mammalian representative, and is very highly specialised, the parts however being developed according to the type foreshadowed in inferior forms. The two main ossifications of the lower part of the Salmon's hyoid arch, the ceratohyal and the epiceratohyal, must be regarded as answering on the whole to the ceratobranchials. The hypohyal of course represents the hypobranchials; the basihyal and the basibranchials are homologous.

798. The homologue of any ossification found in the Mammalia between the incus and the anterior cornu of the hyoid bone must be sought in the cerato- and epiceratohyal of the Salmon. The anterior cornu itself is homologous with the hypohyal. The ossified pieces of the branchial arches which persist in Urodeles and in other types are mostly to be considered as ceratobranchials: there are two segments on either side in many Birds and Newts; they may be cerato- and epibranchials; but it is not very necessary to formulate any precise homology between them and the branchial bones of Fishes.

799. We see then in the ichthyic branchial arches with their four ossifications, the fullest development, according to a simple type, of the lateral rods of the

pharynx; they have a principal region of segmentation, between epi- and ceratobranchials, comparable to the division between sternal and vertebral ribs. The anterior arches, so much modified and specialised, nevertheless agree in being dominated by this segmentation into a supero-lateral and an infero-lateral tract. In almost all cases the mandibular and hyoid arches manifest a distinction into these two regions, this being seen even in Mammals in the ossification of the proximal and the non-ossification and dwindling of the distal part. In the hyoid arch of Anura these two portions become developed more and more independently of one another, as they are specialised, the one for the ear, the other for the service of the tongue; yet after they are definitely formed they grow towards one another, and repeat their primary relations in the Fishes: while in the Mammal the same development is carried to its highest pitch, and yet the whole hyoid arises continuously from the first.

800. The tracing of scales and membrane-bones from their original external and superficial position into their ultimate complete combination with the primitive endoskeletal structures of the skull is a study of high interest. The very numerous superficial bones become diminished in number, certain of them being selected, so to speak, for specialisation, others being dispensed with; and this occurs more and more as the bones are found deeper and deeper, and become modelled on the cartilaginous parts. An intermediate condition is seen in the Sturgeon, where a great number of small bones remain, together with several large ossifications which represent definite membrane-bones of other forms. The regional names of parietal, frontal, &c. have been given to these bones; they are not, however, in the Sturgeon, merely representatives of scales belonging to single segments of the head, but have encroached beyond their primitive morphological territories like the scutes covering the rest of the body. If they were originally segmental ossifications, the precise segment to which they belong cannot be deter-

mined, nor can the true composition of the skull be deduced from them.

801. In those Ganoids which resemble the Teleosteans in their superficial ossifications, as well as in the Teleosteans themselves, while a number of unspecialised membrane-bones remain (in the orbital ring, for instance), we see an advancing specialisation of certain membrane-bones which are recognizable more or less throughout the higher vertebrates. But they often include ossifications of very superficial strata, and are but loosely connected with the cartilaginous parts, which have an independent completeness. The parietals, frontals, nasals, and squamosals (= preopercular + supratemporal) are the bones which become most constant and definite; and if to them we add the ectethmoids (prefrontals) and lachrymals (the latter being specialised from the orbital ring) we have included all the principal membrane elements on the surface and sides of the skull, omitting those which margin the jaws. These bones cannot be assigned to particular segments of the body; only their order and regional position can be predicated; one pair, the frontals, often represented by a single bone, acquire a great predominance; the parietals attain almost a similar extension in many Mammalia, and even in some Amphibia.

802. The ingrowth of epiblastic tissues into the mouth and nares makes it not surprising that membrane-bones should be found therein. But these are even less specialised for segments than the outer bones. The parasphenoid, never found double excepting where the basitemporal wings arise separately at first, as in Birds, is the basal bone underlying the entire cranial cavity; the vomers are a pair related to the internal nares: the premaxillaries are in front of these. At the junction of inner and outer surfaces, the line of the premaxillary is continued by the maxillary, the jugal, and often the quadratojugal. All these bones are frequently found more or less specialised as splints to cartilage, or moulded upon its surface, whether cranial, nasal, prenasal, or palatoptery-

goid; but this does not appear to be the primary condition, and the bones in the higher forms are in more than one of these tracts quite independent of the cartilaginous tracts. The splenial is the corresponding internal bone of the lower jaw.

803. The external series of membrane-bones belonging to the arches only become definite splints in relation to the lower jaw, the angular and the dentary being the two principal bones. In the other tracts the bones are less accurately or not at all moulded on the cartilage. The related tracts and membrane-bones are as follows: upper mandibular region, preopercular and interopercular; upper hyoid, opercular and subopercular; lower hyoid, branchiostegal rays; branchial arches, the denticular series.

804. When we reach the Amphibia these membrane-bones are found much reduced in number and more closely applied to the cartilage; and at the same time there is a diminution of the cranial cartilage, especially above. The squamosal, representing preopercular and supratemporal, becomes accurately moulded on the mandibular suspensorium and on the side of the auditory capsule. The membrane-bones related to the hyoid arch do not occur.

805. In the Reptiles the membrane-bones attain greater perfection and specialisation, with a smaller proportionate development of cartilage; and the external series still form prominent outworks distinct from the cranium proper. The frontals and parietals attain a high degree of perfection, dipping considerably at the sides, forming cranial walls, and even sending laminæ beneath the brain.

806. In Birds and Mammals a much higher condition exists; the membrane-bones are greatly expanded in relation to the increased size of the brain, and more of them are assumed into its walls, while others are closely

engrafted upon them, or unified with the cartilage-bones in the face. The addition of the squamosal to the cranial wall is most important. The frontal retains its old predominance, while the parietals gain a greatly increased influence. Nasals and maxillaries tend more and more to determine the special characters of the face: but the lachrymal is a curiously persistent relic of the circumorbital bones. The squamosal retains its facial relations while becoming engrafted upon the cranial wall; the interopercular element crops up as the tympanic bone. Thus we ultimately get a smoothly compacted skull in which all the principal elements in the heterogeneous skull of the Salmon are used up, but which presents the greatest possible contrast to the Salmon's. Sutures and anchylosis take the place of apposition or overlapping; the braincase is very large proportionately and contains a brain as large. The parts originating in cartilage, and those due to membrane, those which have a history looking back to ancient scales and scutes join harmoniously with the elements related to the vertebral series, or to the protection of sense-capsules, so that a complex mass is produced which seems to defy analysis, and looks as if it had been straightway made as it appears, instead of being the result of a prolonged accumulation of morphological change in the individual, the species, the group, and the class.

807. A few remarks may be made about the segmentation of the bony skull. In the first place, there is no difference caused by ossification in the postoral arches which modifies the views expressed in § 766. Secondly, only one bony segment, the occipital, can be said to be clearly manifest in the skulls of Fishes and Amphibians. And in these forms there are no good grounds for assigning to the cranial bones special names indicating a correspondence to particular parts of vertebræ. From a study of adult structures in the mammalian groups, skull-theories have been devised, lacking the basis of embryology; and granting that they express some of the truth

respecting the highest forms of skull, there is only injury to knowledge in arbitrarily interpreting the lower forms by them. In Reptiles the skull becomes much more perfect, but with wide variations in the different groups, such that they cannot be merely subordinated to and explained by the mammalian type. A careful study of the growth of the Bird's skull, again, will show that it is impossible to express its composition in a simple formula derived from vertebral structure. But from the lower to the higher forms of vertebrates we can discern a growing away from the primordial type of skull, towards and into a loftier development. One feature or another of elevation is manifest in each great group, culminating in the Birds, where a beautifully-finished and compact skull results from the union of elements of very varied primary significance; and in the Mammals, where a stability of skull-construction seems to be reached, the full development of a higher type superimposed on a lower one, and capable of indefinite variation. Here and there only, with various degrees of perfection, may the cranium be said in a certain sense to be made up of segments, enclosing a portion of the neural tube. There are three of these segments, the occipital, the parietal, the frontal; the ethmoidal and nasal tracts cannot be brought within the category of "segments." The squamosal and the periotic bones are not explicable by any simple segmental theory of the skull.

808. Then, if it be granted that in the Mammalia we may perceive with some distinctness three cranial segments, it does not follow that they constitute three cranial vertebræ. The considerations we have advanced previously, in discussing the cartilaginous skull, indicate what difficulties there are in the way of regarding even the basioccipital as equivalent to one vertebral centrum, and the reasons for thinking that it represents ossification in several primitive body-segments. Consequently we cannot admit that our investigations give any reason for describing the skull as constructed by the modification of a

series of vertebræ, still less for viewing it as directly made up of a number of cranial vertebræ. This short and easy road to the comprehension of the skull is really neither short nor easy, for it does not supply the means of grouping together under one conception all the known facts of adult structure; it leads to inconsistency in assigning importance to facts or in framing a nomenclature, and it fails to give a hint of the existence of those changes in growth which we have been tracing; it makes no pretence to supply an explanation of them. The longer and more difficult studies of embryology furnish clues which lead to a conception of the structure of the skull by which a firm grasp of facts can be attained, with a sense that there is no distortion of Nature in the process. The cartilaginous structures are no longer omitted from consideration; the history of the skull is no longer a sealed book.

809. What is the import of these things? What is their place in our conception of Nature?

We find that every form of skull that has been investigated, every stage in development, contributes to one idea, which becomes simpler, more intelligible, more harmonious, by the pursuit of a right process of investigation. There is a unity of structure in the skeleton of the head, a fundamental formal unity which may always be perceived; and an adaptability to the most varied conditions of life in water, on land, in air, which becomes more and not less astonishing as knowledge slowly and surely increases. The illustrations, retrospective to lower conditions, prospective to higher groups, which individual forms present to us, are an assurance that each life-history carries its own abundant lessons, that subjects for investigation are inexhaustible, that each worker may win something for himself towards the comprehension of the creation, to be his own peculiar possession.

810. In these researches the man of sound mind and right spirit walks with bated breath, and is charged with something higher than curiosity as he watches the long-

concealed operations by which the germ, so simple to the eye, so similar in all forms, is in continuance fashioned into the likeness of a vertebrated animal and into the special representation of its ancestry. And he is filled with joy, not merely because he is permitted to discover or to comprehend anything, but because the operation, the plan, the construction that he sees is so beautiful, so infinitely beyond human planning. The little mass of protoplasm and food-material in the egg or the womb is by unseen power, noiselessly, unceasingly, unhastily driven onwards in a growth and differentiation which not merely show us how the individual is built up, but in addition link it with its fellow-creatures. The embryo is not for the sake of the individual only: it expresses a condensed history, a manifest relationship; it is for the sake of those who can study and learn about nature.

811. We are necessarily led to see that this unity of structure, this relationship, includes extinct creatures as well as those now living. And the student cannot but seek for some further light than is involved in the establishment of the fact that there is a unity in the structure of all vertebrate skeletons. An explanation is required; we want to comprehend how this unity in diversity has come about. Morphology studied in the history of embryos reveals to us an evolution by which the skull passes through one grade of structure after another, becoming advanced and changed by almost imperceptible gradations until the adult type is attained, in a certain number of days and weeks. This evolution is continually going on within our experience; and we think little of its marvels. And yet many find it inconceivable that the same process of evolution can have taken place in past ages, so as to produce from small beginnings the varied fauna of the globe. The natural forces which in a few days make a chick out of a little protoplasm and a few teaspoonfuls of yolk, are pronounced incompetent to give rise to a slowly-changing gradually-developing series of creatures under changed conditions of life. Yet to our minds the one is as

great a marvel as the other; in fact, both are but the different phases of one history of organic creation.

812. The genetic connections of all vertebrates, past and present, will not be made out for a long time to come; this is scarcely the place for a full discussion of them; at a future time, after further work, we hope to show what light is thrown upon them by our special study, and how the geographical distribution of animals and past physical geography may be illustrated therefrom. But to indicate the kind of ideas about the past which arise in the mind of the worker in the pursuit of his researches, we append quotations from two of the monographs on which this book has been based.

813. The following passages, written in 1870, are from the Memoir on the Frog, p. 201:

"The mind both of the reader and the writer will be strengthened as well as refreshed by a wider view, and each separate type will be seen in the light of many other types. Indeed thus alone will it be possible to obtain broad views in vertebrate morphology, 'as a man conveniently placed in some eminent station may possibly see, at one view, all the successive parts of a gliding stream; but he that sits by the water's side, not changing his place, sees the same parts only because they succeed, and those that pass make way for them that follow to come under his eye.'

"I must confess to having subjected myself to this mole-like burrowing into so limited a territory that I may obtain fresh material for ratiocination—'that way of attaining the knowledge of things, by comparing one thing with another, considering their mutual relations, connexions, dependencies, and so arguing out what was more doubtful and obscure, from what was more known and evident.'

"To have worked out one single species in this way may seem to be but like the forming of a single track in a primæval forest; yet when well cleared, so perfect is the unity of each subkingdom, by such a narrow path the worker is 'regularly led on through the labyrinths of Nature, when still new discoveries are successively made, every further inquiry ending in a further prospect, and every new scene of things entertaining the mind with fresh delight.'

"Leaving for a while the suggestive morphology of the Frog, it may be worth while for the palæontologist to reflect upon the empty spaces in the great vertebrate circle which are darkly but really revealed by what is seen in both the earliest and the latest stages of the Frog.

"Territories vacant, but larger far than those now occupied by family after family, and order after order, have been suggested to me by my long attention to the growth of the skull in this Amphibian.

"Empty spaces of almost indefinite extent seem, to my mind, to stretch themselves below the Myxinoid prototypes of the Batrachia, and above and beyond the Frogs and Toads, in the direction of the Mammalia.

"This last space is wholly undefined, and no light has yet penetrated its deep abyss, in which lie buried the fundamental Mammalian types. The lowest Mammals known to us, the Platypus and the Echidna, may be fundamental to the Edentata; they are not, they cannot be, to the Marsupials, the Insectivora, and the Rodentia.

"Between the Monotremes and the Batrachia we certainly have the Sauropsida—Reptiles and Birds; but I am bold to say that no Sauropsidan lies in a direct line between, forms any part of a phylum which should connect together the nobler Amphibian forms and the lowest Mammal. On the Mammalian side of this empty space we must suppose a form which should be general to the whole class; I need not say that no such form is extant. The extraordinary and unlooked-for morphological elevation of the adult Anuran, an elevation in very important features attained by no Reptile or Bird, and which brings it almost into contact at certain points with the Mammalian margin, is very suggestive. Such a discovery sheds a certain but feeble light, useful though faint.

"The fact that the higher Batrachia go on metamorphosing until several of their structures are so perfect as to require but the gentlest modification to make them fit for the Mammal, does not require one to suppose that the Toad and the Frog lie in the direct route from the Ichthyic to the Mammalian types. That such power of variation, such aptitude for transformation exists in these essential but metamorphic Fish, suggests the probability that some of the very earliest of the Amphibia, filial perhaps to forms far lower than the Lamprey, did not stop at the last metamorphic stage of an Anuran, but changed still further, and thus laid the foundation of the higher classes.

"We are all looking for further traces of the phylum which shall complete the connexion between the cold-blooded, scaly types of Sauropsida and the feathered, warm-blooded Birds; even should this never be attained to, yet no one will doubt that it has existed.

"An Amphibian, full of latent power of change, need not have taken in its metamorphosis merely the path that leads to the Reptile and the Bird; for the least deflection at first may have sufficed to bring about all the differences which now, in this late human period, we see between the Mammal and the Bird. These warm-blooded groups are huge culminating branches of the tree of vertebrate life; yet it is not a wild fancy to suppose that they may once have existed together in the same common trunk.

"So much for the vacant space *above* the Myxinoids; the lower is much larger and more pathless.

"The lowest existing Fish but one is the Myxinoid (Lamprey, Hag, Bdellostoma); between it and the lowest known Vertebrate, the Lancelet (Amphioxus), there is a gap the extent of which has never been imagined; and yet even the Lancelet itself is not necessarily the actual boundary form.

"I have shown in my comparisons that the larval Lamprey (Ammocœtes) is only a little lower than my third stage of the Frog, whilst my fourth stage answers very closely to the adult Lamprey.

"Let us imagine three families of extinct Fishes below the Lamprey: first, a group arrested as to type at the Ammocœtine stage; secondly, a group which may be morphologically represented by my second stage of the Batrachian embryo; and thirdly, a group no higher than my first stage.

"These three families may have abounded in genera and species, and have been as perfectly in harmony with their surroundings as the highly-specialised and noble Ganoid Fishes. How far these groups would tend to fill up the space between the Amphioxus and the simplest of *their* species, I need not say. Every anatomist will at once see that a creature no higher in type than the unhatched embryo of the Frog is yet an untold distance in advance of the Lancelet, which yet is only the known lowest of the great Vertebrate subkingdom.

814. The second quotation, written in 1868, is from the Memoir on the Fowl, p. 803:

"I would conclude by hinting at the importance of the various isomorphisms displayed by the skull and face of this one type in its stages of growth.

"I have described it *upwards*, but my long and really anxious labour has been in the opposite direction; the stages were traced from that of the old bird *downwards* to that of the chick of the fourth day of incubation.

"Whilst at work I seemed to myself to have been endeavouring to decipher a *palimpsest*, and one not erased and written upon again just once, but five or six times over.

"Having erased, as it were, the characters of the culminating type—those of the gaudy Indian bird—I seemed to be amongst the sombre Grouse; and then, towards incubation, the characters of the Sand-grouse and Hemipod stood out before me. Rubbing these away, in my downward work the form of the Tinamou looked me in the face; then the aberrant Ostrich seemed to be described in large archaic characters; a little while, and these faded into what could just be read off as pertaining to the Sea-turtle; whilst underlying the whole, the Fish in its simplest Myxinoid form could be traced in morphological hieroglyphics."

815. These words express the kind of thoughts which have been developed by and have guided research into the morphology of the skull. The result of this study is to leave upon the mind a strong conviction that the present Vertebrates are but the ultimate twigs of diverging branches of one great tree of life. Some branches are small, others great; some nearer the main stock, others more remote; some bear few twigs and appear isolated, others are so crowded with forms that the branches from which they spring can scarcely be discerned. But happily the growth of every form is capable of revealing to us something of its own relationships, and of the history of a time when the tree of life was confined within narrower limits, when branches or branchlets now extinct were in existence; and by comparing these with fossil remains we understand the meaning of "comprehensive types", which represent in a condensed fashion whole divergent groups of the present day. And we can ask the acceptance of Evolution in past time, without necessarily implying any particular view as to the causes of this Evolution, except that they were slow and continuous.

816. We may be permitted to say in conclusion that in our experience the study of animal morphology leads to continually grander and more reverential views of creation and of a Creator. Each fresh advance shows us further fields for conquest, and at the same time deepens the conviction that, while results and secondary operations may be discoverable by human intelligence, "no man can find out the work that God maketh from the beginning to the end." We live as in a twilight of knowledge, charged with revelations of order and beauty; we stedfastly look for a perfect light which shall reveal perfect order and beauty.

INDEX.

[*The figures refer to the paragraphs.*]

Ægithognathous 599
Aëtomorphæ 597
aliethmoid 520, 546, 575, 636
alinasal 355, 520, 546, 576, 621, 636
aliseptal 354, 520, 636, 690
alisphenoid 158, 174, 449, 455, 469, 479, 483, 518, 531, 542, 556, 581, 620, 634, 647, 658, 691, 722, 771, 787
Amblystoma 293
Amia 210
Amphioxus 217
Amphisbæna 506
amphistylic 742
angular 161, 191, 461, 474, 492, 537, 565, 803
annulus; *see* tympanic annulus
antorbital 72, 84, 96, 110, 147, 269, 309, 312, 395, 485, 520, 546, 719, 777, 795
articular 148, 161, 191, 255, 289, 292, 302, 367, 377, 400, 461, 474, 492, 565, 796
ascending process (of suspensorium) 252, 268, 287, 313, 315, 317, 743
auditory capsule 22, 41, 48, 61, 78, 97, 119, 128, 141, 154, 171, 219, 230, 239, 253, 281, 309, 312, 326, 331, 340, 387, 447, 454, 467, 500, 508, 511, 527, 554, 584, 602, 606, 622, 640, 665, 694, 726, 773, 792
autostylic 742

basibranchial 80, 107, 111, 136, 150, 200, 291, 292, 305, 333, 348, 361, 381, 418, 514, 589, 627, 740, 749, 797

basicranial fontanelle 139, 337, 443, 454, 465, 528, 539, 571, 703
basihyal 65, 80, 106, 124, 149, 163, 193, 333, 361, 402, 514, 797
basilar canal 170
basilar plate 69, 98, 278, 337, 454, 631
basioccipital 155, 168, 279, 454, 465, 478, 539, 553, 646, 678, 690, 788
basipterygoid process (or plate) 504, 536, 556, 685
basisphenoid 158, 174, 456, 465, 480, 498, 504, 532, 541, 571, 584, 647, 658, 678, 691, 789
basitemporal (wing or bone) 184, 270, 364, 374, 391, 533, 541, 556, 594, 685
blastoderm 2
branchial arches 28, 50, 57, 91, 107, 124, 136, 150, 197, 236, 248, 290, 292, 304, 310, 313, 316, 317, 325, 333, 348, 361, 381, 402, 418, 514, 537, 566, 589, 614, 627, 642, 654, 668, 740, 749, 776
branchial clefts 51, 54, 83, 89, 199, 291
branchial rays 91, 94, 108, 110
branchiostegal rays 196, 803
Bufo 411

Callichthys 784
carotid arteries 32, 71, 158, 443, 456, 465, 480, 517, 581, 584
cephalostyle 465
ceratobranchial; *see* branchial
Ceratodus 212
ceratohyal 65, 106, 133, 149, 163, 193, 247, 290, 308, 537, 797

cerebral hemispheres 14
cerebral vesicles 11, 40, 118, 127, 218, 227, 238, 320, 453, 508, 516, 602
Cestracion 216
Chamæleon 506
Chelone 498
Chimæra 215
chorda tympani 360, 665, 757
Clarias 209
clinoid ridge 60, 71, 103, 480, 528, 630, 710
cochlea 23, 448, 454, 511, 527, 530, 584, 640, 682
columella 378, 422, 430, 462, 475, 493, 522, 537, 564, 587, 746, 797
condyle (occipital) 69, 98, 140, 169, 263, 279, 337, 386, 446, 465, 528, 571, 704
cornu (of hyoid) 402
cornu trabeculæ 62, 74, 94, 177, 179, 252, 265, 282, 330, 339, 353, 392, 436, 444, 486, 512, 532, 607, 636, 648, 663, 718, 777
coronoid 461, 474, 492
cranial flexure ; *see* mesocephalic
cranio-facial fenestra 543, 557, 574
cribriform plate 635, 645, 663, 738, 790
Crocodile 507
Crotalus 497

Dactylethra 412
dentary 148, 161, 191, 246, 255, 289, 367, 377, 400, 461, 492, 537, 565, 588, 627, 639, 653, 661, 679, 696, 803
dermal bones 783
desmognathous 596
dromæognathus 594

ear; *see* auditory
ectethmoid 159, 178, 273, 285, 298, 692, 791, 801
ectostosis 779
endostosis 779
epiblast 3
epibranchial; *see* branchial
epiceratohyal 162, 193
epihyal 90
epiotic 156, 168, 279, 467, 478, 555, 659, 694, 792

epiotic fenestra 128, 727
epipterygoid 300, 501, 795
ethmoidal 99, 159, 175, 178, 338, 353, 472, 633, 692, 723, 787; *see* also mesethmoid
ethmopalatine 72, 130, 143, 185, 393
Eustachian tubes 507, 533, 541, 556, 585, 594, 623, 641, 654
evolution 811
exoccipital 155, 168, 263, 279, 294, 362, 371, 386, 454, 466, 478, 529, 539, 553, 646, 657, 665, 677, 786
extrabranchial 82, 109, 778
extrastapedial 379, 523, 587
eyeball 20, 41, 119, 219, 227, 730

facial nerve, 16, 77, 102, 142, 172, 259, 341, 360, 387, 401, 434, 468, 555, 614, 632, 640, 641, 665, 753, 757
fenestra ovalis 253, 267, 331, 401, 448, 522, 727, 729
fenestra rotunda 448, 522, 659, 729
forebrain 13
frontal 146, 182, 255, 274, 286, 299, 363, 373, 390, 457, 470, 485, 547, 560, 634, 649, 656, 670, 683, 801
frontonasal; *see* nasofrontal

Ganoids 210
Gasserian ganglion 231, 234, 258, 341
Gecko 506
gills 43, 52, 68, 83, 89, 108, 223, 229, 236, 238, 291, 322, 329, 336, 348
glossohyal 149, 163, 193
glossopharyngeal nerve 16, 78, 102, 142, 172, 231, 259, 341, 387, 448, 466, 606, 632, 753, 758

hindbrain 15
heart 31, 43, 120, 222, 325
hyoid 50, 55, 65, 80, 90, 111, 124, 149, 163, 193, 221, 225, 236, 247, 290, 303, 322, 333, 347, 361, 381, 402, 418, 451, 461, 566, 589, 614, 625, 627, 642, 654, 668, 696, 741, 798

hyomandibular, 56, 65, 80, 88, 106, 133, 149, 162, 192, 212, 310, 313, 316, 744, 797
hyosuspensorial ligament 290
hypoblast 3
hypobranchial; *see* branchial
hypohyal 90, 133, 149, 163, 193, 247, 290, 303
hyostylic 742

incus 625, 641, 654, 667, 747, 797
infrastapedial 523, 587
interhyal 90, 135, 149, 163, 193, 641, 667
internasal cartilage 62, 73, 84, 159, 177, 252, 282, 439, 444, 512, 716, 735
interopercular 195, 803
interorbital fenestra 544, 731
interorbital septum 143, 157, 173, 498, 504, 518, 532, 543, 570, 715, 731, 774
interstapedial 379
investing mass; *see* parachordal

jugal 161, 189, 298, 502, 505, 535, 564, 650, 674, 802

labial cartilage 82, 92, 94, 100, 105, 110, 147, 180, 326, 334, 345, 360, 377, 417, 458, 487, 599, 663, 734, 778
lachrymal 548, 561, 650, 674, 801
lachrymal canal 21, 611
Lamprey 217
lateral ethmoid 145
Lepidosiren 214
lingula 531, 685, 691, 711

malar; *see* jugal
malleus 624, 641, 653, 666, 747, 796
mandibular arch 50, 55, 63, 79, 87, 105, 123, 131, 147, 221, 225, 284, 243, 322, 332, 342, 360, 450, 602, 613, 624, 742, 796
mandibulo-hyoid cleft 54, 83, 135, 613
—— ligament 290
maxillary 148, 161, 189, 273, 284, 296, 298, 365, 374, 398, 459, 489, 535, 563, 621, 638, 650, 661, 671, 693, 802

maxillopalatine process 29, 221, 228, 442, 509, 602, 611
—— (of bone) 535, 563
meckelian cartilage 63, 79, 123, 131, 138, 147, 191, 235, 243, 255, 274, 289, 332, 344, 377, 400, 417, 450, 461, 474, 513, 521, 565, 639, 741, 796
mediostapedial 379, 493, 587
medullary folds 6
Megalæma 596
membrane bones 779, 800
Menobranchus 317
Menopoma 314
mento-meckelian 377, 400, 796
mesethmoid 389, 498, 557, 570, 573, 581, 621, 790
mesoblast 3, 7, 27
mesocephalic flexure 12, 40, 59, 68, 127, 218, 323, 327, 441, 508, 516, 605, 618, 707, 763
mesonasal cavity 144, 177
mesopterygoid 148, 161, 187
metapterygoid 104, 123, 161, 188, 359, 366, 395
midbrain 15
middle trabecula 45, 59, 71, 224, 437, 605
Mosasaurus 506
mouth 42, 120, 220, 228, 238, 321, 329, 346, 412, 509, 516
Muræuoids 208
muscular segments 230, 239

nasal 183, 273, 298, 363, 373, 397, 457, 470, 486, 548, 561, 649, 656, 670
nasal capsule 18, 41, 74, 84, 119, 121, 159, 179, 219, 227, 244, 252, 266, 283, 339, 354, 393, 441, 445, 456, 486, 499, 508, 546, 574, 602, 609, 635, 718, 734, 772, 801
nasal nerve; *see* orbitonasal
nasal openings, 18, 83, 87, 244, 266, 273, 283, 285, 288, 334, 355, 393, 546, 663
nasal septum 144, 353, 392, 456, 486, 519, 545, 558, 610, 621, 636, 658, 673, 717, 736, 774
nasofrontal process 29, 42, 87, 127, 220, 228, 320, 509, 602
nasoseptal laminæ 74, 717, 734

INDEX. 367

Notidanus 216
notochord 10, 35, 44, 61, 69, 118, 121, 128, 139, 169, 224, 230, 239, 250, 262, 270, 278, 323, 337, 351, 424, 427, 443, 454, 511, 528, 605, 630, 701, 761

olfactory nerve 14, 75, 84, 144, 175, 177, 259, 283, 352, 389, 581, 635, 752
olfactory; *see* nasal
opercular 162, 195, 803
opercular fold 51, 89, 127, 222, 229, 329, 336, 623
ophthalmic nerve; *see* orbitonasal
opisthotic 156, 168, 280, 467, 478, 540, 555, 568, 659, 694, 792
optic foramen 77, 102, 175, 259, 352, 388, 484, 634, 730
optic nerve 13, 753
orbit 21, 732, 773
orbital bones 190
orbital muscles 142, 170
orbital process 123
orbitar process 332, 343, 358
orbitonasal nerve 77, 142, 258, 574, 754
orbitosphenoid 158, 175, 456, 469, 484, 499, 570, 620, 634, 645, 647, 658, 678, 691, 722, 771, 787. See also sphenethmoid
otic; *see* auditory
otic process, 243, 254, 287, 359, 376, 394, 521, 564, 743

palate 29, 180
palatine 131, 148, 161, 187, 245, 257, 272, 288, 296, 332, 343, 356, 375, 395, 459, 472, 490, 536, 587, 563, 638, 652, 661, 671, 693, 741, 795
palatopterygoid 63, 123, 147, 161, 257, 272, 513, 521, 611, 621, 638
palato-quadrate 49, 55, 64, 79, 87, 104, 186
parachordal 46, 60, 69, 121, 128, 139, 250, 263, 324, 330, 337, 443, 511, 517, 528, 606, 619, 702
parasphenoid 146, 160, 170, 184, 257, 270, 284, 292, 295, 338, 364, 374, 391, 422, 429, 432, 456, 471, 481, 802

parietal 160, 181, 256, 274, 286, 299, 363, 373, 390, 457, 470, 482, 499, 547, 559, 649, 670, 801
parietal fontanelle 176, 352, 388
parostosis 779
Parrot 598
pedicle 268, 287, 342, 376, 394
periotic; *see* auditory
Phœnicopterus 596
pharyngobranchial; *see* branchial
Pipa 423
pituitary body and region 26, 45, 71, 118, 158, 174, 224, 232, 323, 531, 584, 633, 685, 706
pneumogastric nerve; *see* vagus
Podargus 596
Polypterus 210
postfrontal 502, 505
postpalatine 269
prefrontal 485, 505. *See* also ect-ethmoid
premaxillary 146, 189, 244, 256, 273, 284, 297, 365, 374, 397, 457, 471, 489, 535, 562, 621, 637, 649, 670, 693, 802
prenasal 62, 73, 99, 110, 456, 486, 532, 545, 558, 562, 576, 607, 716, 791
preopercular 194, 803
presphenoid 573, 633, 658, 678, 685, 691, 790
pretemporal wing 541, 556
prootic 156, 172, 280, 294, 362, 371, 387, 468, 479, 540, 555, 659, 694, 792
prorhinal 355, 392
Proteus 309
protovertebra 12
Pseudis 410
pterotic ridge or bone 88, 97, 141, 156, 168, 171, 281, 340, 387, 623, 728, 792
pterygoid 148, 161, 187, 269, 272, 288, 300, 357, 366, 375, 395, 422, 434, 459, 472, 490, 501, 507, 536, 563, 638, 652, 664, 672, 685, 795
pterygoid plate (external) 645

quadrate 63, 79, 104, 161, 188, 268, 287, 300, 332, 344, 359, 394, 416, 450, 460, 473, 491, 501, 513, 521, 564, 796

quadrato-jugal 366, 535, 564, 802
Rana pipiens, 409
rhinal 354, 373
Rook 599
rostrum 83, 716
rostrum (bone) 534, 557

Salamandrine skull 319
saurognathous 601
schizognathous 595
septo-maxillary 297, 398, 458, 471, 488, 504
Siluroids 209
Siren 312
sphenethmoid 278, 283, 294, 312, 371, 389
Sphenodon 506
sphenoidal 77, 264
sphenotic 86, 96, 156, 170, 542, 568, 792
spiracle, 63, 83, 106
spiracular cartilage 55, 86, 105, 111
splenial 246, 255, 289, 461, 474, 492, 537, 565, 802
squamosal, 254, 274, 287, 301, 367, 367, 396, 460, 473, 491, 547, 559, 651, 656, 675, 683, 694, 801, 804
stapes 267, 281, 331, 340, 378, 401, 475, 583, 587, 622, 641, 682, 729, 797
Struthionidæ 594
Sturgeon 211
stylohyal 462, 475, 493, 523, 587, 641, 665, 668, 676, 695
subocular fenestra 332, 343
subopercular 195, 803
Sula 596
supracranial fontanelle 85, 111, 145, 352, 388, 466, 656, 690, 722
supraethmoid 146, 183
supraoccipital 76, 140, 155, 168, 279, 352, 446, 466, 478, 539, 553, 631, 646, 657, 677, 690
supraorbital band 41, 76, 83, 120
suprastapedial 379, 523, 587
supratemporal 473, 493, 505
surangular 461, 474, 492, 537, 565
suspensorio-stapedial ligament 267, 281, 287

suspensorium 137, 234, 243, 254, 268, 287, 310, 313, 315, 317, 332, 342, 356, 376, 394, 416, 428, 741

tegmen cranii 76, 96, 144, 157, 176, 724
tegmen tympani 340, 387, 415, 529, 623, 640, 667, 728
Tinamus 594
Tortrix 497
trabecula 47, 60, 71, 99, 122, 129, 143, 157, 232, 240, 251, 264, 309, 324, 330, 414, 427, 436, 439, 444, 455, 456, 468, 481, 486, 511, 518, 607, 620, 706, 764
trabecular crest 60, 242, 264
transpalatine 459, 490
trigeminal nerve 16, 77, 102, 142, 172, 258, 341, 387, 449, 511, 555, 634, 753
turbinal 546, 558, 574, 609, 621, 637, 648, 663, 691, 737
Turnix 600
tympanic 654, 660, 676
tympanic annulus 359, 380
tympanic membrane 586, 624
tympanic recess 541, 682
tympanic wing of exoccipital 517, 529, 586
tympano-hyal 668, 676, 683, 695
Typhlops 497

vagus nerve 16, 78, 102, 142, 172, 259, 341, 387, 448, 466, 555, 606, 632, 753, 758
vertebral ossification 36
visceral arches 49, 120, 222, 225, 233, 322, 324, 450, 514, 739, 776
visceral folds 27, 30, 510
visceral clefts 28, 42, 120, 222, 229, 441, 453, 513, 604
vomer 159, 184, 245, 257, 271, 284, 295, 364, 374, 399, 458, 471, 488, 549, 563, 594—601, 621, 636, 648, 673, 685, 802

Woodpecker 601

www.ingramcontent.com/pod-product-compliance
Lightning Source LLC
Chambersburg PA
CBHW030359230426
43664CB00007BB/655